Colchicum

The complete guide

Colchicum

The complete guide

Christopher Grey-Wilson, Rod Leeds & Robert Rolfe

RHS Horticultural Monograph

Inspiring everyone to grow

First published in 2020 by the Royal Horticultural Society:
RHS Media, Churchgate, New Road, Peterborough PE1 1TT, UK

Registered office:
Royal Horticultural Society, 80 Vincent Square, London SW1P 2PE, UK

Registered charity number 222879 / SC038262

rhs.org.uk

© Royal Horticultural Society 2020

The authors assert their moral rights to be identified as the authors of this work.

All rights reserved. No part of this publication may be reproduced, stored in a retrieval system, or transmitted in any form or by any means electronic, mechanical, photocopying, recording, or otherwise, without the prior permission of the publisher.

10 9 8 7 6 5 4 3 2 1

ISBN 9781911666080

A CIP catalogue record for this book is available from the British Library.

Specialist Publications Editors: Mike Grant & Rosalyn Marshall
Specialist Production Editor: Ally Page
Art Editor: Mark Timothy
Designer: Fiona Hood
Colour reproduction: Anthony Masi
Books Publisher: Rae Spencer-Jones
Head of Editorial: Chris Young

Printed and bound by Bell & Bain Ltd, Glasgow, UK

Front jacket: *Colchicum* 'Glory of Heemstede'
Rear jacket: *Colchicum* 'Glory of Threave' (left), *Colchicum* 'Giant' (top right), *Colchicum doerfleri* (bottom right)
Half-title page: *Colchicum* × *agrippinum* painted by Paul Furse
Opposite title page, left to right: *Colchicum* 'Darwin', *Colchicum* 'Glory of Threave', *Colchicum* 'Lilac Bedder'
Opposite: *Colchicum* 'Waterlily'

Contents

Foreword _____ 9

Acknowledgements _____ 11

Introduction _____ 13
   The name of *Colchicum* _____ 15
   Crocus versus *Colchicum* _____ 16

History in cultivation _____ 19
   Gerard's Herbal _____ 19
   Naming of the first species _____ 20
   Early introductions _____ 21
   Table: Dates of introduction of major species to cultivation _____ 22
   The first selections for gardens _____ 23
   Selection in the early to late 20th century _____ 24
   Wild collections in the 1960s _____ 26
   Wild collections in the 1970s and 1980s _____ 30
   Wild collections in the late 20th century _____ 32
   Wild collections and selection in the early 21st century _____ 34
   Parentage of cultivars _____ 37
   Hybrid characteristics _____ 38
   Keeping track of cultivars _____ 39

Classification _____ 41
   Family classification _____ 41
   Family characteristics _____ 42
   Key to genera of the *Colchicaceae* _____ 43
   Characteristics of *Colchicum* _____ 45
   Merendera and *Bulbocodium* _____ 45
   The position of *Androcymbium* _____ 47
   Table: Morphological differences between *Colchicum* and *Androcymbium* _____ 50

   Accounts of *Colchicum* _____ 52
   Chromosome biology _____ 53

Morphology _____ 57
   General description of *Colchicum* _____ 57
   Corms _____ 59
   Cataphylls _____ 62
   Stems, pedicels and bracts _____ 64
   Leaves _____ 64
   Flowers _____ 65
   Tepals _____ 67
   Stamens _____ 68
   Ovary and styles _____ 69
   Fruit capsules and seeds _____ 69
   Flower fragrance and pollination _____ 70

Colchicine _____ 73
   Colchicine in medicine _____ 73
   Colchicum poisoning _____ 75
   Colchicine in plant breeding _____ 76

Cultivation _____ 79
   Selecting plants _____ 79
   Planting _____ 80
   Hardiness _____ 81
   In the open garden _____ 82
   Tidying leaves _____ 83
   Lifting and dividing _____ 83
   Companion plants _____ 85
   As cut flowers _____ 87
   Propagation _____ 87
   Pests and diseases _____ 91
   RHS plant trials _____ 93
   National plant collections _____ 97
   Gothenburg Botanical Garden _____ 98

Species _____ 105
    Habitats _____ 105
    Map: Distribution of the
        genus *Colchicum* _____ 106
    Geographic distribution _____ 108
    Conservation of the genus _____ 108
    Identifying species _____ 109
    Key to *Colchicum* species _____ 110
    Species divisions based on gross
        morphological characters _____ 122
    Hybrids _____ 123
    Table: Identification characteristics
        of *Colchicum* species _____ 124
    Colchicum alpinum -
        Colchicum woronowii _____ 129-429
    Colchicum x *agrippinum* -
        Colchicum x *byzantinum* _____ 431-443

Cultivars _____ 445
    Discovery _____ 445
    Date _____ 445
    Comparable characters _____ 445
    RHS Award of Garden Merit _____ 446
    Identification notes _____ 446
    Botanical descriptions _____ 447
    Table of characters _____ 447
    Identification notes _____ 447
    Table: Characters of *Colchicum*
        cultivars, including AGM _____ 448
    Colchicum x *alberti* cultivars _____ 450
    Colchicum autumnale cultivars _____ 451
    Colchicum bivonae cultivars _____ 456
    Colchicum x *byzantinum* cultivars _____ 459
    Colchicum cilicicum cultivars _____ 460
    Colchicum doerfleri cultivars _____ 461
    Colchicum kesselringii cultivars _____ 462
    Colchicum luteum cultivars _____ 463
    Colchicum macrophyllum cultivars _____ 464
    Colchicum montanum cultivars _____ 466
    Colchicum speciosum cultivars _____ 467
    Colchicum szovitsii cultivars _____ 478
    Hybrid cultivars, A-Z _____ 479-535

Checklist of epithets   536

Glossary   552

Appendix   557

References & bibliography   558

Index   566

Photography credits   574

Publisher's acknowledgements   576

Foreword

In the past the RHS has been pre-eminent in the publication of plant monographs. I am delighted that, with this fourth volume in the series, we are now building on this tradition in our role as a learned society.

Our previous books in the current series have focused on *Kniphofia*, *Hedera* and *Wisteria*. This is the first comprehensive monograph on *Colchicum* for nearly 100 years, since Boris Stefanov's *Monographie der Gattung Colchicum L.* (1926) and E.A. Bowles's *A Handbook of Crocus and Colchicum for Gardeners* (1924). It reflects a renewed interest in colchicums in the UK, a fact that was recognised by the RHS holding a trial of the large-flowered, autumn-flowering types at RHS Garden Hyde Hall in Essex from 2015 to 2019.

In the modern era, following on from horticultural greats such as E.B. Anderson, Richard Nutt and Elizabeth Parker-Jervis, the flame of interest in growing colchicums has been rekindled by Tony Hall, Richard Hobbs and John Morley in the UK, as well as Per Wendelbo and Henrik Zetterlund in Sweden. The extensive botanical studies of Karin Persson in Sweden have been of fundamental importance, as have those conducted by RHS Vice President and former Director General, Chris Brickell.

Colchicums bring a rush of colour to our gardens in the autumn, but there are equally as many spring-flowering species deserving of wider recognition. The recent RHS trial in particular has highlighted these cormous perennials as worthwhile, easily grown an...d mostly readily available plants. They are not just for the collector or other specialists – they add colour and interest to any garden.

This is a valuable monograph that contains a huge amount of work. Beautifully illustrated, it features numerous photographs taken in the wild as well as in cultivation, setting it apart from previous accounts of the genus which were very much prose and no pictures. It includes in its authors a botanist and a taxonomist, and is full of horticultural expertise and gardening excellence. Appealing to botanists, horticulturists and gardeners alike, it contains a detailed overview of the species in the wild, a comprehensive account of cultivars, and much advice on cultivation and using them in the garden.

It has indeed been written with all gardeners in mind and is a monograph which we trust will be an important source of reference for many, many years to come.

Sir Nicholas Bacon Bt. OBE DL
President of the Royal Horticultural Society (2013–2020)
June 2020

Colchicum 'Autumn Queen' is one of the best early-flowering autumn cultivars and holds the RHS Award of Garden Merit.

Acknowledgements

The authors would like to pay due and fulsome recognition to the extensive and intensive detailed research carried out over many years on the genus *Colchicum* (including *Bulbocodium* and *Merendera*) by Christopher Brickell and Karin Persson. They have been instrumental in gathering together and synthesising a huge amount of data, without which this current work would not have been possible.

We have also had considerable help and advice from many others in the preparation of this work, including John Amand, Leonid Bondarenko, Razvan Chisu, Paul Christian, Hagen Engelmann, Jon Evans, John Fielding, John Foster, Alan Gray, Christine Grey-Wilson, Sjaak de Groot, Tony Hall, David Haselgrove, Richard Hobbs, Antoine Hoog, Wim Lemmers, Doug Joyce, Brian Mathew, John Mitchell, John Morley, Johan Nilsson, Jan Pennings, Oron Peri, Jānis Rukšāns, Peter Sheasby, Bob & Rannveig Wallis, Robin White and Henrik Zetterlund.

In addition, we would like to thank the staff of the Herbarium and Library at Royal Botanic Gardens, Kew, and the Royal Horticultural Society's Lindley Library, London. Various members of the staff of the Royal Horticultural Society have also been very supportive, including James Armitage, John David, Mike Grant, Rosalyn Marshall, Andrew McSeveney and Jill Otway.

Grateful thanks also go to Mary Randall for her scrupulous proof-reading of the text, for which all three authors are indebted.

Christopher Grey-Wilson Redgrave, Suffolk
Rod Leeds Sudbury, Suffolk
Robert Rolfe West Bridgford, Nottinghamshire
June 2020

Colchicum 'Rosy Dawn', a watercolour by Christine Grey-Wilson.

Introduction

Most gardeners know colchicums as autumn crocuses, their large flowers pushing through the ground in late summer and autumn, opening their sumptuous goblet-shaped blooms in the sunshine. When much has gone to seed, or is dying down, up they pop to enliven the garden and bring joy towards the close of the year. Yet colchicums are not crocuses (genus *Crocus*), nor are they closely related.

Pink, purple, rose-purple and violet-purple are the predominant flower colours in *Colchicum*. However, a few species are white, or have albino variants, and just one, *C. luteum*, is yellow. Quite a few have tessellated (chequered) tepals, which greatly enhances their overall appeal, while the Central Asian *C. kesselringii* has white flowers in which the tepal reverses often have a bold, central, purple stripe.

While many named cultivars have appeared spontaneously in gardens over the years, others have been produced by careful selecting and breeding. The autumn-flowering, large-flowered cultivars, of which more than 50 distinct ones are available, add interest at a time of year when much in the garden has finished flowering. They are easy to grow and remarkably pest- and disease-free. Furthermore, they will tolerate a wide range of conditions, with the exception of soils that are very acid or waterlogged, and will often multiply with a vigour found in few other corms or bulbs.

LEFT **Autumn-flowering cultivars such as 'Pink Goblet' add much drama to the garden when other flowers are coming to an end.**

RIGHT **Large-flowered cultivars such as 'William Dykes' are tolerant of a wide range of conditions.**

However, their verdant and bold foliage can put some gardeners off growing them, for this may overwhelm nearby plants if consideration is not given to planting location. Yet in the right garden location, for example among shrubs, between clump-forming perennials, among tolerant groundcovering plants, or naturalised in grass, the foliage can be an added

Although sheep grazing helps maintain this population of *Colchicum autumnale* at Hen-allt Common in Powys, Wales, the species is often removed from areas where cattle graze because of its toxicity.

attraction. Besides which, by mid-summer the leaves have usually withered away.

On the other hand, more than half the known species of *Colchicum* are spring-flowering. Many are very beautiful plants, tending to have smaller flowers than their autumn-flowering relatives. They are usually best grown in pots or pans placed in frames, glasshouses or specialised alpine houses. Alternatively, they can be planted outdoors on raised beds, or in troughs and other containers where their beauty can best be admired.

Colchicums are poisonous to both humans and animals, especially cattle. For this reason, in areas where they are a common plant, farmers often eradicate colchicums from their fields. The only native species in the UK is *Colchicum autumnale*, which has a restricted distribution mainly centered on counties bordering the Severn estuary in England and Wales. Richard Mabey in *Flora Britannica* (1996), his compendium of cultural uses of wild plants, records several accounts of farmers' attitudes to colchicum and mentions of fields and woods full of it. Some would forgo spring and autumn grazing, when the leaves or flowers were out, and harvest the corms for the pharmaceutical industry as a source of colchcine. This important pharmaceutical drug is extracted from the corms of colchicums and has been used to treat gout, rheumatic afflictions and other ailments. Colchicine has also been widely used in plant breeding. This is because it disrupts normal cell division, causing the doubling of chromosomes which in turn leads to more vigorous and substantial crops and ornamental plants.

THE NAME OF *COLCHICUM*

The generic name *Colchicum* stems from the Greek *Colchis*, referring to the Caucasus region. William Stearn in his *Dictionary of Plant Names for Gardeners* (1996) notes that the genus is 'Said by the ancient authors to be especially abundant in Colchis, the Black Sea region of Georgia, Caucasus'. Colchis is an ancient region on the east coast of the Black Sea, perhaps more familiar as a location in mythology. However, the centre of diversity for the genus is now known to be nearby, in Greece and Turkey.

The common name 'autumn crocus' alludes to the shape of flowers, like a *Crocus*, and the time of their appearance, although many *Colchicum* species are spring-flowering. The genus has also acquired other common names over the centuries, primarily due to the fact that the flowers of the autumn-flowering species appear without any sign of leaves. These include dainty maidens, meadow saffron, naked ladies, naked virgins and strip-jack-naked. Some names in other languages are quite risqué, with translations including ladies without a chemise and naked arse from the French, and naked whores from the German. Many of these names apply to one species in particular, *Colchicum autumnale*, which is widely distributed in western and central Europe.

Despite the misleading name of meadow saffron, so-called because the flowers superficially resemble those of saffron crocus, colchicums are not the source of saffron. The genuine spice, expensive and much prized in Iberian, North African and Middle Eastern cuisine, comes from a true crocus, *Crocus sativus*, and consists of the red styles that are removed laboriously from the centre of each flower.

Colchicum trigynum is a species native to Colchis, an ancient region in the Caucasus that the genus is named after.

The asymmetric corms of *Colchicum* (left) have a different structure to those of *Crocus* (right).

COLCHICUM VERSUS CROCUS

People might well be confused by the different applications of the common names of colchicums and crocuses. The genera do, however, have very different characters. *Crocus* belongs to the iris family, *Iridaceae*, along with a number of familiar genera such as *Gladiolus* and *Iris*, while *Colchicum* belongs to the *Colchicaceae*. Both *Colchicum* and *Crocus* have a cormous rootstock. That of the former is always asymmetrical, often egg-shaped or oblong, with roots issued from one side of the base as are the flowers and leaves. Occasionally the corm is stoloniferous and horizontal, almost grub-like. The fleshy, whitish *Colchicum* corms are protected by a coat or tunic which is generally brown, yellowish brown or blackish brown and can be membranous, papery or leathery and several-layered. In contrast, the corms of *Crocus* are symmetrical with the roots arising from the centre of the base. In *Crocus* the tunic can be smooth, fibrous or netted. While the *Colchicum* corm is replaced annually to one side of the parent corms, those of *Crocus*, also annually, build one on top of the other. Both may produce a single new corm each year, with the previous corm withering away, or they may produce several new corms, depending on the number of shoots produced. In both, reserve buds may also give rise to additional, small daughter corms.

Colchicum leaves are fleshy and unmarked, although variously edged, pleated or undulate. They can be anything from narrow and strap-like to large and broad. *Crocus* leaves are very different, being linear and grass-like with a distinctive white stripe down the centre.

Both genera have goblet- or funnel-shaped flowers which open in some species to a star shape, and both have a long flower tube (perianth tube) which reaches well below ground level where the ovary is located. The ovary in *Colchicum* is superior, with the perianth tube joining below the ovary, that of *Crocus* in inferior, with the perianth tube attached to the top of the ovary.

The flowers of both genera have six tepals (perianth segments) arranged in two series, with three inner and three outer tepals, which are fused together at their base at the top of the tube. The stamens are attached close to the base of the inner surface of the tepals, but while there are six stamens in *Colchicum* there are only three in *Crocus*, these attached to the three outer tepals. The anthers of *Colchicum* are usually attached in the middle to the filament (dorsifixed) but in *Crocus* the stamens are attached at their base to the filament (basifixed). The styles are also different. In *Colchicum* there are three separate, undivided styles which remain separate for the full length of the perianth tube, with the exception of *C. bulbocodium* which has a three-branched style. In *Crocus* there is a solitary, 3-branched style, whose branches can be entire or further branched, sometimes filigree-like.

While the large-flowered autumn crocuses of gardens are attributable to the genus *Colchicum* it also important to stress there are also some excellent genuine autumn-flowering crocuses. Of these the lovely *Crocus speciosus* and its various cultivars is perhaps best known, often seen in gardens and widely available. Other true autumn-flowering crocuses seen in gardens include species such as *C. hadriaticus, C. kotschyanus, C. longiflorus, C. niveus, C. nudiflorus* and of course saffron crocus, *C. sativus*. To add to the confusion, as with *Crocus*, there are also many species of spring-flowering *Colchicum*.

Colchicums have six stamens, distinguishing them from crocuses, and some have attractively tessellated tepals, as here in 'Autumn Queen'.

Colchicum autumnale.   *Colchique d'automne.*
Var. latifolium.   *Var. à larges feuilles.*

History in cultivation

Colchicums have been cultivated for more than 500 years. *Colchicum autumnale* was the first species to be introduced to European gardens, in 1444. While farmers in Europe were eradicating toxic colchicums from their fields, herbalists took an interest in the same properties for the treatment of gout. Also, early gardeners were intrigued by the autumn flowering of this large-flowered species, coupled with its later production of leaves. It was not until the 19th century that some of the other large, autumn-flowering species were brought into cultivation.

The selection of variants of species and hybrids with novel characters began in the late 19th century. Characters selected for included larger or silkier flowers, variations in colour from paler to darker, increased tessellation, and increases in tepal number. One of the most variable species is *C. speciosum*, and this has yielded the greatest number of cultivars from a single species. Hybridisation has also played a role in cultivar development, but relatively few species are involved, even today. Some cultivars have become confused in cultivation, but plant trials are resolving identification queries and revealing the best garden plants.

GERARD'S HERBAL

The first description in English of the genus *Colchicum* was published in 1597, in John Gerard's *Herball*. He gives a clear account of what he called 'Medow Saffron'. The illustration in the herbal is very clearly of *Colchicum szovitsii* or, as he called it, *Colchicum biflorum*, the 'Twice flouring Mede Saffron'. Gerard was also well aware of 'True Saffron', *Crocus sativus*.

This seasonality of the *Colchicum* that typically flowers

LEFT *Colchicum autumnale* from volume 8 of Pierre-Joseph Redouté's *Les Liliacées* (1802–1816).

RIGHT *Colchicum autumnale* was known to Gerard in the 16th century and was one of the first species to be cultivated.

An engraving by Crispijn de Passe the Younger (1594–1670) in his *Hortus Floridus* (1614) showing several *Colchicum*, the largest one labelled as '*Colchicum Byzantinum multiflorum*'. Now recognised as a hybrid, C. x *byzantinum* was introduced in 1588.

in autumn, produces leaves in spring, and then seeds in summer led to some herbalists, including Gerard, calling it *Filius ante Patrem* 'the son before the father'. Because of this seasonality, Gerard notes in his *Herball* that colchicums are 'cleare contrarie to all other plants whatsoever'. Gerard knew of colonies of *C. autumnale* in Gloucestershire, Somerset and Northamptonshire. He also remarked that gardeners, or florists as he called them, were growing them in their gardens. The herbal uses were quite clearly described. Gout was a great problem in Gerard's time, and he writes that colchicum corms crushed and mixed with milk, whites of egg, barley meal and breadcrumbs applied as a plaster would ease the pain. The ingestion of colchicum or its leaves was known to be potentially toxic, and that milk must be drunk soon after or 'death presently ensueth'. Gerard refers to this death being likened by Dioscorides to choking by mushroom poisoning. It led what we now know as *C. autumnale* being referred to as "Colchicum strangularium".

NAMING OF THE FIRST SPECIES

Colchicum autumnale, widespread throughout Europe and found as far east as Russia, was botanically recognised in 1542 by Leonhart Fuchs in his *De Historia Stirpium* (pp356–359). It is likely to have been in cultivation at that time. The name was later taken up by Carl Linnaeus in *Species Plantarum* (1753), the starting point of botanical nomenclature.

Prior to Linnaeus's (1753) standardisation of nomenclature, *C. byzantinum* (now regarded as of hybrid origin) was named by Carolus Clusius (1601). Corms of this hybrid had been sent from Constantinople (Istanbul) in 1588 to two Viennese women who passed some on to Clusius when he was living in Vienna. The origin of this vigorous plant has not been established, as it has never been found again in Turkey. Over centuries the plant has lost none of its willingness to flower profusely, but it has become infertile. This change in fertility was noted by Edward A. Bowles (1924) about 100 years ago.

In 1771, another colchicum with a mysterious origin first appeared in cultivation (Weston 1771). This was *C. atropurpureum*, although it is discussed in this book under *C. turcicum*.

Two more *Colchicum* species were named in the 18th century by Carl Linnaeus. They were *Colchicum montanum* from Spain and *C. variegatum* from Greece and south-western Turkey. The former was later recognised as *Merendera montana* but is now regarded as a *Colchicum* again.

EARLY INTRODUCTIONS

Among the many consignments of bulbs and corms being imported to western Europe from the Middle East in the 19th and 20th centuries, plants such as tulips, crocuses, cyclamens and colchicums figured prominently. As a result of this, species such as *C. bivonae*, *C. speciosum*, and what were then known as *C. bornmuelleri* and *C. giganteum*, came into the hands of nurseries and plant enthusiasts in the 19th century. The dates of introduction to cultivation for the most commonly cultivated species are given in the table below.

Baker (1879) remarked of yellow-flowered *Colchicum luteum* that 'it would be a great acquisition to our stock of cultivated colchicums and would no doubt be hardy in our English gardens'. Soon afterwards, James Aitchison, a surgeon and botanist, posted corms of the species to Royal Botanic Gardens, Kew, that flowered in January 1875. Yet the species has repeatedly failed in gardens, despite at least one 1980s offering of 'lemon-sized' corms, reputedly repeat-flowering from January to late March. Sometimes a nematode infection is suspected, as with material collected by William van Eeden routinely listed by nurseries towards the end of the 20th century. Bowles (1952), wrote: 'It did well in Mrs Ransome's garden at Ipswich, Suffolk, producing its small but rich yellow flowers in February. Here it suffered the same fate as *C. kesselringii*, the flowers and young leaves being devoured by slugs as they appeared near the surface'. The lack of pest resistance suggests that, away from

Before its introduction, *Colchicum luteum*, being the only yellow-flowered species, was seen as highly desirable, and it still is.

its native lands, pot cultivation under glass is advisable for *C. luteum*.

Bowles was an early champion of spring-flowering colchicums, some of which he called 'these pygmies', noting a wide variation in flower size. This was long before the well-named Turkish *C. minutum* was described. Admitting 'Those that flower during the darkest days of the year... are unsuitable for growing in the open in English gardens', he considered some spring-flowering species 'more robust and in sheltered positions... a welcome addition to snowdrops, *Eranthis* and the early crocuses'.

These spring-flowering colchicums will not approach the affordability, popularity and ease of cultivation of the autumn-flowering hybrids. However, to enthusiasts, anyone wishing to study the genus in depth, or those who have been thrilled by the sight of them growing in countless numbers in the wild, they offer various challenges of cultivation. But more importantly they give undiluted, off-seasonal cheer, brightening an otherwise grey and cold greenhouse, rock garden or raised bed long before the official start of spring.

Dates of introduction of major species to cultivation

C. autumnale	1542
C. × byzantinum	1588
C. atropurpureum	1771
C. alpinum	1820
C. trigynum	1823
C. speciosum	c. 1850
C. robustum	1872
C. speciosum	1874
C. bornmuelleri	1889
C. giganteum	1890
C. decaisnei	1892
C. cilicicum	1896
C. fasciculare	1896
C. laetum	1897
C. szovitsii	1898
C. haynaldii	1902
C. crocifolium	1904
C. stevenii	1904
C. kesselringii	1905
C. hungaricum	1921
C. triphyllum	1938

One of the first hybrid cultivars, 'Glory of Heemstede', was selected by Jacobus Kerbert in the early 20th century. This painting is from when it received an RHS Award of Merit on 27 September 1928.

THE FIRST SELECTIONS FOR GARDENS

The first hybrid cultivars of *Colchicum* were selected in Haarlem in the Netherlands by Jacobus Johannes Kerbert. Kerbert was a breeder of bulbs for Zocher & Co from 1867 until 1919. Kerbert's work was mostly with Triumph Group and late-flowering tulips, but he also crossed large-flowered colchicums using primarily *C. bivonae*, *C. speciosum* and what were then known as *C. bornmuelleri* and *C. giganteum*. His selections included 'Autumn Queen', an early-flowering hybrid which is thought to have *C. bivonae* as one of the parents as the intense tessellation is very apparent. This cultivar has been popular for more than a century and still holds the RHS Award of Garden Merit (AGM). Kerbert also selected 'Daendels', 'Danton', 'Disraeli', 'Giant', 'Glory of Heemstede' (then named 'Conquest'), 'Lilac Wonder', 'President Coolidge', 'Violet Queen', 'Waterlily' and 'W. Kerbert'.

Colchicum 'Waterlily' is still the only pink, double, large-flowered colchicum in commercial cultivation. This unique selection was the result of using pollen from *C. autumnale* 'Album' to pollinate *C. speciosum* 'Album', a seemingly unlikely marriage of two white-flowered cultivars. The true *C.* 'Violet Queen' as selected by Kerbert, is a tessellated, vibrant, dark pink with a contrasting white throat; a highly distinct selection rarely available today. However, Bowles (1924) wrote that 'a great number, perhaps too many, of these Zocher seedlings were named'. Today it seems that 'Danton', 'President Coolidge' and 'W. Kerbert' have disappeared from cultivation.

The Backhouse nursery of York, which existed for most of the 19th and 20th centuries, was best known for its landscaping work and breeding. Their fame in colchicums rests on one single seedling of *C. speciosum*, found in the early 1900s in a seed bed at their York site. It was named *C. speciosum* 'Album' and the first 'roots', as they were called, were sold for five guineas. Bowles wrote that he would rather have been the raiser of *C. speciosum* 'Album' than to have owned a Derby winner. Even today, this cultivar is regarded by many as the finest colchicum. The same nursery, then known as J. Backhouse & Son, also selected and marketed *C. speciosum* 'Atrorubens', which was found in 1900. Both hold the RHS AGM.

Later in the 20th century Kees Visser, of the family bulb nursery Visser near Alkmaar in the Netherlands, selected a number of colchicums, many of which are still available. As with Kerbert, the selections were a minor part of the nursery's output. His most distinct selection was *Colchicum* 'Harlekijn' in 1988. This was the first bicoloured cultivar, having tepals with a pink base and creamy white tips. He also selected *Colchicum* 'Zephyr' in 1985, which is probably a *C. bivonae* hybrid, with quite dark pink, tessellated flowers. One of his earlier selections, in 1977, was *Colchicum* 'Antares', a goblet-shaped flower with mid pink on the outer surface of the tepals and striking ivory-white on the inner surface.

SELECTION IN THE EARLY TO LATE 20TH CENTURY

In the 1930s, Robert Ormston Backhouse, based at Sutton Court in Herefordshire but related to the York nursery dynasty mentioned above, selected at least two outstanding colchicums. *Colchicum* 'Huxley' is the most widely grown example, with large, rounded flowers late in the season. The other is *Colchicum* 'Darwin', one of the most vibrant and late-flowering of all colchicums, but very slow to increase. Both are sometimes listed as cultivars of *C. speciosum*.

The contrasting white and pink tepal surfaces are a dramatic feature of 'Antares', selected by Kees Visser in 1977.

In the 1940s Barr and Sons of Covent Garden were selling colchicums such as 'Rosy Dawn', their own selection, introduced in 1948, and some of the Zocher & Co. hybrids. All these were offered at the modest prices of either one shilling or one shilling and two pence. Nearly 50 years later, Van Tubergen of Haarlem in the Netherlands were selling 16 species and cultivars of autumn-flowering colchicums, including *C. speciosum* 'Album', at 10 times the price, which was still a bargain. Their description of *C.* 'Violet Queen' seems to equate with the original and they were just £8.40 for ten.

In the first half of the 20th century, garden writer Edward A. Bowles trialled and selected many plants at Myddelton House near Enfield in Middlesex. In 1924 he wrote *A Handbook of Crocus and Colchicum for Gardeners*, which was updated in 1952, just two years before his death. Up to now, this was the only book about colchicums available. Thanks to his assiduous research and conclusions, it is still a very useful reference today. He raised the question of the perceived differences between *C. speciosum* and what were then known as *C. bornmuelleri* and *C. giganteum*. He suggested using '*Colchicum speciosum* group' as a naming solution for these plants in cultivation.

Bowles grew all available colchicums and corresponded with other growers, often visiting their gardens and receiving them at Myddelton House. When describing the large-flowered species he mentions the differences between *C. sibthorpii* and *C. bowlesianum* and illustrates the latter with his usual skill. Both are now regarded as synonyms of *C. bivonae*. There is a fine, late-flowering, hybrid cultivar called 'E.A. Bowles' that is still available. A later flowering selection, it seems to have an extra shine to the flowers, especially when seen in sunlight.

A contemporary of Bowles was Richard Durant Trotter, usually known as Dick Trotter. He was a banker and plant-hunter who cherished colchicums. Bowles wrote how well *C. speciosum* 'Album' grew at Leith Vale, the Surrey garden of Trotter, unlike in his own garden. Trotter was lucky enough to regularly obtain seed from this white colchicum which he sowed on his compost heap. This unusual practice paid off when he selected 'Pink Goblet', a fine, truly pink-flowered selection with a rounded shape.

Luckily for later generations of colchicum growers, Dick Trotter passed his love of the genus on to his daughter Elizabeth Parker-Jervis. With her husband Johnnie she ran P-J Nursery, founded in 1970, at Longworth in Oxfordshire. At the time her nursery was an important source of colchicums and snowdrops, in the days when they were less widely available. It was a true nursery, propagating a selection of choice plants and majoring on large-flowered colchicums. Their collection was extensive, with plants from Brin as well as Myddelton House, where Elizabeth had spent many days as a child. This collection was almost a precursor of National Plant Collections overseen by Plant Heritage today. The colchicums were sold at

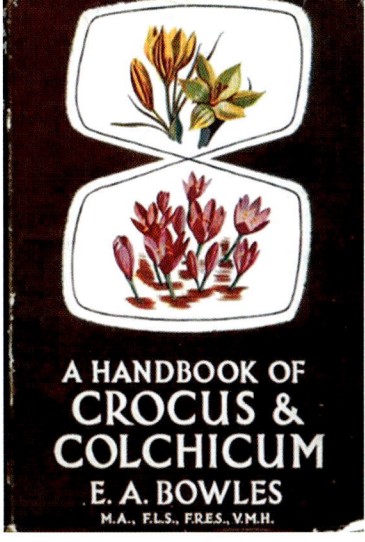

RIGHT **From seed sown on a compost heap, Dick Trotter raised 'Pink Goblet', noted for its rounded flowers.**

FAR RIGHT **Previously, the only book on *Colchicum* was *A Handbook of Crocus & Colchicum* by E.A. Bowles; this is the 1955 edition.**

flowering time so identification was assured. The nursery continued until 1990, when Johnnie died.

Another person with a keen and influential interest in cultivated *Colchicum* during this period was Richard Nutt. He gardened at Great Barfield, near High Wycombe, but for a long time also maintained a second garden, Thornsett in Dore, just south of Sheffield, to which he regularly commuted. He championed many plants, from *Anemone nemorosa* through to *Pulmonaria*, yet his abiding affections were split between *Galanthus* and *Colchicum*. He delivered the Scottish Rock Garden Club Clark Memorial Lecture on these genera in November 1970. In this he dated his interest in *Colchicum* from the early 1960s, declaring that most were grown in a rather large 'cabbage patch', and that while 'they did seem rather expensive per bulb, ... convert it into price per square inch of flower and they are decidedly cheap' (Nutt 1971). He planned to write monographs on both genera but numerous other commitments prevented this. His intellect was easily up to the job but dyslexia was a further hurdle. He gave sage advice from the lectern: 'It is the individual gardener ... who really has the love and patience to keep these plants in cultivation ... if anyone is to try to sort out the genus he needs to know who grows what, under which collector's numbers and, of course, be able to visit'.

WILD COLLECTIONS IN THE 1960s

The 1960s was a fruitful decade for plant introductions from Turkey and nearby countries to the east. It should be noted that nowadays, most countries have regulations concerning the collection of plants and seed from the wild, and permits are required. Introductions in the 1960s were instigated chiefly by Rear-Admiral Paul Furse and his wife Polly, who several times drove overland

ABOVE LEFT **Edward Bowles cultivated a large collection of *Colchicum* at Myddelton House in Middlesex.**

ABOVE RIGHT **Dick Trotter, who grew many colchicums while living in Surrey, seen here in his later garden at Brin House in Inverness-shire.**

to Turkey, and afterwards on to Iran and, at length, Afghanistan. Furse was also a talented amateur botanical artist and painted many of his collections. Around 500 of his paintings are held in the RHS Lindley Library.

In 1962 the Furses spent seven months in Iran and Turkey, collecting plants and herbarium material for RHS Garden Wisley and Royal Botanic Gardens, Kew. This followed a joint trip in 1960 with RHS editor Patrick Synge but 'extending the areas and varying the seasons' (Furse 1963). They relied heavily on their Land Rover, nicknamed 'The Rose of Persia', 'sleeping in her for about 180 nights in all sorts of places, ranging from deep snow drifts to hot deserts'. In March, working down the western side of the Zagros Mountains, near to Kazerun in Iran, they found 'an interesting dwarf colchicum (PF 1276) close to *C. hirsutum*, with up to 15 small, whitish flowers and very long, narrow leaves, almost like a *Crocus*'. Later, between Kermanshah in Iran and the Iraq border, on rocky slopes, they found *C. crocifolium* growing with *Bellevalia glauca*, muscaris, ornithogalums, biarums and dwarf *Prunus*. Above Khorramabad in Iran they collected a spring-flowering

Paul Furse and his wife Polly in the 1970s.

Paul Furse gave advice to the participants of the Bowles Scholarship Expedition to Turkey in 1963. Prior to setting off from RHS Garden Wisley, the participants were (left to right) David Pycraft, Stuart Baker, David Barter and Brian Mathew.

species (PF 1542 and other numbers) similar to *C. szovitsii* 'with bunches of white or colchicum-pink flowers, which grew best in fields of heavy earth, sodden with melting snow when the flowers were out, but baked hard in summer'.

The 1964 Furse expedition to Iran and Afghanistan saw four consignments of material sent back, from April onwards. The first included an undetermined *Colchicum* species from scree at the edge of pine woodland at the Cilician Gates, a pass through the Taurus Mountains, and *C. crocifolium* noted as common in 'fallow earth' at 1,800m on the Paytagh Pass. Both had the ominous observation 'bad corms' under the column headed 'Material'. The third consignment, PF 8020, from Afghanistan, contained plants listed as: *Merendera*, so presumably *C. robustum*, growing with alliums and dwarf tulips at 1,800–3,300m on the Salang Pass; *C. kesselringii* PF 8051, which was growing adjacent to the very distinctive, verticillate, over-arching *Allium regelii* on grit slopes at 2,400m in the same location two days later (24 June); and another collection, as seed, of *C. kesselringii*, PF 8211, from rock crevices at 2,400m in Farqar. Other finds were made through to July, but neither seed nor corms were gathered. Various of the Furses's other finds persist in cultivation, but only a handful of their *Colchicum* introductions have endured.

The Furses advised and encouraged the Bowles Scholarship Botanical Expedition (BSBE) to Turkey in 1963. The participants were Stuart Baker, David Barter, Brian Mathew and David Pycraft, and the trip was detailed in Mathew (1965). They too encountered climatic extremes. In eastern Turkey the snow was 6m deep above the summit road over the Zigana Pass. Shortly after that it was so cold that eggs froze in their frying pan. In north-west Iran, close to Kuh-i-Savalon, west of Ardabil, by late March they found *C. trigynum* in pink and white forms, intermixed with *Iris reticulata* and various *Gagea* in pumice dust screes. A day or two earlier they had come upon, alongside *Iris persica*, 'Merenderas ... rather uninteresting plants, having many flowers to a bulb, with very narrow perianth segments flopping untidily on the ground'.

Still in Iran, at the snow line on Kuh-i-Ushtaran, in early April, *C. szovitsii* was in flower, surrounded by tulips, not yet in flower this early on, by the thousand. That May, in the mountains between Lake Rezaiyeh (now known as Lake Urmia) and the Turkish border, they found *Crocus aerius* in flower with *Merendera trigyna* (now *Colchicum trigynum*) and *Iris bakeriana*. They had already made collections of all three as ripened bulbs 300m lower down the hillside. Later, on granitic Kuh-i-Alwand, above Hamadan in Iran, they headed for the summer pastures where *C. szovitsii* was plentiful, growing in a sward made up of small alchemillas, potentillas and buttercups, with *Primula capitellata* lining the peaty stream banks.

Paul Furse painted many of his finds, including *Colchicum szovitsii* which he found in Iran.

The BSBE plans were successively skewed by a 'minor revolution' in Tehran, an attempted coup in Ankara, and expired visas necessitating a 300km drive to have them renewed. In general, Brian Mathew noted that: 'Bulbs which had been previously dried were usually successfully transported but any green material was in general destroyed before arrival at Wisley'. But even the dried collection was at risk towards the end of the expedition in the autumn of 1963, when having travelled overland back west to the Greek border the 'entire collection was lumped into a huge polythene sack, Customs-sealed, and we were told to go straight through Greece without stopping to collect further'.

Brian Mathew returned frequently to the eastern Mediterranean and beyond in subsequent years. A lengthy 1965 trip to Turkey and Iran with John Tomlinson was very productive. *Colchicum szovitsii* M&T 4530 from high on Mount Erciyes in Turkey was just one of many successful introductions from this.

Paul Christian regards *Colchicum szovitsii* as the best spring-flowering species for outdoor use.

WILD COLLECTIONS IN THE 1970s and 1980s

Martyn Rix completed a doctorate on Greek and Turkish species of *Fritillaria* at the University of Cambridge in 1971, and is a leading authority on that genus. But, while studying these in the wild, he has brought into cultivation many colchicums. This is evident from his book *Bulbs* (Phillips & Rix 1989) which includes images of *C. bivonae* Rix 631, *C. serpentinum* Rix 1591, *C. troodii* Rix 1558 and *C. variegatum* Rix 1262. He elsewhere (Rix 1983) states: 'The finest display of wild bulbs I have seen was in Turkey, in an area close to some cowsheds. The ground was eaten bare of grass, and well fertilised. The *Colchicum speciosum* had thrived to form a carpet of flowers of all shades from palest pink to rich crimson-purple'.

Paul Christian of Rare Plants nursery (formerly Pentre Nurseries) in Wrexham, North Wales, has long championed the genus. His first significant nursery listings were produced in the second half of the 1970s. His subsequent *Varieties List*, first published in 1983, covered the nursery's extant and prospective range of dwarf bulbs and initially included 30 *Colchicum* entries. Some of these were garden hybrids but the list also included species scarce in cultivation such as *C. corsicum*, *C. kotschyi*, *C. micranthum*, *C. szovitsii* ('our best outdoor spring species') and *C. turcicum*. These entries include species he introduced either independently (PJC prefixes), on a 1982 trip to Greece with David Elliott and Antoine Hoog (CE&H), by the latter alone (AH), or involving Paul Christian and Antoine Hoog (CH). This last venture, to Greece in 1987, yielded, among others, *C. haynaldii* CH 871 from north of Konitsa, found at 1,550m in alpine meadows overlying shale. Other offerings over the years have included *C. bivonae* AH 9139 (from Mount Giona), *C. macrophyllum* AH 9806, *C. persicum* (rather surprisingly recommended for outdoor use in a sunny raised bed), *C. sanguicolle* ('reliably hardy here') and *C. tenorei* ('true stock').

The advent of the Brickell (1984) account of *Colchicum* in Peter Davis's *Flora of Turkey* and the Persson (1992) account in Karl Rechinger's *Flora Iranica* made identifying expedition material much easier. Peter Davis, a botanist at the University of Edinburgh, was a scholarly man, somewhat sardonic, a determined explorer and had a trained horticultural eye. He had served an apprenticeship at Ingwersen's Alpine Plant Nursery in East Grinstead, West Sussex, before setting off for the eastern Mediterranean in 1937. This trip served as his introduction to Turkey, to Crete (where he spent all of July) and to malaria, which he developed soon after reaching the island. When he returned to England it was in the company of 'a five-foot sack of uncleaned seeds, boxes of bulbs and three fat flower-presses' (Davis 1939). On a trip two years later, he made the type collection of *C. peloponnesiacum*. However, it was the Turkish rather than the Greek flora that truly enthralled him. His tenure in the botany department at the University of Edinburgh from 1950, when he was appointed as a lecturer, facilitated numerous extended visits to that country. He lent his name to many species, but in *Colchicum* he is remembered by *C. davisii*, a species from Adana in Turkey.

The research team that worked on the *Flora of Turkey* from 1961 to 1985 under Davis's editorship included the similarly enterprising Oleg Polunin, a teacher and botanist perhaps best known for writing various field guides. Others who came within Davis's orbit included John Watson, mastermind of the botanically and horticulturally ground-breaking AC&W (Albury, Cheese and Watson) trip to

Colchicum davisii was named for Peter Davis, editor of the *Flora of Turkey* published in the 1980s.

Jim and Jenny Archibald undertook extensive seed-collecting expeditions to many parts of the world. Their seed lists ensured wide distribution of their collections.

Turkey in 1966. *Fritillaria alburyana* tops the list of their bulbous discoveries, but in the context of this book *C. boissieri* AC&W 2352 was a significant collection, seen in September on top of Mount Ida, flowering around the fringes of spiny *Dianthus webbianus* cushions. Chris Brickell, a former director general of the RHS who wrote the *Colchicum* account in Davis's *Flora of Turkey*, also undertook notable fieldwork. One of his collections, made with Brian Mathew in late 1980 in Crete, was *C. cupanii* B&M 10149.

WILD COLLECTIONS IN THE LATE 20TH CENTURY

Mike Salmon ran Monocot Nursery from a couple of Somerset locations, most enduringly from Jacklands Bridge in Tickenham. From the 1970s to the 1990s he collected throughout the Mediterranean, including from North Africa to Israel. These expeditions were either on his own or with Jim Archibald, Peter Bird, John Blanchard, Mark Fillan or Chris Lovell. He annually sent out both seed and plant lists, and among the bulbs offered in his 1997–1998 list were: *C. autumnale* var. *algeriense* (now *C. lusitanum*) AB&S 4353 (Ifrane, Morocco), *C. cilicicum* MS&CL 541 (Yakapinar, Turkey), *C. corsicum* Watt s.n. (Vizzavona Pass, Corsica), *C. levieri* (now *C. lusitanum*) MS 937 (Esterel, France), *C. neapolitanum* var. *micranthum* (now *C. longifolium*) AB&S 4522 (Tischka, Morocco), *C. peloponnesiacum* MS&CL 195 (southern Greece), *C. procurrens* (now *C. boissieri*) M&T 1862 (Turkey), *C. psaridis* MS&CL 198 (Limeni, Greece), *C. pusillum* (three Cretan collections, all his own) and *C. variegatum* PB 408 (Samos). His 1999–2000 seed list included most of these, but was also noteworthy for its five listings of *C. lusitanum*, one from Puerto de Opacua, Spain, and four from Morocco; *C. neapolitanum* from Larderello, Italy, and what was then Yugoslavia, and two Moroccan offerings of *C. neapolitanum* var. *micranthum* (now *C. longifolium*) AB&S 4422 and AB&S 4493.

Mike Salmon's longstanding friend Jim Archibald developed a fondness for *Colchicum*. On his landmark 1966 odyssey, during which he travelled overland from southern Spain to Iran, he made collections from Morocco (*C. triphyllum* JCA 825), Libya (*C. ?ritchii* JCA 903 and JCA 973) and Iran (*C. szovitsii* JCA 1147 (white), JCA 1360, JCA 1362, JCA 1741, JCA 1796 and four others, through to JCA 2760C from near Esfahan).

Jim Archibald and his second wife Jenny subsequently sent out seed lists, invariably using the prefix J&JA, once or twice a year from 1983 to 2010. These were far and away the best commercial source of accurately named material over several decades. The lists included rare items that they and their close circle of friends gleaned from the Mediterranean through to Iranian and Central Asian localities. The August 2005 list offered a bumper crop, running to several pages and offering, among others: *C. decaisnei* J&JA 313.708 (Jebel Nusairi, Syria), *C. feinbruniae* J&JA 314.250 (Laqlouq, Syria), *C. kurdicum* J&JA 314.789 (Siabishe, Mazandaran Province, Iran), *C. persicum* J&JA 316.707 (Kurdistan province, Iran, among oak scrub on shale, and noted as 'widespread in central Iran and extending westwards across the Syrian desert to Lebanon'), *C. polyphyllum* (above Hasanbeyli, Nur Mountains, Turkey, at 1,100m 'in heavy red clay among deciduous *Quercus* and *Styrax* scrub') and *C. varians* J&JA 3178.700 (Isfahan, south-east of Khonsar, Iran, at 2,600m on steep shale slopes, 'close to *C. szovitsii* but a plant of drier habitats with more, narrower leaves)'.

In June 2005 Archibald had travelled to north-west Iran with Norman Stevens, having previously gone to Uzbekistan and Kazakhstan with him three years earlier. Here they garnered, in addition to the species listed above, *C. kotschyi*, *C. speciosum* (overwhelmingly those in cultivation are from north-east Turkey, but this was from Gilan province in Iran, south-west of Asalem, at 2,000m 'among bracken in *Carpinus* woodland'), *C. szovitsii*, *C. wendelboi*. These were all included in the 2005 Alpine Garden Society Seed Exchange, along with *C. trigynum* from open, stony slopes at 1,900m north-east of Tabriz. The last of these was privately circulated in an addendum, along with diverse collections of *Iris reticulata* and others, under the handwritten subtitle: 'These are the collections they did not get'.

Rare species such as *Colchicum antilibanoticum* have been successfully grown by Bob and Rannveig Wallis.

WILD COLLECTIONS AND SELECTION IN THE EARLY 21ST CENTURY

Bob and Rannveig Wallis joined Archibald on his last trip, to the Georgian Caucasus in 2010. The Wallises met as undergraduates at the University of Bristol, moved to Sussex and later relocated to Carmarthen in Wales. They have established themselves in the top rank of exhibitors at Alpine Garden Society and Scottish Rock Garden Club shows. Travelling widely over the past 40 years, they have made introductions of *Colchicum* from Morocco through to Syria, Lebanon, Turkey, and more recently Kyrgyzstan. Rarely grown species such as autumnal *C. antilibanoticum* from Syria (it also occurs in Lebanon, Jordan and Israel) have settled down with them, and Rannveig has annually issued a late-summer list of bulbs, *Buried Treasure*, making available a select range of the genus.

In the 21st century there has been a resurgence of interest in colchicums in Europe. Most selections are now coming from Germany and eastern Baltic states such as Latvia and Lithuania. However, Antoine Hoog's *C. hungaricum* 'Velebit Star' is a rare example of a Dutch selection, albeit collected in the Velebit range of mountains in Croatia. It is seed-raised rather than clonally propagated and varies from white through to pale pink. Also from the Netherlands, *C. doerfleri* 'Valentine' (often incorrectly listed as a cultivar of *C. hungaricum*) is pale rose-pink. Named for its propensity to flower around St Valentine's Day on 14 February, this does depend on where you garden. Jānis Rukšāns, a nurseryman and bulb expert based in Latvia, has raised and offers a number of selections. In particular, he has named some superb selections of *C. szovitsii* collected by Arnis Seisums from Armenia, such as 'Snowwhite', 'Tivi' and 'Vardahovit'.

Jānis Rukšāns also participated in the 2010 Archibald and Wallis trip to Georgia. Long before then he established his credentials as a nurseryman specialising in *Colchicum* and other genera of interest to bulb growers. His Latvian business, at first known as Bulb Nursery, is more recently called Rare Bulb Nursery. This rebranding was fully justified, for he has single-handedly pioneered the distribution of all manner of hitherto unavailable bulbs, those sourced from Turkey through to Central Asia in particular. Many have been introduced as a result of his far-ranging travels, and also collaborations with Dimitri Zubov from Ukraine and Václav Jošt and Jiří Bydžovský from the Czech Republic. He has distributed accurately identified material of notoriously misidentified species, such as *C. laetum* from Tbilisi, Georgia. He has also pioneered superior selections of *C. szovitsii* from Armenia and the Nakhchivan Autonomous Republic, along with rarities such as *C. hirsutum* (near Arzpli, Turkey), *C. minutum* HZ 8819 (Aydos Dagi, Turkey),

'Vardahovit' is a cultivar of *C. szovitsii* found by Arnis Seisums in Armenia.

C. munzurense KPPZ 208 (Munzur, Turkey) and *C. serpentinum* (Tortum, Turkey).

Jānis Rukšāns has also worked with fellow Latvian Arnis Seisums, whose expertise is principally in *Allium*, *Corydalis* and juno irises. In 1989 Seisums discovered plants deduced to represent *Colchicum kesselringii* × *C. luteum* near the airfield of Tavildara in Tajikistan. Several clones were selected, two named 'Jānis' and 'Jeanne', and have been passed around. They have also proven fertile, if not spectacularly so, and F_2 seedlings come in every shade from cream to yellow, some with just a faint trace of the median external perianth stripe often seen in *C. kesselringii*. Karin Persson (2007) discusses these hybrids and refers them to *C. albertii* (styled *C.* × *alberti* in this book), a taxon first noted from Kyrgyzstan by Regel in June 1880 'having violet flower tubes and whitish to yellow limb segments'. She also states that the Kyrgyzstan and Tajikistan plants are the only two examples of this hybrid she is aware of. Unfortunately, the epithet *alberti* has been widely treated as a synonym of *C. luteum*. Because there is good reason to regard these plants as very occasional natural hybrids, it would surely be more sensible to agree upon a new hybrid epithet – preferably one that recognises the fieldwork of Arnis Seisums three decades ago. The hybrid has also been recreated in cultivation by Ian Young in Aberdeen and perhaps others.

'Fuller's Mill' was named by RHS trial judges in 2017 when they spotted it at the garden of the same name belonging to the late Bernard Tickner.

In Lithuania, Leonid Bondarenko has crossed *C. kesselringii* 'Snow of Highland' with *C. luteum* to give *C.* × *alberti* 'Lucky Selfmade'. He also pollinated *C. luteum* with pollen from *C. kesselringii* 'Yeti' to produce *C.* × *alberti* 'Moonlight', giving flowers of pure, pale yellow. It is rumoured that Bondarenko has succeeded in storing *C. luteum* pollen under refrigeration and has used it to pollinate autumn-flowering hybrids, resulting in successful seedlings.

Jan Jilich from Bludov in the Czech Republic has offered an astounding range of *Corydalis*, *Eranthis*, *Fritillaria* and *Iris* especially, but with a distinguished list of *Colchicum*. Some are of Afghan provenance, echoing the notable flourish of bulb discoveries made in that country in the 1960s and early 1970s. His 2019 online catalogue listed six accessions for *Colchicum kesselringii* from across its distribution, Afghanistan included, while under *Merendera*, three collections of *C. robustum* were offered, along with an unidentified species from Kandahār province, which by implication is either new to the Afghan flora, or to science.

Several cultivars were named in 2017 as a result of an RHS trial of large-flowered *Colchicum* at RHS Garden Hyde Hall in Essex. These plants had been submitted to the trial without established names, but were recognised by the judges as outstanding and deserving of a cultivar name. They include 'Boxford', 'Chequers', 'Felbrigg', 'Fuller's Mill' and 'Redgrave'.

PARENTAGE OF CULTIVARS

While there are many cultivars selected from single species, such as *C. autumnale* 'Album' or *C. macrophyllum* 'Cretan White', the parentage of the majority of hybrid cultivars is obscure. Most of the crossing that led to these hybrid cultivars happened spontaneously on nurseries, and few records have been kept as to the origin of new seedlings when they appeared. Suggestions of parentage have often been speculative, primarily concentrating on which species were growing nearby. However, what is certain is that relatively few species have been involved. These include *C. autumnale* in its various forms, *C. bivonae*, *C. cilicicum* and *C. speciosum*. Of these it is Turkish-Caucasian *C. speciosum* that has had probably the greatest input. *Colchicum speciosum* is very variable in the wild and several variants from the wild have been introduced to gardens. These readily cross with one another, and with other species in cultivation. Added to this is the fact that many hybrids have had the chance in cultivation to cross with one another, or backcross to one or other parent. In these situations a very complicated picture of origin soon builds up.

Another species that should not be overlooked in this connection is *C. turcicum*. Although it has rather smaller flowers than most of those mentioned in the previous paragraph, they are of an intense reddish purple colour. This shade may have been imparted by chance to some of the darker-flowered cultivars such as 'Benton End' and 'E.A. Bowles', although it could be equally well argued that some of the darker-flowered forms of *C. speciosum* could have had a similar and perhaps more significant influence.

Interestingly, Patrick Synge (1961) comments that 'A number of hybrids and garden forms have been raised from crosses made between this sp. [*C. speciosum*], *C. giganteum* and the large flowered tessellated species from Greece, *C. sibthorpii* and *C. bowlesianum* [both now recognised as *C. bivonae*]. They have, however, inherited the good garden constitution of *C. speciosum* and have varying, although rarely very conspicuous tessellated petals. It is these forms that are found most commonly in gardens. Several of them, however, are so close to each other in colour, season of flowering and habit that it is hardly possible to tell them apart, far less to describe on paper the differences.'

Today, the only way to resolve some of these issues is through careful morphological analysis linked to genetic studies. As molecular analyses of the majority of species have been published, it should be possible to examine the cultivars and at least try and understand their origins.

Cultivars such as 'Huxley' have a silky sheen to the surface of the tepals.

HYBRID CHARACTERISTICS

It is possible to make various comments on some of the cultivars found in gardens today. Below are comments on hybrids and hybrid cultivars of known or putative parentage, none of which are of wild origin.

1. *Colchcium × agrippinum* (2n=45) = *C. autumnale* (2n=36) × *C. variegatum* (2n=54).
2. *Colchicum × alberti* (2n=54) = *C. kesselringii* (2n=54) × *C. luteum* (2n=54).
3. *Colchicum × byzantinum* (2n=45) = *C. cilicicum* (2n=54) × ?*C. autumnale* (2n=36). The white-flowered cultivar 'Innocence' belongs here.
4. *Colchicum* 'Lysimachus' (2n unknown) = purportedly *C. autumnale* (2n=36) × *C. haynaldii* (2n=96).
5. *Colchicum × ambiguum* (2n=72) = *C. cilicicum* (2n=54) × ?*C. neapolitanum* (2n=90).

Cultivars with pronounced chequering on the tepals, such as 'Autumn Queen', 'Disraeli' and 'Glory of Heemstede', are likely to have *C. bivonae* and *C. speciosum* in their ancestry.

Cultivars with intense colouring and goblet-shaped flowers, such as 'Benton End', 'E.A. Bowles' and 'Glory of Threave', are derived from *C. speciosum* and possibly also have *C. turcicum* in their ancestry.

Cultivars with pale flowers and a silky surface, such as 'Dick Trotter', 'Huxley' and 'Pink Goblet', have *C. speciosum* in their constitution, which has been crossed or backcrossed with other cultivars.

Many of the remaining cultivars are of such complex or mysterious origins that it would be misguided to speculate without further research.

KEEPING TRACK OF CULTIVARS

Numerous *Colchicum* selections have become muddled in cultivation. There is often no definitive description of cultivars as there would be for a species. Buying *Colchicum* cultivars, which all appear very similar as corms, can therefore be a bit of a lottery.

This is also because the identification of colchicums is fraught with difficulties. The colour of the tepals varies depending upon the amount of sunlight received, and the stigmas often darken as they age. Artists tended to ignore colchicums, probably because spring has been their inspiration and not autumn. Even Jan Brueghel the Elder (1568–1625), who mixed seasons hugely in his paintings of exuberant flower vases, did not seem to include any colchicums. The Dutch nursery of G.C. Meeuwen, established in 1856, published a good coloured plate depicting 'Glory of Heemstede', 'Lilac Wonder', 'Violet Queen' and 'Waterlily' in the early 20th century. In a 2018 exhibition at the Garden Museum in London of Sir Cedric Morris's (1889–1982) work there was a painting, *Flowers in a Vase* (1930), which included *Colchicum autumnale*. Morris was an artist and plantsman who grew many of the selections of *Colchicum* available in the 20th century in his garden at Benton End near Hadleigh in Suffolk.

There are some good descriptions of cultivars and selections of species written by E.A. Bowles in his books (1915, 1924). Mentioned earlier, Patrick Synge's *Collins Guide to Bulbs* (1961) was an excellent and erudite volume for gardeners. Of the coloured plates painted by Paul Furse, one included an assortment of colchicums, including the amazing, tessellated 'Disraeli'.

A number of cultivars seem to have disappeared from cultivation. The original descriptions of them were often poor and it has been impossible to assign current plants to those original names. Most were raised in the first half of the 20th century, although some a little later, and were primarily of British or Dutch origin. They include the following: 'Danton', 'Ferndown Beauty', 'Glorie van Holland', 'Guizot', 'Hidegkut', 'James Pringle', 'Klondike', 'Naeisanum', 'Petrovac', 'President Coolidge', 'Purity', 'Ruby Queen', 'Surprise' and 'W. Kerbert'. It is possible that some of these linger on in gardens unrecorded. What is certain is that they are no longer commercially available.

Classification

Colchicum contains 104 species as currently understood by most modern researchers. This figure includes species that have previously been assigned to other genera, such as *Bulbocodium* (1 species) and *Merendera* (11 species). *Androcymbium* is maintained as a separate genus here, as explained later in this chapter. This chapter also summarises the relationship of *Colchicum* to other genera and its family placement.

FAMILY CLASSIFICATION

For many years the genus *Colchicum*, along with its related genera *Androcymbium*, *Bulbocodium* and *Merendera*, was included within the large and diverse lily family, *Liliaceae*. In the latter part of the 20th century intensive anatomical studies saw the division of the *Liliaceae* into a number of separate, though closely related, families. These include *Alliaceae*, *Alstroemeriaceae*, *Asparagaceae*, *Asphodelaceae*, *Colchicaceae* (which includes *Androcymbium*, *Bulbocodium*, *Colchicum* and *Merendera*), *Hyacinthaceae*, *Liliaceae*, *Melanthiaceae*, *Philesiaceae* and *Smilacaceae*. Subsequently, further reorganisation and repositioning of these families and genera has taken place, based extensively on DNA sequencing information.

The *Colchicaceae* is now included in the *Liliales* order, which contains the following families: *Alstroemeriaceae*, *Campynemataceae*, *Colchicaceae*, *Corsiaceae*, *Liliaceae*, *Melanthiaceae*, *Petermanniaceae*, *Philesiaceae*, *Ripogonaceae* and *Smilacaceae*. The other former *Liliaceae* families mentioned above (*Asparagaceae*, which now includes *Hyacinthaceae*, and *Asphodelaceae*) are now included in the *Asparagales* order, which

LEFT **A coloured engraved plate from volume 4 of Pierre-Joseph Redouté's *Les Liliacées* depicting *Colchicum autumnale*.**

RIGHT **Although now in its own family, *Colchicaceae*, the genus *Colchicum* was once assigned to the *Liliaceae* when it was a much larger family.**

also includes other large families such as *Iridaceae* and *Orchidaceae*.

The *Colchicaceae* contains around 15 genera and almost 300 species, many of which are geophytes. The two largest genera are *Colchicum* itself (with or without the inclusion of *Androcymbium*, *Bulbocodium* and *Merendera*) and the Afro-Australasian genus *Wurmbea*. These two genera account for more than half the species in the family. The other genera are *Baeometra*, *Burchardia*, *Camptorrhiza*, *Disporum*, *Gloriosa*, *Hexacyrtis*, *Iphigenia*, *Kuntheria*, *Ornithoglossum*, *Sandersonia*, *Schelhammera*, *Tripladenia* and *Uvularia*. The family has a wide distribution, including Europe and the Middle East to central Asia, China and Japan, Africa (mostly tropical east and southern Africa), to south-east Asia, Australia (including Tasmania) and New Zealand.

FAMILY CHARACTERISTICS

So diverse are the genera of *Colchicaceae* that it is difficult to see what holds them together morphologically, whatever gene sequencing might indicate. The characters that unite the genera are:

1. Simple leaves with parallel, generally arched veins, with sessile blades that are sometimes amplexicaul or auriculate at the base.
2. Flowers bisexual (rarely unisexual in *Wurmbea*), mostly actinomorphic, rarely slightly zygomorphic.
3. Tepals nearly always 6 (in two series of 3), occasionally up to 12, more or less similar in size and shape, free or partially fused.
4. Stamens 6, free or fused at the base to the tepals, with dorsifixed anthers (basifixed in some species of *Colchicum*) that open by longitudinal slits.
5. Ovary superior, 3-locular, the three carpels fused or partially fused.
6. Styles 3, free or partly fused.
7. Fruit a septicidal or loculicidal capsule, but a berry in *Disporum*.

Wurmbea is the second largest genus in the *Colchicaceae*, with about 50 species. This is *Wurmbea stricta*, native to South Africa.

KEY TO GENERA OF THE *COLCHICACEAE*

The following key enables identification of all the genera within the family.

1a Plants acaulescent, flowering in advance or with part-developed leaves; flowers with a long perianth tube; ovaries located well below ground   *Colchicum* (including *Bulbocodium* and *Merendera*)
1b Plants caulescent (sometimes with a very reduced stem in some *Androcymbium* species); flowers with a short perianth tube or no tube; ovaries located at ground level or above   2
2a Flowers borne in the axils of the uppermost stem leaves   3
2b Flowers borne in clusters or spikes, or scorpioid cymes, rarely solitary   9
3a Flowers large (tepals at least 6cm long), with reflexed tepals and exposed stamens, ovary and styles; styles bent at right angles to the ovary; leaf tips tendrillate   *Gloriosa*
3b Flowers medium to small (tepals less than 5cm long), tubular-campanulate to campanulate, with stamens, ovary and styles concealed; styles straight; leaf tips tendrillate or not   4
4a Tepals fully united into a lantern-shaped, pendent corolla; leaf tips tendrillate   *Sandersonia*
4b Tepals free; leaf tips tendrillate or not   5
5a Flowers campanulate to rotate, solitary or several at the upper leaf-axils; fruit a capsule   6
5b Flowers tubular or tubular-campanulate, generally 2 or more at the leaf-axils; leaf tips not tendrillate; fruit a berry or capsule   8
6a Perianth campanulate, yellow; leaf-tips tendrillate   *Littonia* (sometimes included in *Gloriosa*)
6b Perianth rotate, white; leaf-tips not tendrillate   7
7a Anthers versatile; leaves elliptical   *Tripladenia*
7b Anthers basifixed; leaves linear   *Schelhammera*
8a Fruit a berry; flowers solitary, paired or clustered; leaves neither amplexicaule or perfoliate, sometimes shortly petiolate   *Disporum**
8b Fruit a capsule; flowers solitary; leaves amplexicaule or perfoliate   *Uvularia*
9a Bracts well-developed, leaf-like, broad, subtending all the flowers   10
9b Bracts absent or reduced, small, slender, not leaf-like and only subtending the lowermost flowers if present   11
10a Flowers several in a raceme, actinomorphic or zygomorphic, often nodding, yellow, green, purple or brown, often bicoloured; bracts small and green, not surrounding the flowers; styles slender, curved and spreading   *Ornithoglossum*
10b Flowers tightly clustered, white, often pink-lined, greenish or brownish and surrounded by bracts; bracts often enlarged, leaf-like or petal-like, white or greenish; styles short and straight   *Androcymbium*

11a	Flowers borne in distinct spikes or spike-like scorpioid cymes, with or without bracts	12
11b	Flowers borne in umbels or corymbs, sometimes in a compound inflorescence, bractless	15
12a	Flowers yellow or orange, short-stalked, the lower subtended by a slender bract; styles short and hooked	*Baeometra*
12b	Flowers not yellow or orange, generally white, pink or maroon, sessile, unbracted or the lowermost subtended by a large leaf-like bract; styles not as above	13
13a	Tepals united at the base, not or only slightly auriculate above the claw, persistent; flowers sometimes unisexual, bractless	*Wurmbea*
13b	Tepals free, often auriculate above the claw, persistent or not; flowers always hermaphrodite	14
14a	Tepals oval to obovate with pouch-like nectaries; styles slender, terminal on the ovary	*Onixotis*
14b	Tepals attenuate, wispy, with a pair of auriculate nectaries towards the base; styles short and hooked, arising laterally near top of ovary lobes	*Neodregea*
15a	Filaments greatly expanded (bulging) close to the base	*Camptorrhiza*
15b	Filaments filiform or slightly expanded at the base	16
16a	Flowers borne in a compound inflorescence, nodding, with reflexed tepals and a solitary style, tepals fused at the base	*Hexacyrtis*
16b	Flowers borne in a simple umbel or corymb, not nodding, star-shaped, with 3 short styles and spreading, free tepals	17
17a	Flowers borne in umbels; filaments terete	*Kuntheria*
17b	Flowers borne in corymbs; filaments flattened	*Iphigenia*

* An Old World genus; North American species formerly in *Disporum* are now transferred to the genus *Prosartes* in the *Liliaceae*.

This key below covers just *Colchicum* and *Androcymbium*. The groups formerly included in *Bulbocodium* and *Merendera* are keyed out separately.

1a	Tepals united into a short or long perianth tube; plants flowering in advance of the leaves or with leaves partly developed	*Colchicum* (excluding species formerly included in *Bulbocodium* and *Merendera*)
1b	Tepals free from one another, flowers tubeless; plants flowering when leaves well-developed	2
2a	Leaves bract-like, arising from the top of a stem, present at flowering time, forming a distinct collar surrounding the flowers; tepals with	

	conspicuous swollen yellow glands at the base of the blade; ovary at or slightly below ground level	*Androcymbium*
2b	Leaves not as above, arising from the top of the corm; swollen yellow glands absent from tepals; ovary well below ground level	3
3a	Styles 1, 3-lobed at apex; tepals free but knitted together near the base by small hooked teeth; flowering with the leaves shortly developed; anthers versatile	*Colchicum* (species formerly in *Bulbocodium*)
3b	Styles 3, free; tepals free; flowering in advance of the leaves or with the young leaves showing; anthers versatile or basifixed	*Colchicum* (species formerly in *Merendera*)

CHARACTERISTICS OF *COLCHICUM*

Within *Colchicaceae*, only *Colchicum* (including *Merendera* and *Bulbocodium*) has a much reduced stem and basal leaves, the underground portion of the leaves contributing to the corm tunics. All the other genera in the family have distinct stems with several or more numerous, alternate leaves. The tepals of those species that have always belonged to *Colchicum* unite into a long perianth tube reaching almost to the base of the subterranean corm. Those of the species previously included in *Merendera* and *Bulbocodium* are not united, but instead possess a long claw which also reaches well below ground level. The styles of all *Colchicum* are very long, extending from the flower to well below ground. This feature is unique in the family, dictating that fertilisation takes places below ground level. After fertilisation, a pedicel or peduncle, very short and obscure at flowering time, extends rapidly, pushing the developing fruit capsules to ground level. In some species the capsules extend well above ground level, up to 30cm in some instances. Moreover, only in *Colchicum* and *Androcymbium* are the stamens fused at the base of the filaments to the corresponding tepal.

Colchicum is a European-Mediterranean genus extending into North Africa and eastwards into the eastern Mediterranean, Turkey and Central Asia as far as the western Himalaya and the Tien Shan. *Androcymbium* is a Mediterranean genus (including the southern and eastern Mediterranean region) that extends as far south as southern Africa. All the other genera in the *Colchicaceae* occur in North America, eastern Asia (including Japan), south-east Asia and Australia and southern Africa, with no presence in either Europe or the Mediterranean region.

MERENDERA AND *BULBOCODIUM*

The genus *Merendera* has always been closely associated with *Colchicum*. Different authorities have been at odds over the years over whether the two genera should be merged or kept separate. They both share a very similar distribution and often similar habitats, and both have autumn- and spring-flowering species.

Species hitherto segregated in *Merendera* have chromosome counts of $2n=20$, 22, 24 and 54. *Colchicum* has a huge range between $2n=14$ and $2n=216$. However,

LEFT *Colchicum raddeanum* was originally described as a *Merendera*, a genus that is now included in *Colchicum*.

BOTTOM LEFT *Colchicum bulbocodium* was previously known as *Bulbocodium vernum* and was the only species in the genus *Bulbocodium*.

both have a greater number of species with 2n=54.

Morphological differences between the two genera are scant. The only prime difference is the lack of a perianth tube in *Merendera*, with all six tepals free to the base, and the tepals narrowed below into a long claw. There is a tendency in *Merendera* species for the flowers to fall apart as they age, but two devices often stop this from happening. Firstly, small fimbriae at the base of the tepal lobes act like Velcro, knitting the tepals together and holding them in place so that the claws form a split, pseudoperianth tube. Secondly, in several of the spring-flowering species of *Merendera* the young, partly developed leaves surround the flowers in a collar-like fashion, preventing the tepals from falling into disarray.

It has been noted that in some cultivated specimens of *Colchicum szovitsii*,

partial splitting of the perianth tube occurs and Persson (1992) adds that this was once observed in the wild. *Colchicum szovitsii* has a distinctly furrowed perianth tube that can appear split. In fact, chromosome and DNA evidence suggests that it is closely related to *C. kurdicum* and *C. raddeanum*, both formerly included in *Merendera*. In addition, further evidence suggests that *Merendera robusta* is most closely allied to *Colchicum luteum* and *C. kesselringii*, these three species representing the easternmost extension of *Colchicum* and *Merendera*. In light of this it is illogical not to include *Merendera* within *Colchicum*.

The other onetime genus, *Bulbocodium*, presents a similar problem. It contains a single species, *B. vernum*, with a wide distribution in the mountains of Europe and southern Russia. In leaf and flower *Bulbocodium* looks very like some species of *Colchicum*. It possesses separate tepals (i.e. with free claws) like those species formerly included in *Merendera*, and is only distinguished by the solitary, 3-branched style instead of 3 separate, un-united styles. However, two species of *Colchicum*, *C. luteum* and *C. kesselringii*, also have united styles. This is not obvious from a casual glance, but if the flowers are prized apart carefully it can be seen that the styles unite midway down the perianth tube.

Thus, the prime distinguishing features of both *Merendera* (split perianth) and *Bulbocodium* (united styles) are to be found in species otherwise assigned to the far larger genus *Colchicum* and this is supported by DNA evidence (Persson *et al.* 2011). *Bulbocodium vernum* is now known as *Colchicum bulbocodium*.

With a genus of this size it might possible that *Colchicum* in the broad sense could be divided into subgenera or sections. The most obvious grouping would be to make species formerly in the genus *Merendera* a subgenus. However, that would make key characters such as split perianth tube and stylar features in the adjacent genus *Bulbocodium* difficult to uphold. Instead, the species are generally divided more arbitrarily on whether they flower in advance of the leaves emerging (hysteranthous species) or with the leaves partly developed (synanthous species), thus separating the majority of autumn-flowering species from those that flower during winter and spring. However, there some species that can fit into either grouping, depending on the stage of development, and this can be season-related.

THE POSITION OF *ANDROCYMBIUM*

Androcymbium has had a chequered history since its establishment in 1808, with various authorities either maintaining it as a separate genus or uniting it with *Colchicum*. In recent times Manning *et al.* (2007) redefined *Androcymbium* based on molecular evidence. They came to the conclusion that the genus should be absorbed within an expanded circumscription of *Colchicum*. With this in view they made

Androcymbium palaestinum exhibits some features that distinguish the genus from *Colchicum*, including short styles and purple-striped tepals.

LEFT *Androcymbium vanjaarsveldii* is native to Western Cape in South Africa.

ABOVE *Androcymbium psammophilum* on Fuerteventura in the Canary islands.

the necessary recombinations under *Colchicum*, publishing 61 new nomenclatural combinations. While their studies are clearly detailed and thorough in a limited field, they did not conduct a wider and more detailed analysis of morphology, which we think must play an important role. While no one would argue that *Colchicum* and *Androcymbium* are not closely related, and indeed this was long-established before molecular genetics started to dominate taxonomy, we feel this merger to be unwarranted.

The position of *Androcymbium*, with its 60 or so species, should not be controversial, for the genus clearly differs from *Colchicum* on morphological criteria. Persson (2007) makes the following pertinent observations:

'Another closely related genus, *Androcymbium*, was united with *Colchicum* s. lat. to the tribe *Colchiceae* by Buxbaum (1925, 1936), Nordenstam (1982, 1998) and Dahlgren *et al.* (1985). Molecular analysis of the two genera by Vinnersten & Reeves (2003) and Vinnersten & Manning (2007), showed *Colchicum* (including *Bulbocodium* and *Merendera*) to be nested within a paraphyletic *Androcymbium*, and the conclusion was therefore that the generic status of the two genera needs to be addressed. In another recent molecular phylogeny of *Androcymbium* made with cpDNA non-coding sequences (Del Hoyo & Pedrola-Monfort 2006), it was also found that *Androcymbium* was not monophyletic, and the necessity of a nomenclatural revision of the genera belonging to the *Colchicaceae* was suggested.

However, the material was in both cases rather small regarding the number of taxa studied from one or both genera, particularly for *Colchicum* (incl. *Bulbocodium* and *Merendera*), and the clades were in many cases not very well supported. Furthermore, one of the species of *Merendera* included in the Vinnersten articles, *M. schimperiana* (with its synonym *M. longifolia*), was removed already by Stefanov (1926) from the *Colchicum* alliance into *Androcymbium*, an inference that is supported on morphological grounds [...] This finding is also corroborated by our molecular investigation, the first part of which considered DNA sequences from 6 regions of the plastid genome (Persson *et al.* [2011]). More importantly, this investigation. which comprises c.35 species of *Androcymbium* and next to all *Colchicum* species hitherto recognized, has also shown that taxonomic decisions to change the delimitation of the genus *Colchicum* to include *Androcymbium* were premature. Also, more genes, preferably nuclear genes, will have to be included, particularly as it is reasonable to assume that evolution in *Colchicum* in large parts is reticulate, to judge from the high percentages of polyploid, probably alloploid species: c.75% are polyploid, from 4× up to 24×, 2n=54 being the most frequent number (Persson 1993a, b). Lastly, results will have to be considered in combination with morphological characters.'

It is clear that although *Androcymbium* and *Colchicum* share a common ancestry they have, in the intervening years, diverged and taken different evolutionary pathways, with centres of diversification in both southern Africa and the Mediterranean and Middle East.

It is reasonably straightforward to conduct a molecular study and come to the conclusion that two genera, or more, are similar enough to be merged. However, in most cases the only way to accommodate such mergers is to greatly expand the circumscription of the one genus to accommodate the other. However, in the case of *Androcymbium* and *Colchicum* this does not take into account the important fact that the two genera

TOP *Androcymbium europaeum* is native to southern Spain and northern Morocco.

BOTTOM **Further east, and also with a Europe-Africa distribution,** *Androcymbium rechingeri* **can be found on Crete and in northern Libya.**

do not look at all alike in their morphological profiles. In fact they differ in a number of important criteria which are listed in the table opposite. These differences amount to far more than can be reasonably expected if the genera are to be combined satisfactorily.

Rather than combining *Colchicum* and *Androcymbium*, morphological evidence would appear to point in the opposite direction. Within *Androcymbium* there are two main types:

1. Species with plain tepals (perianth segments) and leaf-like bracts (sections *Dregeocymbium*, *Erythrostictus* and *Marlothiocymbium*).
2. Species with cucullate tepals and elaborated floral bracts (sections *Androcymbium* and *Kunkeliocymbium*).

Viewed in this light, both *Colchicum* and *Androcymbium* are best maintained as separate, though allied, genera. The classification of species in *Androcymbium* is not within the scope of this work and this is up to others to resolve. The purpose of this discussion is to outline the clear differences between *Colchicum* and *Androcymbium* and to firmly reject the inclusion of the latter into the former.

To sum up, *Androcymbium* has several unique characters. Some of the species look superficially like *Colchicum* but plants bear a short to long stem, sometimes rather obscure, with a basal tuft of leaves and a whorl of large, leaf-like, green or white bracts which are sometimes very petal-like in some of the South African species. The flowers are sessile and without a tube and, as a result, the styles are short, although separate. Unlike *Colchicum*, the ovary is located at or above ground level. Another important difference is the presence of idioblasts on the leaves and tepals of *Androcymbium*, which are not present in *Colchicum*, nor in species formerly in *Bulbocodium* and *Merendera*. The idioblasts are present in the leaves and flowers and appear as small, blackish-purple dots, especially visible in dried material, more particularly on the tepals (Stefanov 1926, Buxbaum 1936). Morphologically at least, it would be nonsense to include *Androcymbium* in an expanded concept of *Colchicum*, since the genus has clearly taken a different evolutionary route.

ACCOUNTS OF *COLCHICUM*

One of the first major accounts of the genus was provided by John G. Baker (1879), which provided a solid foundation for future studies. It is a detailed and descriptive morphological study with identification keys. Being a wide-ranging account, it covers 63 genera, including *Colchicum* for which he recognised 29 species. He recognised *Bulbocodium* and *Merendera* as seperate genera.

The first detailed monograph of the *Colchicum* genus alone was published almost 50 years later, in Bulgarian, by Boris Stefanov (1926). Stefanov included *Bulbocodium* and *Merendera* in *Colchicum*, a position that had been taken by earlier authors such as Brotero (1816), Ker Gawler (1821), Bentham & Hooker (1883) and Baillon (1894). Sprengel (1825) took a different view, including

Morphological differences between *Colchicum* and *Androcymbium*

Morphological feature	*Colchicum*	*Androcymbium*
Corms	Hypopogia foot-like	Hypopogia fan-like
Idioblasts*	Absent from leaves and tepals	Present on leaves and tepals
Stems	Absent	Present (sometimes very short)
Pseudostems	Present	Absent
Bracts	Minute, hidden below ground	Large and conspicuous above ground, leaf-like or petal-like
Flowers	Solitary or fascicled	Clustered in bracted heads
Peduncle	Very short, elongating markedly in some species	Absent or very condensed
Pedicels	Undeveloped at flowering	Well-developed at flowering
Perianth tube	Very long and cylindrical	Very short or absent, not cylindrical
Tepals	Variable in size, often large and showy, brightly coloured, flat	Small, whitish (often purple-striped), greenish or brownish, flat or cucullate
Styles	Very long (i.e. 6–30cm or more)	Very short (less than 3.5cm)
Ovary	Located well below ground	At ground level or above
Pollinators	Bees, butterflies, occasionally hoverflies (some flies and ants are casual visitors)	Beetles, flies, rodents**

* Idioblasts are isolated plant cells that differ from neighbouring tissues. They have various functions such as storage of reserves, excretory materials, pigments, and minerals. They contain oil, latex, gum, resin, tannin or pigments etc. Some can contain mineral crystals which may be acrid-tasting and poisonous.

** Many of the South African species of *Androcymbium* (sections *Androcymbium* and *Kunkeliocymbium*; the vast majority of species) have well-defined nectar reservoirs. Strong evidence points to the fact that these have evolved specifically for pollination by rodents (Kleizen *et al.* 2008).

Merendera in *Colchicum* but recognising *Bulbocodium* as distinct, a position shared by few others.

Thereafter, most revisions to the genus appeared in regional floras. These contain a great deal of useful factual and analytical data. Major regional accounts include D'Amato's (1955, 1957a & b) revision of the Italian species, Feinbrun's (1953, 1958) account of Middle Eastern species which also included an overall survey of the genus as then known, and Brickell's detailed accounts of European (1980) and Turkish (1984) species.

Other significant regional accounts include those by Velenovsky (1891), Maire (1958), Pignatti (1982), Meikle (1985), Feinbrun-Dothan (1986), Persson (1992), Franco & Afonso (1994) and Arrigoni (2015).

In recent years, Karin Persson, based at Gothenburg Botanical Garden at the University of Gothenburg, Sweden, has done much for our overall understanding of this complex genus and its allies. Her papers, alone and in partnership with other authors, have taken in a wide range of disciplines, primarily chromosome and pollen studies, but including geography, morphology, pollen and chromosome studies. They were published mainly in the period 1988–2009, culminating in Persson *et al.* (2011).

To explore the wider classification surrounding *Colchicum* and related genera, refer to the Angiosperm Phylogeny Group website (www.mobot.org/mobot/research/apweb).

In this book we have relied heavily on the work of Chris Brickell and Karin Persson to guide our approach to the classification of the genus.

CHROMOSOME BIOLOGY

Persson and colleagues have done a great deal to increase our understanding of the chromosome biology and evolution in *Colchicum*, indeed of *Colchicaceae* in general. While additional species have been described by other authors in more recent years, no further major advancement on Persson's studies have been made since, and it is not intended to repeat that already published information here. Those wishing to delve deeper into this complex subject should refer to the papers by Persson and Persson *et al.*, with Persson (2007) being the best starting point.

Another important paper (Chacón *et al.* 2014) focusses on chromosome numbers. They point out the extraordinary karyological variation within the *Colchicaceae* from 2n=14 to 2n=216. Furthermore, they stress the extremely complex situation found in *Colchicum*, with different species having variable chromosome numbers as well as ploidy levels. Nordenstam (1998) has argued that the presence of colchicine and its effect on chromosome separation during mitosis must be a major consideration. Chacón *et al.* (2014) also point out that, apart from polyploidy that is so prevalent in the *Colchicaceae*, chromosome fission or fusion can take place, processes known respectively as ascending or descending dysploidy. High levels of polyploidy can have a marked evolutionary effect, both on adaptive success in populations but also in the creation of new species (Levin 1983, Abbott *et al.* 2013).

The inferred ancestral number in the *Colchicaceae* is n=7 (2n=14). This base number seems to have been maintained in divergent genera in North America (e.g. *Uvularia*) and Australia (e.g. *Kuntheria*, *Schelhammera* and *Tripladenia*). Several species of *Colchicum* (e.g. *C. antilibanoticum*, *C. crocifolium*, *C. fasciculare*, *C. schimperi* and *C. tuviae*), interestingly all from the eastern Mediterranean region, have this chromosome count. Other genera in *Colchicaceae* have higher basic numbers.

Colchicum records an extraordinary range: 2n=14, 18, 20, 22, 24, 28, 32, 36, 38, 40, 42-44, 46, 48, 50, 54, 58, 90, 96, 108, 140, 144, 162, 180 and 216. Species corresponding to these numbers are listed below, noting that although the majority of species have had their chromosomes counted, a few remain to be assessed. Species marked in **bold** were formerly included in *Merendera*.

- 2n=14 *C. antilibanoticum*, *C. crocifolium*, *C. erdalii*, *C. fasciculare*, *C. hirsutum*, *C. ritchii*, *C. schimperi*, *C. tuviae*
- 2n=18 *C. lagotum*, *C. leptanthum*, *C. serpentinum*, *C. szovitsii* (incl. subsp. *brachyphyllum*) (also 2n=36), **C. trigynum** (also 2n=22)
- 2n=20 *C. kotschyi*, **C. kurdicum**, **C. raddeanum**
- 2n=22 *C. feinbruniae*, **C. ignescens**, *C. polyphyllum*, *C. sanguicolle*, **C. trigynum** (also 2n=18)

Colchicum ritchii, here growing wild in Jordan, is one of only eight species with the ancestral chromosome number of 2n=14.

Here growing wild in Turkey, Colchicum szovitsii is one of the species with two recorded chromosome numbers, 2n=18 and 2n=36.

2n=24 C. munzurense, C. umbrosum, **C. wendelboi**, C. woronowii
2n=28 C. antepense
2n=32 C. chimonanthum
2n=36 C. asteranthum, C. autumnale (also 2n=38), C. cretense, C. szovitsii (also 2n=18)
2n=38 C. autumnale (also 2n=36), C. speciosum (also 2n=40, 42 & 44)
2n=40 C. confusum, C. speciosum (also 2n=38, 42 & 44), C. triphyllum
2n=42 C. graecum (also 2n=44), C. laetum, C. speciosum (also 2n=38, 40 & 44), C. speciosum Bornmuelleri Group
2n=44 C. graecum (also 2n=42), C. minutum, C. peloponnesiacum, C. speciosum (also 2n= 38, 40 & 42)
2n=46 C. davisii, C. varians
2n=48 C. bivonae, C. paschei, C. woronowii
2n=54 C. alpinum (also 2n=56), C. arenarium, **C. atticum**, C. balansae, C. boissieri, C. burtii, C. chalcedonicum (subsp. punctatum 2n=50), C. chlorobasis, C. cilicicum, C. cupanii (incl. subsp. glossophyllum), C. decaisnei, C. doerfleri, C. dolichantherum, C. euboeum, C. figlalii, **C. filifolium**, C. freynii, C. heldreichii, C. hungaricum,

C. imperatori-friderici, C. inundatum, C. kesselringii, C. lingulatum (and subsp. *rigescens*), *C. luteum, C. macedonicum, C. macrophyllum, C. manisadjianii, C. micaceum, C. micranthum, C. montanum, C. nanum, C. osmaniyense, C. parlatoris, C. parnassicum, C. persicum, C. psaridis, C. pulchellum, C. pusillum, C. rausii,* **C. robustum**, *C. sfikasianum,* **C. soboliferum***, C. stevenii, C. troodii, C. tunicatum, C. turcicum, C. variegatum*

- 2n=56 *C. alpinum* (also 2n=54)
- 2n=90 *C. neapolitanum* (also 2n=140)
- 2n=96 *C. haynaldii*
- 2n=108 *C. balansae, C. lusitanum*
- 2n=140 *C. neapolitanum* (also 2n=90)
- 2n=144 *C. longifolium, C. multiflorum*
- 2n=162 *C. arenasii, C. lusitanum*
- 2n=180 *C. gonarei*
- 2n=216 *C. corsicum*

Colchicum atticum, in common with around half the genus, has a chromosome number of 2n=54.

While there is a wide span of chromosome numbers it is quite clear that 2n=54 is by far the commonest, being found in about half of all known species.

Morphology

Colchicums are geophytes that grow from underground, bulb-like corms. Although they are superficially similar to crocuses, the two genera belong to different plant families and differ in fundamental aspects of corm, leaf, flower structure and in the position of the ovary.

All *Colchicum* species have stemless flowers held on long perianth tubes or with free tepals right to the base. In the wild the flowers are produced from late summer to early winter, before the leaves develop or late winter to spring at the same time as the leaves. Tepal colour varies from purple and pink to white and sometimes yellow and tessellations are a feature of some of the autumn-flowering species. Identifying species is difficult because morphological characters overlap and some species are extremely variable (see key in Species chapter). The desert or semi-desert species are particularly difficult to distinguish. They mostly have white to pale pink flowers, occasionally darker pink, with the leaves part-developed at flowering time. However, in exceptionally dry seasons the flowers may appear in advance of the leaves or with the leaf tips only just showing.

GENERAL DESCRIPTION OF *COLCHICUM*

Cormous perennials without well-developed stems. **Rootstock** a fleshy tunicated corm with numerous slender fleshy roots issuing from a small basal plate, the corm vertical, often ovoid to ovoid-oblong, slightly to markedly asymmetrically concave on one side, flattened on the other, sometimes more elongated and twisted, or inclined or horizontal and soboliferous (i.e. the rootstock sometimes with narrow, horizontal, rhizome-like outgrowths), with or without a hypopodium, a one-sided, drop-like, downward–directed appendage or 'foot', with or without a cylindrical, tunicated neck (sometimes referred to as a collar); tunic membranous

LEFT **The larger morphological features of the genus are apparent in flowering corms of *Colchicum* 'Autumn Queen'.**

RIGHT ***Colchicum* x *agrippinum*, a hybrid with relatively small corms, showing fully developed leaves.**

or coriaceous, entire, split or shredded, pale to dark brown or blackish brown, occasionally yellowish brown. **Stem** very reduced, often obscure and appearing stemless, but expanding to form the replacement corm that develops to one side of the existing corm. Leaves and flowers developing within a whitish, membranous, cylindrical sheath or cataphyll that reaches the surface of the ground or rises shortly above it. **Leaves** all basal, developing after flowering (hysteranthous) or partly or wholly developed at flowering time (synanthous), occasionally developing as the flowers fade (subsynanthous); leaves often 2–9 but as many as 20 per corm in some instances, linear to linear-lanceolate, elliptic, ovate, oblanceolate or ligulate, entire, parallel-veined, smooth or undulate, channelled or flat, longitudinally ridged or pleated, sometimes keeled beneath, glabrous or occasionally pubescent or puberulous in whole or part, smooth, scabrid or ciliate at the margin; leaf base narrowed to cuneate or sometimes auriculate, wrapped around one another to form a short to long tube (pseudostem) which may be wholly closed or split for a short distance from the top. **Flowers** hermaphrodite (one exception), solitary through to 12, occasionally more per corm, fascicled, opening at the same time or in succession, each subtended by a small, generally inconspicuous, bract located well below ground level, shortly pedunculate or pedicellate, the peduncle or pedicels elongating as the fruit capsules develop, the pedicels at flowering generally longest in synanthous species. **Perianth** consisting of 6 more or less equal tepals in two series of 3, campanulate to infundibuliform, cupped or rotate, pink, lilac-pink, lavender-pink, mauve-pink, purple, rose-purple, rose-magenta, white, rarely yellow or bicoloured (white with a purple stripe down each tepal), sometimes tessellated; tepals ascending to spreading, equal or the inner somewhat smaller, rarely auriculate below, fused into a short to long cylindrical tube at the base, the tube circular or more or less triangular in cross-section, or tepals free and with a long claw, often with tooth-like auricles at the base of the limb; perianth tube sometimes ridged at the top within where the filament bases form a ridged channel, the ridges glabrous or downy. **Stamens** 6 in two series of 3, the series often of uneven length, inserted near the base of the tepals or at the top of the perianth tube; filaments slender, often thickened towards the base; anthers dorsifixed and versatile, occasionally basifixed and rigid, with introrse dehiscence, yellow, orange-yellow, greenish, greenish brown, grey-brown, purple-brown or blackish brown; pollen cream, yellow or orange-yellow, occasionally brownish. **Styles** 3, free, long and slender, rarely fused, the stigmas punctiform or decurrent along one side of the tip, the tip straight, curved or hooked, shorter than, equalling or longer than the stamens. **Ovary** superior, subterranean, usually located within the corm tunics, 3-locular with numerous ovules; placentation central. **Fruit** usually at ground level or above, occasionally partially subterranean, 3-locular, the capsules septicidal, each loculus containing few to numerous rounded or subrounded seeds, the seed-coats brown or reddish brown; endosperm hard.

CORMS

Colchicum corms are swollen stem bases. They are solid, fleshy, starch- and sugar-filled, unlike bulbs which are multi-layered like an onion, with their growing points hidden at the base, encased in the fleshy, swollen leaf bases that form the main part of a bulb.

The corms of colchicums are very variable both in shape and dimensions, ranging in size from that of a quail's egg to that of a goose's egg. In many species and the majority of cultivars, the corms are asymmetric. They are concave on one side but with a slightly flattened face on the other, and at the bottom is located a small basal plate. It gives rise to a cluster of slender fleshy roots and a solitary (rarely several) tube-like, generally white, membranous cataphyll. In a few instances the corms are more elongated and horizontal, and behave like rhizomes or stolons. In these instances the corms are quite small and the base plate is often located a short distance from one end, but they may produce several horizontal outgrowths or soboles that become independent corms in due course. Soboles can be smooth, uneven, nobbly, variously lobed or branched. In a few instances (as in *C. davisii*), perfectly ordinary corms may give rise to one or several soboliferous outgrowths. To date, eight soboliferous species are known. *Colchicum boissieri*, *C. minutum*, *C. psaridis*, and *C. soboliferum* produce corms that spread by narrow, horizontal, more or less cylindrical soboles, while *C. baytopiorum*, *C. munzurense*, *C. rausii* and *C. sieheanum* produce more intermittent, oblique soboles that are sometimes short and tooth-like.

The cataphyll encloses the young leaves and flowers and at maturity expands to the soil surface or stands slightly proud of it. In the spring-flowering species leaves and flowers develop together, expand up through the cataphyll and appear above the soil surface. In the autumn-flowering species, and indeed many of the cultivars grown in gardens, the cataphyll

Here in *Colchicum* 'Chequers' a new corm is forming to the right of the old one which has had its tunic removed.

and flowers push up to the surface well in advance of the leaves. At the base of the corm, immediately subtending the base plate, there may be a small or very pronounced 'foot' or hypopodium, sometimes referred to incorrectly as a 'dropper'. The hypopodium is a vertical, downward-directed portion of the fleshy corm, not always present and sometimes represented by only a small bump. Corms may produce one or several hypopodia which have a bud near the tip, each developing in the following season to form a new corm. Hypopodia serve two purposes. Firstly, they adjust corms to a deeper level in the soil, otherwise they gradually get closer to the surface. Hypopodia do not contract to pull the corms down into the soil, as droppers do in some other bulbs and corms. However, as the growing point is close to the base they can position new corms deeper in the soil. It is especially noticeable in seedlings and young plants that hypopodia can act like vertical soboles, delving deeper into the ground. In cultivation corms will sometimes push themselves above the surface of the soil, especially when they become crowded. The second purpose of hypopodia is that where several are produced, they aid vegetative duplication of the parent corm.

In all species the fleshy corms are encased in a pale to deep brown or blackish brown, papery or rather leathery tunic or coat. In many instances the tunic is permanent, but in species with very thin, papery tunics it may disappear before the corms mature. Tunics are usually several layered and the inner layers may be differently coloured to the outer, being often paler brown or yellowish brown, thinner and more membranous. The average number of layers of the tunic is directly related to the average number of leaves per corm, as it is the leaf bases that transition into the tunics. However, the smaller, uppermost leaves may not contribute to the tunics.. The tunic may extend above the top of the corm as a long

Hypopodia extending from the base of corms in *Colchicum macrophyllum*. One clearly shows where the new growing point is.

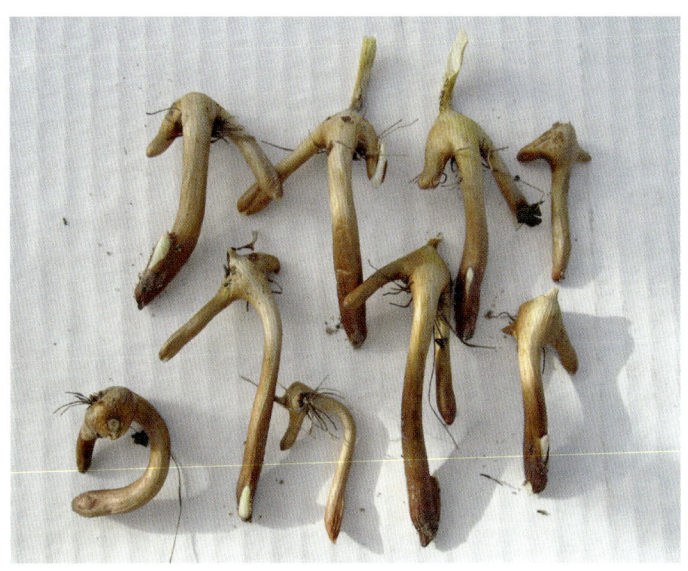

ABOVE **The soboles of species such as *Colchicum soboliferum* will eventually separate from the parent corm, making a new corm.**

BELOW **On the right the tunic has been removed to show the replacement corm on the right and a reserve bud on the left.**

tunicated neck; this can sometimes be as much as 30cm long, occasionally more, and reach the soil surface, although generally stopping short of this surface. Tunics can sometimes become quite fibrous upon aging, with the neck splitting up into a lot of parallel fibres. The corms, along with the tunics, are generally fully developed by the time the leaves wither away in late spring or summer.

Unlike bulbs, corms are replaced annually. New corms are produced to the side of the parent corm from the base plate. As the new corms develop within the tunic of the parent corm, so the old corm is gradually exhausted and withers away, although it may be partly present at the beginning of the following season. The replacement corm develops from the axillary bud of the lowermost leaf, often referred to as the regular renewal bud. In addition, a reserve bud may also develop into a corm. On rare occasions, and more especially in cultivated plants, several reserve buds, usually in the axil of the second or even third leaf, may produce a cluster of new, often small corms. They appear piggy-back on the side of the mother corm. While the base of the stem (which is obscure and below ground) swells to form the replacement corm, so the leaf bases swell and stretch around it as it develops to form the tunic. Soft and fleshy at first, the leaf bases eventually dry out, harden and darken, becoming leathery, papery or membranous, depending on the species. As corms multiply they may separate in the soil or remain closely associated. They can share the same tunicated neck for several seasons, each season adding an additional tunic, so that tunicated necks can be very leathery and persistent. This can happen in a number of species and is obvious in plants such as *C.* × *agrippinum*. In some instances, especially in young plants or under less favourable conditions,

MORPHOLOGY **61**

the developing corms may be slim and small with proportionately larger and longer hypopodia which can be vertical or horizontal. These can act as soboles, in effect stolons, and may terminate in small cormlets seen as small knobs. They will eventually separate from the parent corm and act as independent plants; at first they produce a single leaf in a similar manner to that of a young seedling.

CATAPHYLLS

Cataphylls are vertical, white, fleshy, cylindrical tubes that develop from the base plate and generally reach the soil surface or rise just above it. While most are white throughout at first, the tip that appears at the soil surface can be greenish, purplish, crimson or reddish. Cataphylls are tough and fleshy, even leathery, ensheathing and protecting the developing leaves and flowers as they emerge. Often housed within the tunic neck of the corm, they usually extend beyond it to the soil surface. In those species with long-necked corms, the cataphylls often extend up the inside of the neck or they may break through it to the surface. As they develop from close to the base of the corm, and as corms can lie deep in the soil, cataphylls to 25cm long are not exceptional for some of the larger species and cultivars, and even longer ones have been recorded. For example, *C. balansae* has been recorded with cataphylls to 50cm long.

Many corms produce a single cataphyll, others may produce two, three or four from the area of the base plate. Occasionally a cataphyll will develop away from the base plate, particularly in species that have soboliferous corms. Cataphylls stay in place until both flowers and leaves have emerged, in both synanthous

FAR LEFT **On the right the tunic has been removed to show the shoots developing from axillary buds which will eventually become new corms. Developing from the base plate, these cataphylls are now emerging from the tunic.**

LEFT **The red cataphylls of *Colchicum sanguicolle* are a dramatic feature of the species.**

RIGHT **The pale cataphylls of *Colchicum szovitsii* ensheathing the leaves and flower.**

BELOW RIGHT **If these corms had been planted the cataphylls would eventually extend to the soil surface.**

and hysteranthous species. As the new corms begin to expand, the cataphylls eventually rupture and begin to break down. However, elements of the cataphylls may still be present when the new corms have matured.

All things being well, the number of cataphylls indicates the number of new corms that will develop. Vigorous species and cultivars that produce a number of cataphylls per corm are quick to multiply. At the other extreme, some species and indeed one or two cultivars ('Darwin', for instance) only produce a single cataphyll and therefore just manage to reproduce themselves annually. For gardeners, seed, when produced, is the best option for these 'slow' species. As outlined above, new corms develop by expansion of the base of the stem located directly on the base plate. As the new corms develop the cataphylls usually wither away while the leaf-bases expand, and eventually dry out to form the corm tunic. This can be in various shades of brown, yellowish-brown or blackish-brown, occasionally orange-brown, maturing at the same time from fleshy to become papery, membranous or coriaceous.

STEMS, PEDICELS AND BRACTS

Colchicums possess much reduced stems. In fact, most of the stem eventually swells to produce the replacement corm. The stem is very condensed and from it arise from two to eight, occasionally more, close-set, alternate, spirally-arranged nodes. The leaves with the cataphylls ensheath the nodes in the young state. Immediately above the nodes one finds very reduced and insignificant bracts in whose axils arise the flowers. After fertilisation the developing fruits are pushed up to the soil surface, or above it, by rapid expansion of the pedicels. In some cases there may be several fruits on a common peduncle, as in some of the large-flowered species such as *C. autumnale* and *C. speciosum*. However, in most of the smaller-flowered species each fruit is borne on a separate, expanded pedicel. By the time the fruit capsules reach the soil surface, or expand above it, they are more or less fully grown but take some months after that to fully mature.

LEAVES

Colchicum leaves vary in number from two to 12 (exceptionally up to 30 in *C. polyphyllum*), although three to seven is more usual. In species whose corms produce several cataphylls, plants produce a dense cluster of leaves. The leaves develop through the cataphyll and expand markedly once above ground. Whether or not the species are autumn- or spring-flowering, the leaves begin to appear in late winter and early spring, are fully developed by late spring, and persist well into summer. Leaf development is often triggered by the onset of winter rains.

The leaves are spirally arranged and have short to long ensheathing bases. These enfold each other to form a short or long tube (pseudostem) which is inconspicuous in the smallest species but well-defined and stem-like in the large, autumn-flowering species. In species such as *C. autumnale*, *C. bivonae* and *C. speciosum* the pseudostem can reach 15–20cm in length. In these latter species and their cultivars the pseudostem can be entirely enclosed, or it may be open for a short distance, depending on the species. Leaf position can be a diagnostic character, the

The pseudostem, formed of ensheathing leaf bases, can be an obvious feature in species such as *Colchicum autumnale*.

A variety of leaf shapes exhibited by (clockwise from top left) *Colchicum cretense*, *C. variegatum*, *C. macrophyllum* and *C. peloponnesiacum*.

leaves ranging from strictly upright to ascending or spreading, or recurving to some extent, depending on the particular species.

Leaf shape varies and can be linear to linear-lanceolate, ligulate, elliptic to oval, ovate, oblanceolate or obovate. The leaves of all species have entire margins. The leaf base can be narrow, cuneate, rounded or variously auriculated.

Leaf colour varies from pale to deep green, blue-green or yellowish green, and matt to variously glossy. While the majority of species have glabrous leaves, in a few the surface can be finely hispid, shortly pilose (as in *C. burtii*) or variously scaberulous. The leaf surface can be smooth, slightly ridged or folded along the length, or in some instances distinctly pleated. While the margin is always entire it can be flat, channelled or undulate, sometimes hispid or ciliate, although the vast majority are glabrous. The leaf margins are often also thinly cartilaginous, although this is sometimes difficult to observe.

The leaves of *C. macrophyllum* are large, elliptical and pleated, resembling those of *Veratrum*. At the other extreme are species such as *C. serpentinum* which has a cluster of three to six linear, channelled leaves up to 12–15cm long.

FLOWERS

Colchicum flowers appear either from late summer through to early winter or in late winter and spring, depending on the species. However, several species, especially in cultivation, bridge the winter gap, influenced by variable weather conditions from one year to another. Each cataphyll can carry one or as many as 12 flowers, rarely more, which usually open in quick succession. As some corms produce several cataphylls, there can be a considerable cluster of flowers open

FAR LEFT **The narrow, star-like flowers of *Colchicum cupanii*.**

LEFT **The globose flowers of *C. triphyllum*.**

BELOW LEFT **Some species, such as *C. trigynum*, bear many flowers per cluster. Note also the split perianth tube typical of species formerly in *Merendera*.**

at the same time. In some, corms build up in close, multi-flowered groups which can be mistaken for the product of a single, large corm.

The flowers are always upright but vary in shape from that of a narrow wine glass to goblet- or bowl-shaped or funnel-shaped, or more spreading and star-like, all borne on a short to long tube.

Several species bear rather unusual floral features that may reflect the harsh environment to which they have become adapted. *Colchicum tuviae* has either male or female flowers, the only species in the genus with this trait, while the species with the proposed name of *C. ramonensis* has hermaphrodite flowers with a solitary style.

TEPALS

The six tepals that make up the perianth of each flower are arranged in an inner and an outer series of three, with the exception of *C. gonarei* which has only four to five tepals. The tepals in the inner series can be very similar to those in the outer one, or they can be broader and slightly shorter. Tepals vary in shape from linear to linear-lanceolate to lanceolate, or elliptical, oval or obovate. The apex can be acute or obtuse and the margin can be flat or undulate, glabrous or occasionally ciliate. In most of the species the tepals are fused together at their base into a short or sometimes very long tube (perianth tube) which reaches down below ground to near the base of the cataphylls, where the ovary is located. On the outside the perianth tube is grooved down its entire length by three pairs of parallel lines corresponding to the fusion of the tepals, and these may be faint or pronounced. In cross-section the perianth tube can be round or more or less triangular.

In contrast, in those species formerly included in the genus *Merendera* the tepals are free from one another (there is no perianth tube) and the tepal limb extends below into a long claw. This means that the tepals can readily fall apart. In some of these species this is prevented by the presence of small appendages often referred to as auricles, tooth- or hair-like, that lock the tepals together. In other species the flowers are held together by the young encircling leaves.

The predominant flower colour in the genus is pink or purple-pink. The only species with yellow flowers is *C. luteum*, while *C. kesselringii* in most of its forms has white flowers with a bold pink or reddish purple stripe down each tepal. In addition, quite a few species have a chequered pattern on the tepals. These tessellations are more often a feature of the autumn-flowering species than of the spring-flowering ones, and can be a useful diagnostic character. Tessellation is more pronounced in some species (such as *C. bivonae*, *C. macrophyllum* and *C. variegatum*, except in the rare albinos), while other species have fainter or less

RIGHT **The tesselations of *Colchicum variegatum* are some of the most dramatic in the genus.**

FAR RIGHT **In species formerly in *Merendera*, such as *C. atticum* there is no perianth tube and the tepals are free from one another.**

obvious tessellations only lightly evident (such as *C. cilicicum*, *C. lingulatum* and *C. turcicum*). The tessellations may cover most of the tepal, or else be confined to the upper half or the marginal area. Occasionally, tessellations may be more like a delicate, reticulate pattern.

STAMENS

Colchicums have six stamens which, like the tepals, are arranged in an inner and an outer series. In *C. gonarei* the number of stamens is reduced to four or five. They are attached close to the base of the tepals, generally shortly above where the tepals unite, or they may sometimes be attached in the top of the perianth tube. They can be attached all at more or less the same level, or the outer three may be set slightly lower and, as a result, the outer stamens may appear to be shorter than the inner. The attachment points of the filaments run down into the top of the perianth tube, sometimes forming a pronounced ridge either side of the filament base, generally referred to as the filament channels. These ridges may be glabrous, pubescent or adorned with teeth-like or thread-like lamellae that can be quite pronounced in some species. The presence of teeth or hair-like growths at the base of the stamens is particular to several of the desert species, such as *C. ritchii* and *C. tuviae*.

The filaments, occasionally stout but generally quite slender, are typically whitish, yellowish-green or pale greenish, sometimes swollen at the base, occasionally markedly so. These bases, often coloured, yellow or green predominately, ooze a slightly oily, sweet nectar. The anthers are dorsifixed and versatile in the majority of species, although they can be basifixed, such as in *C. kesselringii*, *C. luteum* and *C. robustum*. They vary in colour from species to species, from cream to pale or deep yellow, to orange, brown, blackish brown, purple-brown or greenish black. In contrast, the pollen, released from the inner side of the anthers, is nearly always yellow, varying from pale creamy-yellow to rich dark yellow or orange-yellow, and very occasionally cream.

BELOW **Ridges either side of the filaments can be seen here in *Colchicum autumnale*.**

BOTTOM **The slender, white styles of *Colchicum pulchellum*.**

All Colchicum have three styles, seen clearly here in 'Autumn Herald'.

OVARY AND STYLES

The ovary is superior, with the perianth tube joining below the ovary. It is found close to the base plate of the corm, well below ground, but can be elevated to some extent on a short, indistinct pedicel, in which case the location is partway up the cataphyll. The base of the perianth tube surrounds it. The ovary has three locules, each compartment containing numerous ovules.

Each flower has three distinct, linear, filiform styles which stretch the length of the perianth tube. In *C. bulbocodium* the three style arms unite below into one, close to the point of attachment of the stamens. In *C. kesselringii* and *C. luteum* the style arms unite inside the perianth tube. In all other species the three styles remain separate throughout their length. The length of the styles is quite remarkable and is one of the most fascinating features of the genus. It is not unusual for the styles to reach 15–25cm long. In *C. balansae* styles more than 50cm long have been recorded. As the ovary is right at the base of the corm, upon pollination the pollen tube has to extend the full length of the style if fertilisation is to be achieved, a quite remarkable journey, virtually unsurpassed in any other plant except in a few species of *Iris*, such as. *I. lazica* and *I. unguicularis*.

The portion of style arms above the point of attachment of the tepals may be shorter than the stamens, the same length, or longer. In quite a few of the large-flowered species at least, the styles elongate as the flowers mature, in some instances eventually pushing them above the stamens.

The styles may be white, greenish or yellowish white, or tipped with purple, pink or red, and occasionally the whole style may be flushed with colour. The tips can be straight, somewhat kinked or curved, punctiform, or dot-like with the stigma only at the very tip. Alternatively, the end of the style may be flattened on one side with the stigma decurrent, running for a short distance from the tip, often for 2–6mm. The style details are a useful aid to accurate species identification.

FRUIT CAPSULES AND SEEDS

The fruit capsules are green at first but pale to dark brown at maturity, sometimes with various darker spots or blotches. They are often glabrous but occasionally scabrid or pilose overall or only in the upper part. The capsules are often more or less ovoid but they can be oblong, ellipsoid or rounded and the apex can be blunt, pointed, acuminate or beaked. They split eventually into three sections along the inner sutures, in the upper third or half usually, to form 3-compartmented cups that release the seeds. After fertilisation, the short pedicels begin to lengthen, pushing the developing fruits to the soil surface or beyond. The mature capsules may reside at or above ground level or half below ground, nestling in the centre of the leaves. In many of the large-flowered species the capsules are located well above ground, sometimes as much as 30cm or more, on pronounced pedicels that are closely enveloped by the sheathing leaf bases. The capsules may be solitary or several can be clustered tightly together. In species such as *C. bivonae* and *C. speciosum* each capsule in turn is shortly pedicellate. The seeds are generally numerous, ovoid to globose, pale to deep brown or blackish brown.

FLOWER FRAGRANCE AND POLLINATION

Only a few *Colchicum* species are noticeably scented. *Colchicum kesselringii* and *C. luteum* both have fragrant flowers, while a number bear a faint, rather *Crocus*-like fragrance resembling that of saffron (some forms of *Crocus speciosus*). However, many of the smaller spring-flowering species seem to be scented, although this requires more careful observation. Instead of pollinator-attracting nectaries or specialised nectar-secreting organs, colchicums have swollen filament bases that exude oily nectar secretions that collect at the top of the perianth tube. The swellings are accompanied on either side by filament grooves that act as guides for insects. Nectar is never as apparent as it is in the spurs and

ABOVE **Each fruit, and there are two visible on the foreground plant of 'Glory of Threave', splits into three compartments along sutures.**

FAR LEFT **The seeds are visible in each of the three compartments of *Colchicum macrophyllum*.**

LEFT **An empty fruit capsule of *Colchicum arenarium*.**

nectaries of many other flowers, and it is presumed that the action of visiting bees somehow stimulates the flowers to exude small amounts from the swollen filament bases. In the related genus *Androcymbium* small pools of nectar glisten at the base of the flowers.

For species with goblet- or funnel-shaped flowers the number of different visiting insects is rather limited. Observations clearly show that both in the wild and in cultivation the prime visitors are honey bees, while bumblebees are rarer visitors that in most cases seem often to shun the flowers. The reason for this is unclear. Others visitors include various small flies which seem to use the flowers more for shelter than act as agents of pollination. Hoverflies are also frequent visitors, not seeking the base of the flowers but the anthers, attracted by the pollen. Various small beetles, such as pollen beetles, have also been observed in the flowers but they do not appear to be prime pollinators.

Butterflies may certainly be important pollinators in the hot, dry Mediterranean regions where many of the smaller flowered species are found. Some of the species with rotate flowers that open out widely, such as *C. boissieri*, *C. lingulatum*, *C. stevenii* and *C. variegatum*, also attract butterflies. Unfortunately, few other such observations have been made on colchicums in the wild and many pollinators may remain unrecorded.

ABOVE RIGHT **Depending on species, seeds can be pale brown to very dark brown.**

RIGHT **Swollen filament bases secrete an oily nectar, apparent here in a mature flower of *Colchicum lagotum*.**

FAR RIGHT **Honey bees are frequent pollinators of cultivated *Colchicum* such as 'Nancy Lindsay'.**

Colchicine

All parts of colchicums – the corms, leaves, flowers and fruits – are poisonous. The principle toxic compound in colchicums is the alkaloid colchicine. Its toxicity is well known to farmers, who avoided grazing animals on pastures where colchicums grow. Colchicine also has anti-inflammatory properties that have been exploited medicinally since at least 1500BC. Nowadays it is sometimes used to treat gout and other inflammatory disorders such as Behçet's disease, pericarditis and familial Mediterranean fever.

Colchicine is also used in plant breeding. It acts by inducing polyploidy through the inhibition of mitosis by disrupting the microtubules that pull the chromosomes apart during cell division. Polyploid plants are often larger, or have more thickly textured flowers, or are faster growing.

COLCHICINE IN MEDICINE

Despite its toxic properties, colchicine has a long history of use in medicine. The Ebers Papyrus of Ancient Egypt dates back to 1500BC. One of the oldest papyri in existence, it is thought to have been discovered in a tomb in Thebes, Egypt. It was purchased in 1873–74 by Georg Ebers from an American farmer based in Luxor, and now resides in the library of the University of Leipzig, Germany. Much of the papyrus is copied from earlier texts, and there is one reference to the use of colchicum in the treatment of rheumatism and swelling.

The first mention of the use of colchicum for treating gout is probably by Pedanius Dioscorides in *De Materia*

LEFT *Colchicum autumnale* was once the main source of colchicine.

RIGHT The Ebers Papyrus of 1500BC is one of the first known documents mentioning the medicinal use of *Colchicum*.

Medica, dating from the first century AD. Dioscorides distinguished between the consumable 'small colchicum' (now associated with widow iris, *Iris tuberosa*, also known as *Hermodactylus tuberosus*) which had various nourishing qualities, as well as therapeutic and aphrodisiac properties, and the 'wild bulb' or *colchicos* that grew in the ancient country of Colchis (Caucasus) and was dangerous. It has been argued that the latter was referable to *Colchicum autumnale* but that species is not known from the Caucasus region; the species was probably *C. speciosum*. Later mentions include the 'hermodactyl' of the physician Alexander of Tralles (c.525–605), which was almost certainly a colchicum, who also recommended the use of the corms in the treatment of gout in one of his medical texts. *Colchicum* corms were also used medicinally by Avicenna (980–1037), a Persian physician and polymath. Avicenna lived through the Golden Age of Islam and is often referred to as the 'father of modern medicine'.

Herbalist John Gerard (1545–1612) was well aware of the toxic properties, commenting that 'it must be mixed with other substances such as bread crumbs, egg white and barley meal to make a poultice to treat gout. This same poultice will increase sperm production. If eaten it must be mixed with ginger, anise, pepper and cumin. If taken alone it is essential to drink plenty of cow's milk to prevent death. It is very hurtful to the stomach and kills by choking so some call it colchicum strangulatorum.' The use of colchicum in the treatment of gout was described in the *London Pharmacopeia* of 1618 and by the 1820s many medical texts included colchicum.

John Gerard wrote of using a poultice of colchicum to treat gout.

Colchicum bulbs are thought to have first been introduced to the US from France in 1798 by Benjamin Franklin (1706–1790), one of the founding fathers of the USA, who suffered from gout.

In 1820 two chemists, Pelletier and Caventou, managed to isolate an alkaloid substance from the corms of *Colchicum autumnale*. A few years later, in 1833, Geiger and Hesse succeeded in purifying this as colchicine. By 1884, Laborde and Houdé were able to obtain crystallised colchicine. It was not until 1945, however, that its chemical formula ($C_{22}H_{25}NO_6$) was established.

Commercially, colchicine was extracted from the seeds and corms of *Colchicum*, principally *C. autumnale* but also *C. × byzantinum*. However, in Greece

Colchicine can now be manufactured, using amino acid precursors.

and Turkey other species would have been used. Today, colchicine can be biosynthesised, using the amino acids phenylalanine and tyrosine as precursors.

Colchicine is sometimes used today to treat inflammation and pain related to gout, Behçets disease, pericarditis and familial Mediterranean fever, but it has mainly been superseded by other anti-inflammatory drugs. The mechanism of colchicine action has not been fully established. It reduces inflammation by inhibiting mechanisms that lead to this condition, including inhibiting cell division while preventing cell migration at the same time. Although it might be thought to have potential as a cancer treatment, its side effects and the fact that is transported out of cancer cells mitigates against this.

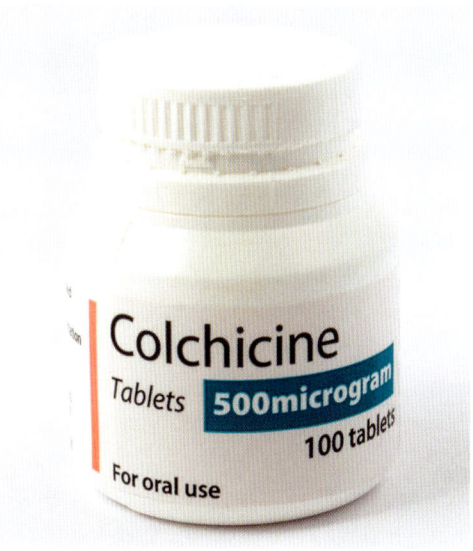

COLCHICUM POISONING

Cases of poisoning and death from consuming colchicums have been reported. In several cases the leaves have been mistaken for those of wild garlic, *Allium ursinum*. The symptoms of *Colchicum* poisoning are similar to those of cholera and appear within 24 hours of ingestion. They include convulsions, cardiovascular collapse, multi-organ failure, blood clots anywhere in the body and respiratory arrest. It should be noted that colchicine can be toxic if inhaled, ingested or absorbed through the eyes. There is no specific antidote for colchicine poisoning.

Its use in medicine needed to be carefully controlled because of its toxicity. Adverse effects of overdosing when used medicinally include gastrointestinal upsets, fever and muscle pains, seizures, delirium, low white blood cell counts and, in the worst scenarios, organ failure.

It is wise for anyone handling *Colchicum* corms or any part of the plant to do so with care. Never lick your fingers and preferably wear gloves when handling the corms or any part of the plant, especially if you have any cuts on your hands. The large-flowered cultivars make excellent cut flowers and if picked carefully will cause no harm.

In the past, farmers have often had a policy of eradicating colchicum (usually *Colchicum autumnale*) from meadows and pastures to protect grazing animals from poisoning. This led to a great decline in the species in many areas where it was formerly found. However, today a number of sites are included in local nature reserves. The same species is also found wild in open woodland, woodland rides and clearings where its presence is less likely to cause harm.

COLCHICINE IN PLANT BREEDING

The effect that colchicine has on mitosis (normal cell division) was discovered as long ago as 1889 by Italian pathologist B. Pernice. He injected dogs with colchicine and observed a vast number of dividing cells, due to mitotic disruption, in their intestinal mucosa.

The discovery of the effect of colchicine on plants took place when Havas (1937) found that an application of colchicine affected the growth of wheat seedlings. In the same year, Blakeslee & Avery (1937) obtained diploid and tetraploid pollen grains from plants of *Cucurbita*, *Datura* and *Portulaca* which had been treated with colchicine. These and subsequent findings revolutionised cytogenetics because the technique allowed for the experimental doubling of the entire chromosome complement of a cell.

Colchicums make good cut flowers, but wash your hands after handling them.

Colchicine works by disrupting the separation of chromosomes when cells divide. Replicated chromosomes are normally pulled apart during cell division by structures called microtubules. However, colchicine inhibits microtubule function by binding to its constitutive protein, tubulin. So, when colchicine is applied to dividing cells, the chromosomes still replicate but the two sets do not migrate to opposite ends of the cell. Normal division is prevented, resulting in a cell with twice the number of chromosomes in its nucleus. Thereafter, these cells multiply as normal but with double the number of chromosomes in subsequent daughter cells.

Colchicine treatment does not always go to plan. Some chromosome material may be lost and treatment may cause gene mutation. Alternatively, only part of the tissue may be polyploidised, such as the epidermis or subepidermal cells. Colchicine works best on material that is actively growing, but it can also be used on seeds and dormant bulbs where the influence of the colchicine can last long enough to have some effect.

The increase in the number of chromosomes often increases the size of the affected cells and may alter their function and enhance the plant's ornamental value. Although it is difficult to predict the degree of change, polyploidy may cause increased height, increased thickness and size of stems, leaves and petals, and changes in the texture of leaves, petals and fruits, as well as various physiological responses. Some polyploids have greater tolerance to growing conditions, soil type and drought.

A wide range of plants has been subject to colchicine treatment to improve various qualities. These include *Gladiolus, Hemerocallis, Lilium*, various woody plants, and crops such as grape and potato. *Magnolia* has been a favourite target of breeders using colchicine. In the 1940s plant scientist E.K. Janaki Ammal, working at RHS Garden Wisley, experimented with applying colchicine to a number of different plants, with her best-known raising being *Magnolia kobus* 'Janaki Ammal'. Later in the 20th century, magnolia breeder August Kehr of North Carolina, USA, raised *Magnolia* 'Sun Ray' and *M. stellata* 'Two Stones' using colchicine.

Colchicine-induced polyploidy is most useful for allowing the potential of hybridisation between species that have different chromosome numbers. Sterile triploid plants can also be rendered fertile through chromosome doubling. A well-known example of this is in wheat breeding. Wheat (*Triticum* species) exists in a diploid state in the wild, but over thousands of years of cultivation tetraploid wheats have arisen naturally and been selected for their vigour and higher yields. Breeders have long predicted the benefits of crossing wheat with rye (*Secale* species) for disease resistance and tolerance of colder climates. Tetraploid wheats, often the durum type, when crossed with diploid rye, result in infertile triploid hybrids (× *Triticale*, triticale). Chromosome doubling was achieved by colchicine treatment, often of anther-culture-derived plants, and this led to major advances in triticale breeding.

Colchicine can also be useful to make plants less fertile or even sterile. This can be useful for horticultural or agricultural crops grown for their foliage, or to make flowers last longer. They do, of course, have to be propagated vegetatively.

To induce polyploidy, one of the most common methods is to apply colchicine cream to growing points, where active cell multiplication is taking place. This may be an apical shoot, root or sometimes a sucker. In addition, seeds can be soaked in a colchicine solution before planting to induce polyploidy.

Colchicine treatment was used to raise *Magnolia stellata* 'Two Stones', resulting in a tetraploid cultivar with larger leaves and flowers than the species.

Cultivation

Colchicums that are readily available from nurseries and garden centres are not difficult to grow and make excellent, eye-catching garden plants. They are generally undemanding and the majority are suitable for most soil types and conditions in the garden.

The great diversity in habitat and flowering time across the genus has an influence on their cultivation and place in the garden. The larger species and cultivars, particularly the autumn-flowering ones, create the greatest impact. Most are free-flowering and the corms build up into sizeable colonies in the garden without much maintenance. Large groups flowering in borders or naturalised in grass can be a most arresting sight. They bring flowering interest to the garden when many of the flowers of summer are fading away.

Most of the spring-flowering species and cultivars tend to be smaller and trickier to maintain. These are better suited to bulb frames, alpine houses, rock gardens or raised beds. Some specialist growers cultivate the smaller ones in permanent beds in bulb frames where their growing conditions can be carefully regulated. Advice on growing these types of colchicums is given at the end of this chapter in the section on Gothenburg Botanical Garden

SELECTING PLANTS

Corms are generally offered in nurseries and garden centres in colourful packaging in late summer and autumn. They can be planted immediately. Autumn plant fairs and bulb sales are also good sources, often with specialist growers in attendance, who sell corms individually in most instances. At these you are likely to find a wider range of colchicums, including some more difficult to obtain and rarer kinds. It is also worth searching online since some bulb outlets offer

LEFT **Autumn-flowering *Colchicum* hybrids, such as 'Fuller's Mill' here, combine well with ground-covering plants, particularly if there is a colour contrast.**

RIGHT **Commercial growers in the Netherlands supply many of the colchicums offered in garden centres.**

a good range of different colchicums in this way, both in the UK and abroad. Nurseries run by growers such as Jānis Rukšāns in Latvia and Leonid Bondarenko in Lithuania offer a wide range.

If selecting corms in person then choose only undamaged and firm corms with a clean tunic. Being poisonous, it is wise to wash your hands after handling them or use protective gloves. Beware though, as plants are not always correctly labelled and misidentifications are frequent in the commercial bulb trade. Some nurseries sell them in flower, when you can make a more confident identification.

LEFT **Free-draining soils that do not dry out quickly in summer are ideal. 'Poseidon' in the foreground and 'Autumn Queen' are thriving here.**

BELOW **Plant corms at a depth of twice their height.**

PLANTING

Colchicums prefer free-draining soils. These should be neither too acid nor in any way waterlogged. The ideal is a rich, deep loam that does not dry out too rapidly as summer arrives. Coarse horticultral grit should be added to heavy, poorly drained soils, especially clay ones. Most colchicums thrive best in an open, airy position in full sun, but many are tolerant of some shade, especially *C. autumnale* and its hybrids. This British native autumn crocus, is more tolerant of heavier, moister soils than most and can be found in the wild in wet, not waterlogged, meadows.

Planting depth is important but not critical. However, corms planted too shallowly will often eventually find their way to the surface after a couple of years, or even stand slightly above the surface. They are then more prone to damage and will not flower so prolifically as a consequence. Shallow planting may also invite more slug damage, as many slugs live in the soil close to the surface. The best planting depth for the majority of colchicums is about twice the height of the corm, excluding the necks. Specifically, a large corm of *C. × byzantinum* or *C. speciosum*, or of the large-flowered cultivars which can be 7.5–12.5cm long, will need to be planted at a depth of 15–25cm. Ensure that corms are placed upright. Most will have a neck or a pointed extension at the top, while the base plate, from where the roots emerge, should be obvious at the bottom of the corm.

Exceptions to the depth rule are those colchicums that have rhizome-like corms, *C. boissieri* and *C. psaridis*, for instance. These need to be planted at 5–7.5cm depth and laid flat. These kinds are seldom planted in the open garden, being generally grown in raised beds or pots.

HARDINESS

The large-flowered, autumn-flowering species and cultivars are fully hardy in the open garden. *Colchicum autumnale*, which is widely grown as various selections, is hardy to -20°C. Many of the species, both autumn- and spring-flowering, are hardy to -15°C. In contrast, a few of the smaller, autumn-flowering species such as *C. lingulatum* and *C. variegatum*, and some of the lower-altitude, spring-flowering species will not survive freezing much below -5°C for extended periods, although an occasional low of -10°C should not affect them adversely. While it is true that a number of mid-winter- and spring-flowering colchicums are resilient in the face of wintry weather, the flowers will not open properly unless the sun shines and temperatures rise.

Translating this to RHS hardiness ratings, most of the larger *Colchicum* species can be assigned to H5 (-15°C to -10°C) and some can be rated H6 (-20°C to -15°C). Some of the smaller species, especially those from the semi-desert regions of the eastern Mediterranean, are probably best considered as H4 (-10°C to -5°C). For those gardening in more extreme climates with prevailing low winter temperatures the protection of a glasshouse or cold frame may be required.

Having said this, young foliage can be damaged by heavy frost or bitter winds. Such conditions generally affect only the tips of the leaves, but when the whole

The larger autumn-flowering cultivars, such as *Colchicum speciosum* 'Album' here, are hardy to -20°C.

leaf expands this damage soon appears insignificant. Mature foliage can be damaged by hail storms which have been known both to puncture and lacerate the leaves, but these are rare occurrences.

IN THE OPEN GARDEN

The majority of colchicums grown in the open garden are large-flowered, autumn-flowering species, hybrids and cultivars. The most widely-grown smaller ones include *Colchicum × agrippinum*, *C. laetum* and *C.* 'Pink Star', all of which flower prolifically. All of these are tolerant of a wide range of conditions and a variety of free-draining soils.

Choosing the best position in the garden does require some planning, mainly because the leaves that develop some time after flowering are large and can swamp neighbouring plants. The foliage can, however, be very handsome and certainly provides good groundcover. On the whole they are not suitable for herbaceous borders for the reasons given above, but also because by early to mid-summer the leaves have died away, leaving a gap which is only filled later in the year when the new flowers emerge. However, they are perfect for infilling in shrub borders. A mixture of species and cultivars can give a flowering display from mid-August until November, adding great interest as summer flowers wane.

Alternatively, colchicums look extremely good naturalised in grass. Their pinks, mauves and purples look especially resplendent against the green of the turf. *Colchicum autumnale*, *C. giganteum* and *C. speciosum* are naturally meadow plants so take to grass cultivation very readily. Some of the cultivars whose flowers tend to flop, such as 'Lilac Wonder', look better in grass as it supports them to some extent. It is a matter of trial and error to see which ones do best in your own garden. As corms often multiply quite rapidly, it is easy to put a few aside for naturalising in grass and watch what happens. The grass can only be cut once the foliage of the colchicums has yellowed and begun to die down. Moreover it is vital to complete the last mowing before the plants start start to flower, otherwise they will be damaged as they push through the sward. In most instances the last mowing should take place around late August, while the first mowing cannot take place until the leaves yellow, usually in May or June, otherwise the new corms will not have fully developed.

Colchicums look at their best when planted in groups or drifts, and used in this way will make a real statement when they come into flower. They can also look particularly effective planted around tree bases.

Many of the larger cultivars such as 'Glory of Threave' do well in grass.

The larger, autumn-flowering colchicums look good in a relatively formal planting around the base of a tree.

The different kinds can be isolated from one another in the garden, or they can be highly effective when mixed, especially if they flower at the same time. *Colchicum speciosum* 'Album', for instance, can look spectacular mixed with one of the deep pink or purple cultivars such as *C. speciosum* 'Atrorubens', *C.* 'Benton End' or *C.* 'E.A. Bowles'.

TIDYING LEAVES

Colchicum leaves are fully developed by late spring or early summer. In most types the leaves begin to yellow, collapse and wither away in early to mid-summer, although some linger on into July and August. They can be cut off while still slightly green without detriment to the plants, as by this time the new corms will be fully developed. Otherwise, sharply pull the withered foliage horizontally while pressing down on the ground above the corms with your foot. This stops dormant corms being inadvertently pulled out. If keeled slugs are a problem then immediately sprinkle some dry, sharp sand down into the dry tunic necks from the soil surface. This makes access less easy for these damaging molluscs.

This is quite a good time to lift and divide colchicums as it is easy to see where the plants are and there is less likelihood of damaging the corms.

LIFTING AND DIVIDING

It is best to lift and divide colchicums in mid-summer. Try and avoid damaging the corms, but if you do then leave them above ground until the scars dry before replanting them. Divisions can be replanted at the same time or stored for later.

OPPOSITE:
FAR LEFT The gleaming white of *Colchicum speciosum* 'Album' contrasts with the dark ground cover of *Ajuga reptans* 'Atropurpurea'.

LEFT The pink of autumn-flowering cultivars such as 'Rosy Dawn' can be set off against the fading browns of grasses.

BELOW FAR LEFT The silvery green of an ornamental grass sets off the colour of *Colchicum* 'Glory of Threave'.

BELOW LEFT Dark-leaved heucheras such as 'Palace Purple' make a good foil for *Colchicum* x *byzantinum* 'Innocence', but also for pink cultivars as well.

Corms will survive well when out of ground as long as they are kept cool, dry and shaded. They can then be planted any time up until mid-autumn, although such late planting will usually delay the onset of flowering.

Colchicums naturalised in grass can also be lifted, separated and replanted in a similar way. Unless there is no alternative, colchicums should not be dug up 'in the green', unlike snowdrops and some other bulbs. This is because the corms are only fully developed when the leaves have begun to yellow and die down. Lifting before then will not kill plants, but it does mean that the corms will be immature and probably will not flower well, and may take a couple of years to recover.

COMPANION PLANTS

The plantsman, explorer and writer Reginald Farrer had mixed feelings about colchicums, remarking that '*Colchicum* has an evil colour but a valuable blooming season, even though the coarse handsomeness of the contemporaneous or subsequent leaves unfits them for the choicer associations that otherwise their cups of magenta-lilac would ask' (Farrer 1928). However, modern gardeners are not generally dismissive of the genus. With a bit of knowledge and imagination, good companion plants for colchicums can be found.

Many plants can be associated with colchicums in flower, but later the large leaves tend to overwhelm companion plantings. There are exceptions, however, and some plants can tolerate the dense canopy provided by the foliage. As the leaves generally die away by mid-summer this gives the rest of the summer for congeners to recover. On the whole, the best companions are vigorous carpeting perennials such as bugles (e.g. *Ajuga pyramidalis* and *A. reptans* and their cultivars), New Zealand burrs (e.g. *Acaena inermis*, *A. microphylla*, *A. novae-zelandiae* and their cultivars) and brass button (*Cotula hispida* and *Leptinella pusilla*). Certain plants, such as some polypody ferns (*Polypodium* species and their cultivars), are dormant in early summer and as the *Colchicum* leaves die down the polypodium fronds begin to grow, disguising the gap left in the border. The fronds make an attractive background to the *Colchicum* flowers in autumn. For open beds in full sun, colchicums can even be grown with low-growing grasses and sun-loving, carpeting perennials.

In addition, ground-carpeting shrubs such as *Cotoneaster dammeri*, *C. procumbens* and small-leaved ivies (*Hedera*) make perfect foils. Both the flowers and, later, the leaves, can push up through such shrubs, and their foliage also protects the *Colchicum* flowers from being disfigured by soil splash. This problem can affect colchicums when planted in bare soil.

Other useful companions are lungworts (*Pulmonaria*), *Brunnera* and *Heuchera*, although avoid tall cultivars of the last two. When using lungworts, you can shear their leaves off in July, often when they are afflicted by mildew. This will initiate fresh, non-mildewed leaves in autumn which form an attractive backdrop for the naked *Colchicum* flowers.

AS CUT FLOWERS

The large-flowered colchicums, especially the autumn-flowering ones, make good cut flowers. Many have sturdy stalks which are easily snapped off close to ground level. The best time to cut them is when the flowers have coloured up but are still in the closed goblet phase, or just as they open. They will then last for up to a week in a vase. Although usually placed in jars of water, this is not necessary for colchicums and crocuses as perianth tubes have no means to take up water. In fact, they last surprisingly well once cut without water. However, if placed in a mixed bunch with other autumn bulbs that have scapes that can take up water, sternbergias for instance, then water will be necessary. A medley of different colours placed in a jar or vase can be very eye-catching and will enhance any table or windowsill in the house. For *Colchicum* aficionados they can be a talking point, and visitors can be teased into trying to identify them accurately. As colchicums are poisonous it is wise not to lick your fingers after picking them and to wash your hands.

ABOVE LEFT **The foliage of 'Glory of Threave' in grass with fritillaries.**

ABOVE RIGHT **The large leaves of cultivars such as 'Benton End' can be handsome in the right setting.**

BELOW **Colchicums make good cut flowers.**

PROPAGATION

Colchicums can be increased by allowing them to clump up in the garden and then dividing them, or from seed. Unlike many ornamental bulbs, the corms do not lend themselves to slicing or chipping and commercially they are allowed to increase by natural division.

Division

Most of the species and hybrids grown in the open garden multiply freely by division of the mother corm. In fact, a single flowering-sized corm can increase three-, four- or five-fold in a single season. In this manner substantial clumps can soon build up. Such clump-formers are best lifted and divided every three or four years by digging them up and carefully pulling the clusters apart. The best time to do this is when the foliage withers and dies down in mid-summer. It is then easy to locate the position of the plants in the ground. Even then, care should be taken when lifting, preferably with a garden fork, not to damage the corms which can be located quite deep in the soil.

In some soils the corms build up to such an extent that they begin to push themselves above the surface, a sure sign that lifting and separation is required. Corms can be replanted immediately in your chosen spot or left for later planting, although it is best to plant them by mid-autumn. However, larger ones can be dried and stored for up to six months without detriment, provided they are kept cool, dry and in the dark.

Many of the large-flowered colchicums grown commercially are lifted on an annual basis and divided for sale after careful drying and packaging. Corms of some of the smaller species and cultivars can be treated in the same way. However, with some of the rarer species, which can have quite small corms, these are best not stored for too long as they can deteriorate. Stored corms often do not settle down to a regular routine for the first year or two after being planted out, often flowering out of season or producing malformed flowers.

Some of the smaller species can be slow to increase, scarcely multiplying from one year to the next. For these, the best option is to grow new stock from seed.

The best time to lift corms for replanting is when the foliage has just died back in mid-summer.

Sourcing and collecting seed

Seed can come from one of two sources, either collected from the wild (for which permits are needed) or from plants in cultivation. Seed for sale is usually available late in the year or in early spring. There is an undoubted thrill to be had from raising plants from seed, with the added fun of using seed of hybrid origin which gives unpredictable results. There is always the chance that an exceptionally fine plant may arise. It is wise to be highly selective when seedlings come into flower and to eliminate any that are poor in either form or colour.

Some species regularly set seed while others only do so reluctantly. It is probably true to say that the rarer and more difficult the species is in cultivation, the less the chance of it setting fruit and producing seed. In the garden many of the large-flowered species and cultivars develop fruit capsules, although their productivity in this respect varies from year to year.

Garden-collected seed sometimes produces hybrid seedlings, especially if a number of different species are grown in close proximity in the garden and flower at the same time. Most cultivars will not come true from seed as hybridisation occurs quite easily and many are hybrids in the first instance, often of complex or unknown parentage. Seed cannot be relied upon to replicate the parent plant, so that seed-raised plants from such origins can be expected to, and will almost certainly, vary. This can be an rewarding way of raising new and novel plants. The best way to prevent hybridisation is to isolate plants to rule out cross-pollination, but this is not always possible.

The fruit capsules are usually ripe by early summer. Generally, as the leaves begin to yellow the fruit capsules, if set, will have ripened. The sign that seed is ready to harvest is when the tips of the fruit capsules begin to part, revealing the seeds within. Anticipating this point and acting promptly ensures that the maximum number of seeds can be gathered.

Seed can be stored in dry containers or seed packets and kept in a refrigerator if it is not to be sown immediately after harvest. On the whole, however, seed is best sown as soon as it has been collected, generally around mid-summer.

These fruit capsules will start to split in a few days, at which point the seed can be harvested.

Sowing seed

Seeds of colchicums are generally quite large, mostly 1.5–3mm across, occasionally larger. They are therefore simple to handle and it is easy to avoid sowing them too densely. Crowded seedlings are susceptible to fungal infections such as damping off. Seed of the smaller species is best sown in pots or pans using a well-drained seed compost, preferably with a little extra added grit. A compost that maintains moisture but is free-draining at the same time is ideal.

Sow seed thinly and cover with 2–3mm of compost, then an equal layer of horticultural grit. The grit helps prevent damage to the young foliage and protects the compost from over-vigorous watering, as well as being a good weed suppressant. Remember to label each batch carefully with the name of the plant, origin and date of sowing. Sown pans and pots are best placed outdoors in a cool, shady place to await germination, which usually takes place in late winter or early spring. Alternatively, they can be placed in a shaded cold frame which also has the benefit of preventing bird damage.

Seed of large-flowered garden colchicums can also be sown directly in the garden. An area set aside in a vegetable plot or similar is ideal. The soil needs to be carefully prepared and free of weeds, especially perennial ones, and dug and raked to a fine tilth. Make straight drills 15cm apart, marked at both ends with a cane. After sowing and infilling the drills with soil, the area should be covered with a fine layer of horticultural grit and wire netting secured over the top to prevent birds, cats or other animals disturbing the plot. Each row should be carefully labelled to avoid confusion. Ensure that colchicums are situated well away from edible crops, especially salads, to prevent the toxic leaves being harvested accidentally.

Growing on

Seed mostly germinates in late winter or early spring and produces a single grass-like leaf (cotyledon) initially. Some develop two or three true leaves in their first year but many do not do so until the second year. Thereafter they gradually build up to the average number of leaves for that particular species or cultivar, usually between three and seven, as the corms develop and swell. Regular liquid feeds will help to build strong, young corms.

Colchicum seed is usually large enough to handle relatively easily.

A high germination rate exhibited by seeds of *Colchicum speciosum* 'Album'.

    Whether in pots or in the open ground, seedlings should be kept watered until mid-summer when the foliage begins to die down. Seedlings in pans and pots, especially those of the larger species and cultivars, can be transplanted to a nursery bed after two years. Colchicums are not especially quick to flower from seed, three to five years being the norm, but it is sometimes longer, depending on the growing conditions. Young bulbs can be kept growing vigorously by fortnightly applications of a weak liquid feed during the growing season. Liquid tomato feed or similar is fine.

    Most of the above comments are relevant to the large-flowered, autumn-flowering species and cultivars, most of which can be grown in the open garden. In contrast, the many fine, usually smaller-flowered winter- and spring-flowering species are more suited to growing in pots and pans, in bulb frames or in raised beds. Their cultivation needs greater consideration and, in some instances, more exacting requirements if they are to thrive. Many enthusiasts have good collections of these smaller species, and they regularly appear at early spring flower shows. The finest and most comprehensive collection by far is to be seen at Gothenburg Botanical Garden in Sweden. The collection there, and methods of cultivation of these types, is described at the end of this chapter.

PESTS AND DISEASES

On the whole colchicums are remarkably free of most pests and diseases. Being poisonous in all their parts they tend not to be eaten by grazing or browsing animals such as rabbits, hares and deer.

Slugs

Colchicum flowers and leaves are fairly resilient to slugs and snails, but sometimes these can damage the flowers. Small black keel slugs can sometimes get between the corm and its tunic and eat away at the growing points near the corm base. This either kills them, or sets them back for a year or two before they can reach flowering size once again. A broader range of slugs can also attack seedlings and young foliage, especially that of some of the smaller species.

Slugs are more common in some gardens than others and damage to colchicums can vary greatly. On the whole, gardens on wetter soils will be prone to greater predation by slugs. This, however, also varies from season to season, with more damage apparent to flowers during wet autumn periods. Slugs can be kept at bay using various products on the market. However, slug pellets, particularly those containing metaldehyde, can pose a risk to wildlife in the garden so should only be used sparingly, if at all.

Biological control with parasitic nematodes can be highly effective but the treatment should start early in the season, well in advance of flowering, otherwise damage will still occur. These nematodes are living organisms and require specific conditions to survive and be effective, such as soil temperatures above 5°C and high moisture levels.

Various tapes and shields are also available but are not practical for larger plantings and are lacking in scientific evidence for their effectiveness. Torchlight

Snails have attacked the young foliage of this *Colchicum speciosum* 'Album'.

ABOVE LEFT **Colchicum smut is a rare fungus that is not widely recorded on garden plants.**

ABOVE RIGHT **Virus infection of colchicums manifests as streaking on the flowers.**

inspection of plants in order to remove slugs is the most targeted control option. There is no long-term solution to slug prevention but vigilance and good husbandry will greatly alleviate the problem.

Aphids

Aphids can sometimes colonise emerging flower buds or leaves. This applies particularly to colchicums grown under glass. Aphids are best dealt with as soon as they are seen, otherwise the flowers and foliage can end up being marked and distorted. They can be removed by hand or by using organic or synthetic insecticides.

Grey mould

The fungus grey mould (*Botrytis cinerea*) can infect dying flowers. It is visible as a grey fuzzy presence, especially during periods of dull and damp weather. Left unchecked it can extend down below ground and ruin the corms. Grey mould infection is a particular threat to the smaller species, especially those that flower in late winter and early spring, which is another reason for growing them under glass where conditions are easier to regulate. It is good practise to remove all dead flowers carefully, especially those showing any sign of fungal infection. By the time the flowers have withered they should have been pollinated and fertilisation below ground will have taken place, so removing dead flowers will not hamper the production of fruits and seed. Grey mould can, on rare occasions, afflict autumn-flowering species, affecting flower buds and dying flowers.

Grey mould can also infect pans of young seedlings, particularly if they are sown too densely. Diligence and good hygiene should prevent it happening. There are no fungicides available for gardeners to control grey mould.

Colchicum smut

A smut fungus, *Urocystis colchici*, has been recorded on *Colchicum autumnale* leaves in the UK. It manifests as black streaks that rupture to release black spores. However, it is extremely rare and has been assessed as Critically Endangered in the *Red Data List of Threatened British Fungi* (Evans *et al.* 2006). Its rarity suggests that it is unlikely to cause significant harm to affected plants.

Viruses

Little is known about viruses that infect colchicums. However, some commercial stocks of cultivars such as 'Autumn Queen', 'Giant' and 'Rosy Dawn' are known to be affected. The infection reveals itself as streaking on the flowers, sometimes accompanied by some distortion. Leaves can, on occasion, be similarly affected. Virused stock is best destroyed. The symptoms of virus infection are very similar to those seen in other unrelated cormous plants, such as irises and crocuses.

RHS PLANT TRIALS

There have been two RHS trials of *Colchicum*, both concentrating on the large-flowered, autumn-flowering species and their cultivars.

The first took place at Felbrigg Hall in Norfolk during 1996 and 1997. What was then a National Plant Collection of *Colchicum* held by the National Trust at Felbrigg Hall from 1995 to 2010 was used as a basis for this assessment. One of the aims was identifying accurately the different accessions present. Plants were also judged for the Award of Garden Merit (AGM), for both species and cultivars. Nearly 100 different accessions were acquired by the head gardener Edward (Ted) Bullock from various sources. These were primarily from individuals but a few were from commercial outlets, particularly Dutch companies. The trial was first set out in the walled garden, with the various accessions laid out in squares, and the whole site was dressed with coarse gravel.

Three visits were made in autumn 1996 to view the flowering season. Initially the trial was deemed a great success and a lot of information was gathered and discussed by the trial panel which consisted of Joy Bishop, Chris Brickell, Kath Dryden, Alan Edwards, Peter Erskine, Alf Evans, Sibylle Kreutzberger, Rod Leeds, Alan Leslie (RHS trial secretary), Tony Lord, John Morley, Richard Nutt, Elizabeth Parker-Jervis, Mary Randall and Pam Schwerdt.

Although a great deal of information was gleaned from the trial, few awards

The RHS trial at Felbrigg Hall in Norfolk resulted in cultivars such as 'Glory of Threave' being named and distributed.

were made and nothing substantial was put in writing other than a report in the minutes of the Joint Rock Garden Plant Committee (Anon. 1998). Perhaps the most important aspect was the close scrutiny of the various entries, establishing and confirming the names of the species and cultivars, and applying names to those that were deemed to be new or novel.

From these activities came cultivars such as *C.* × *byzantinum* 'Innocence' (previously *C. byzantinum* 'Album'), *C.* 'Felbrigg' (thought at the time to be a selection of *C. cilicicum*) and *C.* 'Glory of Threave' (received from Threave Garden in Scotland as *C.* 'E.A. Bowles') all of which have subsequently earned the accolade of an AGM. Another that has since won popular appeal is *C.* 'Pink Star', which had been mistaken for a Russian species, *C. laetum*, for many years.

Interestingly, Elizabeth Parker-Jervis supplied the rare and late-flowering C. 'Darwin' for the assessment, but unfortunately it was still not showing any flowers at the time of the last visit in October 1997.

Subsequently, a change of management at Felbrigg Hall led to a decline of interest in colchicums. The collection was moved to a new site outside the walled garden where it thrived for a couple of years, but thereafter a decline set in. As a result, the trial never achieved all its objectives.

The cessation of the Felbrigg Hall trial led eventually to the setting up of a new trial at RHS Garden Hyde Hall. This ran for a three-year period with the initial planting taking place in summer 2014 and its eventual dissolution in 2019. The new judging panel consisted of John Amand, Keith Bankier, Chris Brickell, Alan Gray, Christopher Grey-Wilson, Tony Hall, Richard Hobbs, Doug Joyce, Rod Leeds, Andrew McSeveney (RHS trial secretary), John Morley and Mary Randall. The Hyde Hall trial again focussed primarily on the large-flowered, autumn-flowering species and cultivars which find most favour in the open garden.

The site chosen for the trial was an open sunny, bank adjacent to the dry garden at RHS Garden Hyde Hall. Additional grit was added to the existing soil to ensure perfect drainage, essential for successful cultivation. Once more, plants were elicited from both commercial and private sources. The number of accessions for the trial eventually reached 132, with some later additions for comparison purposes. Each accession consisted of five corms and each was photographed and measured at the time of planting, which took place in the summer of 2014. Corms were planted in squares, with one corm at each corner and one in the middle, and each individual was given a number starting at one.

The site was visited by the judging panel on specified dates in late summer and

The trial of *Colchicum* at RHS Garden Hyde Hall being judged in 2015.

The trial site at RHS Garden Hyde Hall was a sunny bank with a rounded profile.

autumn to assess the flowers and the general performance of the plants, and then in late spring to review the foliage. This pattern was repeated for the three years of the trial, which was then extended for a fourth year into 2019. For the first two years the panel concentrated on ensuring that the accessions were correctly named, finding initially that 40% were not. Staff from the herbarium at RHS Garden Wisley took samples of flowers in autumn and mature leaves in late spring to provide specimens for future reference. Any nomenclatural and identification problems were resolved as far as possible. The panel also recorded the performance of each accession. In the third year, having gathered all this information from records and photographs, AGMs were given to those deemed to be of exceptional garden value (Grey-Wilson 2019). This assessment was aided by visits to see and study a National Plant Collection of *Colchicum* established in 2016 at East Ruston Old Vicarage in Norfolk and collections at other gardens including those of Rod and Jane Leeds, John Morley, Christopher and Christine Grey-Wilson, and Fuller's Mill, the garden of the late Bernard Tickner now owned by Perennial, all in Suffolk. Colchicums do particularly well in East Anglia, hence many gardens where they grow well are in the eastern counties of the UK.

The trial was a great success and much admired by all those who saw it at Hyde Hall. Although there were a few failures, many accessions proved extremely vigorous, increasing in a number of cases from the initial five corms to more than 40 during the trial period.

NATIONAL PLANT COLLECTIONS

In the UK, National Plant Collections (NPCs) are administered by Plant Heritage, a cultivated plant conservation charity. The first NPC of *Colchicum* was held by the RHS at RHS Garden Wisley from 1982 to 1994. This collection was almost certainly the legacy of Frederick J. Chittenden, a director of the garden from 1919 to 1931, who initiated the planting of all available *Colchicum* types in the then Azalea Garden. Little remains of this collection today, but RHS Garden Wisley still has excellent displays of colchicums, especially as circular underplantings around tree bases. The next NPC to be designated was that held by the National Trust at Felbrigg Hall, Norfolk, from 1995 to 2010. The only current NPC is that held by Alan Gray at East Ruston Old Vicarage in Norfolk, designated in November 2016. It is possible that the current collection at RHS Garden Hyde Hall, a legacy of the recent trial, will be designated as an NPC of large-flowered, autumn-flowering *Colchicum* in the future.

Other countries also have national collection schemes. In 2014 national collection status was granted to a gathering of 81 taxa of *Colchicum* assembled by Paweł Kaźmierski in the Unieście district of Mielno, Poland. The collection (http://arborator.pl/zimowit-colchicum-l) is based largely on the nursery catalogue listings of Leonid Bondarenko and Jānis Rukšāns but with several novel inclusions.

BELOW **The only current National Plant Collection of *Colchicum* in the UK is held at East Ruston Old Vicarage Garden in Norfolk.**

BELOW RIGHT **The extensive plantings of *Colchicum* at RHS Garden Wisley, such as 'E.A. Bowles' here, are the legacy of a former director and a National Plant Collection.**

The spectacular collection of spring-flowering colchicums at Gothenburg Botanical Garden.

GOTHENBURG BOTANICAL GARDEN

The collection of winter- through to spring-flowering colchicums at Gothenburg Botanical Garden in Sweden is spectacular and surely the most extensive presently cultivated. It has been rigorously appraised by botanists and horticulturists for more than four decades.

A former director of the garden, Norwegian botanist Per Wendelbo, was instrumental in building up the collection during his tenure from 1965 until his premature death in 1981. He has a spring-flowering Iranian *Colchicum*, *C. wendelboi*, named after him. An authority on the flora of southwest Asia, he made field trips to Iran, Afghanistan, Turkey and Israel between 1950 and 1978, greatly enriching the garden's bulb accessions. This work has been substantially built upon by Henrik Zetterlund, horticultural supervisor at the garden, who has ventured even more widely. He has joined or instigated plant-collecting expeditions to botanically under-explored areas across several continents. Of these, and focusing only on those countries where *Colchicum* is represented,

Collection number BATMAN 223-06 from north-east Turkey was described as a new species, Colchicum lagotum.

the following collection number prefixes apply: S&Z (with U. Strindberg; Macedonia, 1988), KPPZ (with M. Kammerlander, E. Pasche & J. Persson; Turkey, 1990), T4Z (with G. Tjeerdsma, D. Zschummel & R. Zschummel; Iran, 2002), BATMAN (with J. Rukšāns and A. Seisums; central and eastern Turkey, 2004) and LST (with J. Rukšāns; Turkey, 2005). Some, such as *C. lagotum* BATMAN 223-06 and *C. munzurense* KPPZ 90-193, proved new to science. Zetterlund's long-time colleague Jimmy Persson also has a great interest in the genus, while Jimmy's wife Karin Persson has conducted numerous pivotal studies of their taxonomy, publishing many new taxa. Type material established from all the above, and from other similarly enterprising collectors, forms the core of the present living collection at the garden.

Some unusual hybrids are also grown at the garden. One is a highly ornamental plant labelled '*C. eichleri* f. *rosea* aff. hybrid'. It dates from 1998 and is thought to have arisen from a spontaneous cross. There is also a plant of *C. szovitsii* × *C. luteum*.

Colchicum growing areas at Gothenburg

Around half a dozen broadly speaking spring-flowering colchicums are grown outdoors, some on the rock garden (such as *Colchicum doerfleri*, *C. hungaricum*, *C. szovitsii*), but one of the latest-flowering, *C. bulbocodium*, is naturalised in the sloping bulb meadow that overlooks the main glasshouse complex. Here it typically blooms alongside crocuses, dwarf daffodils and reticulate irises in mid-April. An annual top-dressing of bone meal and wood ash is applied every autumn, doubtless boosting the vigour of this sometimes difficult to establish European species. The stock here at Gothenburg is far superior to those with rather ragged, flopping tepals that are often seen in gardens.

However, the vast majority of the collection is grown under cover, either in pots or planted out in the Per Wendelbo Memorial Garden. This latter area (with projected development likely in the future) is an interlocking series of raised beds, covered by a cantilevered glass roof and open on three sides, catering largely for dryland geophytes. The temperature at ground level can drop to -15°C, or even lower, but thermal screens are only used in extreme circumstances.

A few pots are positioned in a side branch of the main undercover complex, in raised, sand plunge-beds at waist-height or more, in the equivalent of an alpine house. This makes a handy, comfortable perch when it comes to the lengthy and difficult job of hand-pollinating the plants, weeding and general maintenance. Anybody aiming to grow bulbs under cover in pots would do well to follow suit,

Numerous accessions of *Colchicum serpentinum* at Gothenburg, just about to flower.

creating raised beds at the height of a similar bed to fit their particular circumstances. Too low and you will spend much time on your knees and detailed observation of the plants is made more difficult.

Nearby neighbours to the colchicums have included recently introduced material from Iran, *Corydalis* of section *Leonticoides*, scillas and a wide range of unusual alpine plants. Their placement has been fluid from year to year.

A couple of the accessions cannot satisfactorily be ascribed to any of the existing species and are grown under their collectors' numbers (such as *Colchicum* LST 040 from Afyon province in west-central Turkey), pending further study and possible description as new taxa. The garden has generously distributed seed of otherwise virtually unobtainable species and unclassified taxa via their annual *Index Seminum*. In the most productive years (2011–2014) there were 25 or more listings in the seed list, including several collections of *C. triphyllum* and *C. trigynum*, and *Colchicum* sp. nov. aff. *trigynum* EP 02/36, presently catalogued as *C. trigynum* Palandöken Group.

There is an ambitious plan to rebuild the entire glasshouse area and incorporate a visitors' centre. Up until now, the main areas where the alpine and hardy dwarf bulb collections are kept have not been accessible to visitors, aside from pre-arranged visits by educational establishments or researchers. Occasionally, bulb displays are transferred and mounted in the areas open to the public. Whatever the outcome, the structure in which the main *Colchicum* collection is kept will very likely need demolishing. Clear polycarbonate ridged sheeting was used instead of glass and, while this has the advantages of strength and excellent light transmission, only those versions that are combined with fire-retardant materials are nowadays permissible for such constructions.

Cultivation at Gothenburg

The equivalent of crop rotation is practised. In one year the collections of autumn-, winter- and spring-flowering species were in adjacent beds. This allowed direct comparison of their varied types of foliage in mid-spring. By 2019 they had been moved much further apart, and those flowering in early March did so in unison with a scattering of hyacinthellas nearby. The sand of the plunge-bed is slightly mounded to form a very low dome along the centre of the bed, partly for purposes of display but also to assist airflow and prevent surface water from lingering.

The colchicums are grown in 12.5cm-diameter clay pots, between three and six together depending on corm size. With new material, where perhaps only a single corm is to hand, a smaller pot is used. There are also a few slightly larger pots, used where material is in generous supply, but in general there is not sufficient space for many larger pots. In any case, widely differing sizes would disrupt the pattern of close-packed, parallel, transverse rows in which the pots are arranged up to 12 deep, typically by species name in alphabetical order, sometimes by taxonomic allegiance, as with the mutable *Colchicum szovitsii* and its close allies.

A compost comprising equal parts of sterilised loam, peat, coarse sand and grit is made less acidic by the addition of Dolomitic lime and bone meal. Repotting of the winter- and spring-flowering species, but not their autumnal relatives, is deliberately delayed until October, long after most other bulbs, the pots having

The close proximity of plantings at Gothenburg allow useful comparisons of foliage as well as flowers.

The peak display of *Colchicum* at Gothenburg is in early March.

been left unwatered from May onwards. *Erythronium* and *Narcissus* are dealt with first, in August or early September, and *Allium* and *Tulipa* are repotted after these colchicums.

Colchicums requiring a cooler, less arid summer are kept in the alpine house and part-shaded, the compost being topped off with a 1–1.5cm topdressing of washed granite chips. Some corms are pronouncedly long-necked and the depth has to be adjusted in accordance. The first watering, using a spray lance, is often not until mid-November. A weak solution of a liquid fertiliser is given with every subsequent watering and the plunge bed remains moist, with supplementary watering applied through to dormancy.

Any time from November, and throughout the following four months, there will be at least some species in bloom, with some habitually flowering around Christmas time and before. The earliest species to flower, from November to December, are *Colchicum antepense*, *C. atticum*, *C. chimonanthum* and *C. szovitsii* subsp. *brachyphyllum*. The last of these has always been an early flowerer in glasshouses and is often grown for that reason. However, the peak display occurs from middle to late February to earliest March, when hundreds of potfuls come into bloom. At this very early stage in the Swedish spring, or more accurately late winter, there can be up to 30cm of snow the other side of the greenhouse wall, with temperatures remaining well below zero throughout the hours of daylight. On even the greyest of days, turning on an industrial fan heater indoors near the entrance transforms the scene within 20 minutes, inducing the flowers to open en masse until around 4pm, when falling light levels cause them to close again. The honeyed scent of species such as *C. munzurense* is very pronounced under cover on a warm afternoon, with the occasional bee in evidence, although few if any other insects are on the wing so early. Prolonged freezing when the plants are in growth is to be avoided, particularly for the more southerly-occurring species from low altitudes.

The aim overall is to limit the minimum temperature by late winter to -5°C. But as a rule of thumb they are markedly hardier than *Crocus* and *Fritillaria*, for example, kept under the same conditions. Some of the Turkish and Central Asian snow-melt species can cope with much colder weather and lose condition in gardens where milder winters are routinely experienced. As an example, at the Alpine Garden Society Caerleon Show in Wales in February 2018, Bob and Rannveig Wallis exhibited a fine pan of *C. soboliferum* whose floriferousness they attributed to the enduringly cold weather over the previous three months.

Regular vigilance is required at flowering time to guard against outbreaks of *Botrytis* and a build-up of insect pests and incursions by slugs and snails. Staff at Gothenburg inspect the collection at least weekly, at which time spent flowers are removed along with any damaged foliage. This is also when controlled cross-pollination is carried out. While some species will set seed without much encouragement, in such conditions and with so few natural pollinators present, cross-pollination with a paintbrush helps the more reluctant species.

Cold weather can encourage flowering, as in this exhibit of *Colchicum soboliferum* displayed at an Alpine Garden Society show in February 2018.

CULTIVATION **103**

Species

The genus *Colchicum* contains 104 species and four hybrids with binomial names. Colchicums have many features that make it difficult to define the differences between species using good, clear-cut characters. This has led to divergence among botanists and taxonomists; some preferring a broad approach, while others have adopted a narrower vision in which a greater number of species is recognised. This chapter covers all species currently accepted by the authors of this book, including those transferred from *Bulbocodium* and *Merendera*. We have also listed all the widely used synonyms here and in the Checklist, but for a full synonymy see Persson (2007).

HABITATS

The different species of *Colchicum* have widely differing habitat requirements and occur over a broad altitudinal range. The types of places that they can be found in range from meadows, woodlands and rocky places in the lowlands up to exposed, rocky areas on mountains, including screes and alpine meadows.

The common, widespread and highly variable *Colchicum autumnale* is a denizen of moist habitats, meadows, open woodland and woodland margins, rides and waysides. It prefers cooler habitats away from the Mediterranean. In countries such as Italy and Greece it is only found in the north, in the cooler, higher mountains, while in central and western Europe it can be found at lower altitudes.

Quite a few species, such as *C. cupanii*, *C. psaridis* and *C. pusillum* favour the Mediterranean region, where

LEFT *Colchicum pusillum* is found in rocky and open places in Greece and some of its islands.

RIGHT *Colchicum speciosum* is a variable species with a wide distribution across the Caucasus, Turkey and north-west Iran.

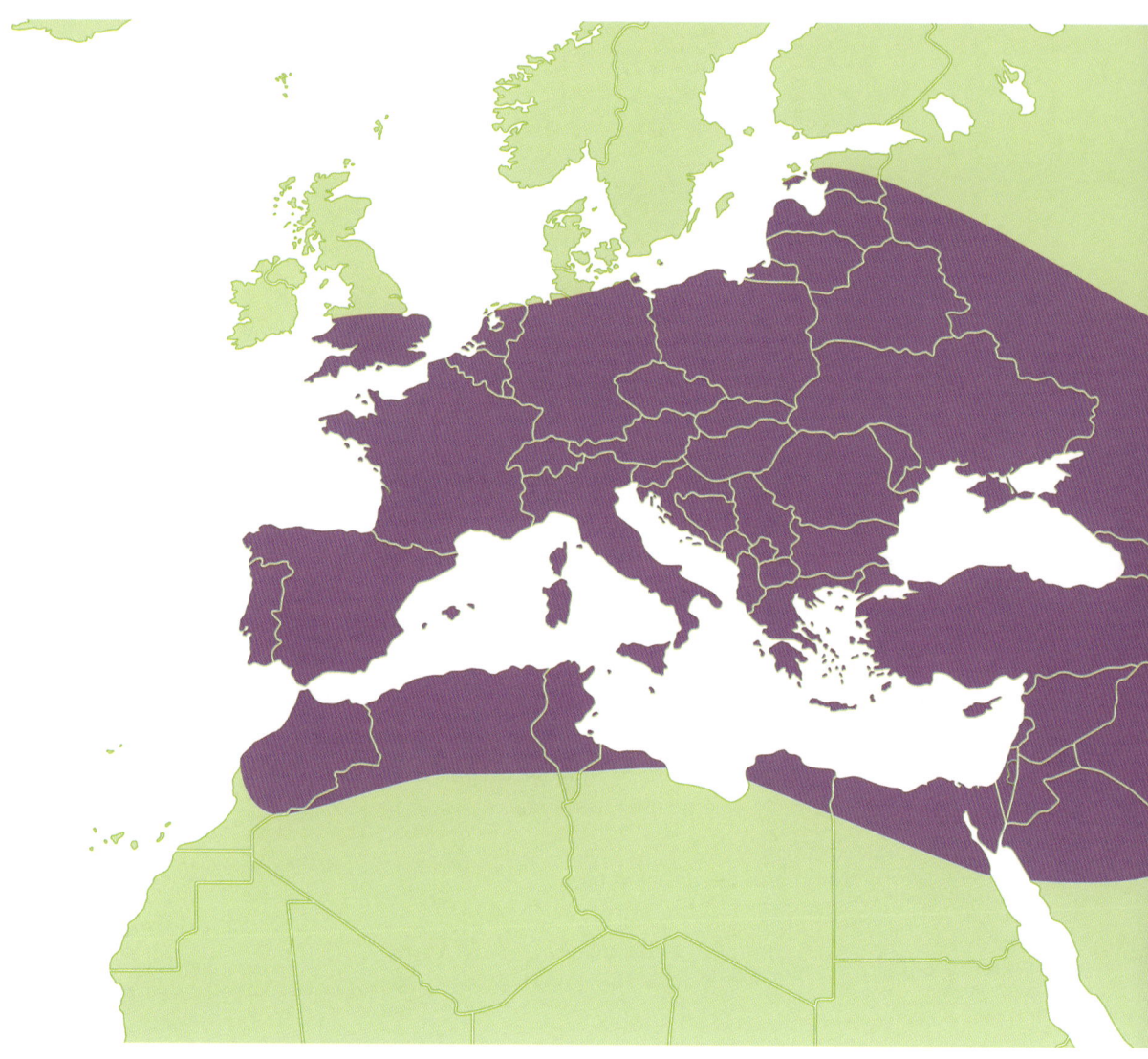

they experience hot, dry summers and cool, wet winters. Other Mediterranean species, such as *C. antilibanoticum*, *C. cilicicum* and *C. guessfeldtianum*, are found in mountain habitats, often between 1,000m and 2,500m, where they inhabit north-facing, rocky slopes or open forests where they get some relief from the fierce summer sun.

Others, such as *C. fasciculare*, *C. "ramonensis"*, *C. ritchii* and *C. tuviae*, are true desert inhabitants, growing on sandy or rocky terrain in some of the hottest areas close to the Mediterranean. Many of these species grow in deep terra rossa or basaltic soils and can be seen in habitats such as fields, orchards and olive groves or open woodland, typically of pine or oak.

A further group of species, including *C. kesselringii* and *C. luteum*, are found only at high elevations, up to 4,000m, where they occupy mountain habitats,

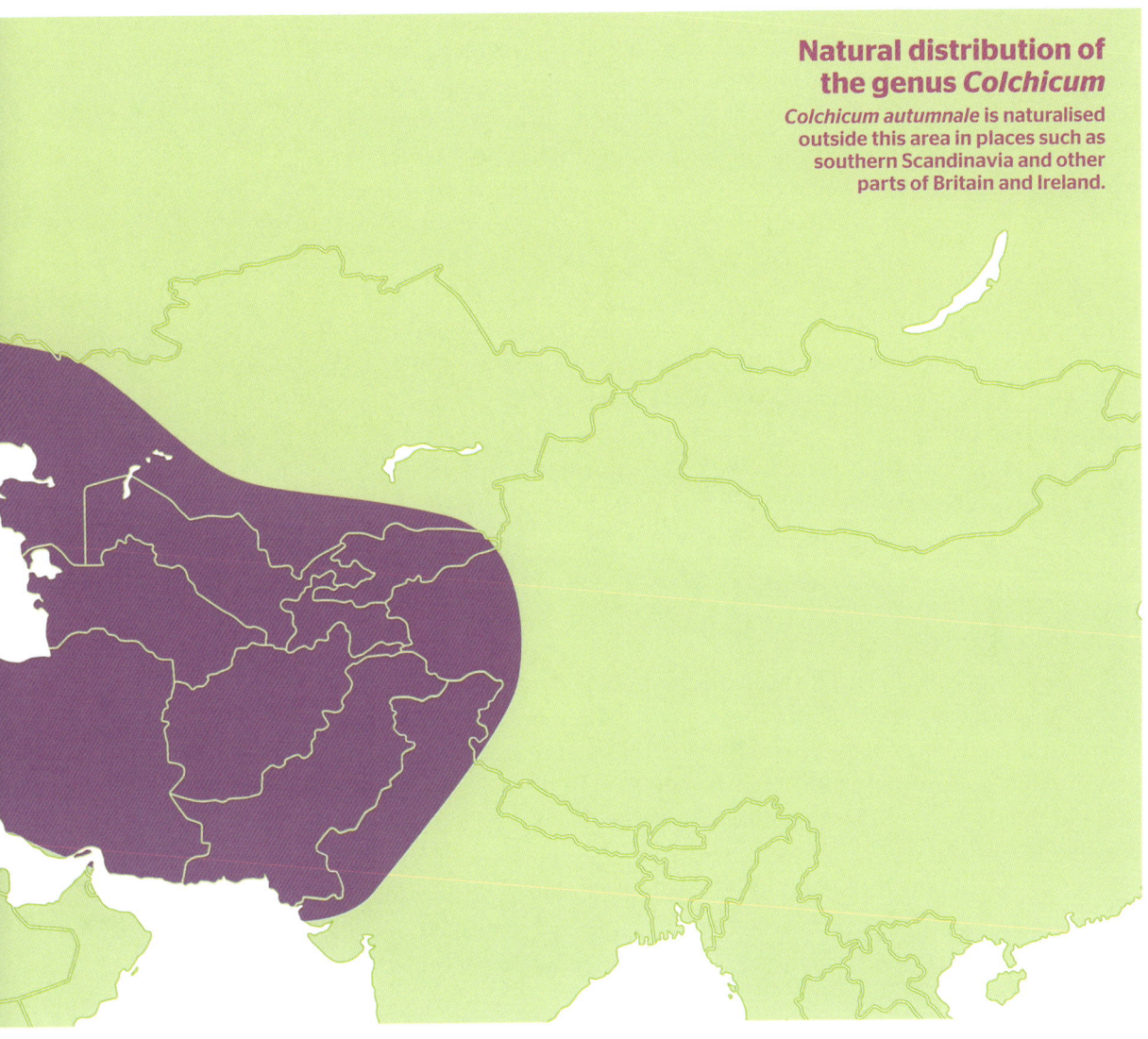

Natural distribution of the genus *Colchicum*
Colchicum autumnale is naturalised outside this area in places such as southern Scandinavia and other parts of Britain and Ireland.

meadows and pastures, stream banks, moist depressions and snow hollows. They typically flower in spring, shortly after snow melt. A few montane species, such as *C. kurdicum* and *C. szovitsii*, are even able to withstand seasonally flooded ground when the snow melts in the spring on quite heavy clay soils.

Colchicum morphology is surprisingly uniform, with diagnostic characters ill-defined or difficult to circumscribe. Despite this, some species, such as *C. autumnale* and *C. speciosum*, are inherently extremely variable. The desert species found particularly in the Middle East are especially problematic, with wide variations in flower size and colour, and in the shape of the tepals. Glabrous species can also occasionally have individuals with partly pubescent or ciliate-margined leaves. Also, within a species, the corms may have no hypopodium in some plants and a very pronounced one in others.

Habitat can also affect plant vigour considerably, especially with regard to the number of leaves and flower size. Flowering time is also quite variable. While the majority of species can be divided into spring- or autumn-flowering species, others fall midway, flowering in late autumn or winter with leaves partly developed. A few may flower before their leaves develop in most years but with their leaves in others, particularly if the rains arrive early in the autumn.

GEOGRAPHIC DISTRIBUTION

The genus is distributed from western Europe (including Britain) eastward, encompassing much of the Mediterranean region as well as North Africa, to the Middle East. Its eastern distribution extends to Kazakhstan, Kyrgyzstan, Tajikistan and northwest India.

Greece and Turkey are the two main regions of diversity and speciation for the genus. From Greece, 31 species have been recorded, 14 of them endemic, and there are 35 species in the whole Balkan Peninsula. Turkey has 37 species, 16 of which are endemic.

CONSERVATION OF THE GENUS

Colchicum is not listed as an endangered genus by the Convention on International Trade in Endangered Species of Wild Fauna and Flora (CITES), an international agreement between governments that aims to ensure that the trade of wild animals and plants does not threaten their survival. However, some of the more localised species of *Colchicum* are very restricted in the wild, particularly in Turkey and Iran, and several have been targeted by unscrupulous collectors. Of the species assessed for the IUCN Red List (International Union for Conservation of Nature's Red List of Threatened Species), two are critically endangered (*C. leptanthum* and *C. trigynum* (listed as *C. greuteri*)) and two are vulnerable (*C. corsicum* and *C. macedonicum*).

Many countries have regulations that prohibit the digging up of plants in the wild, indeed the collection of any plant material, including seeds. Those wishing to carry out research into a particular plant group must seek permission to do so within the country concerned. This is often done by applying through a specific botanical institute or botanic garden. Reserves and national parks often have more stringent laws regarding collecting.

Today, a great deal of information is available on the localities of rare and endangered species of plants and animals, with their occurrences pinpointed by GPS technology. Fortunately, the vast majority of those seeking plants in the wild do so to enjoy their beauty and appeal where they grow, not to dig them up. For them the camera is their prime tool, not the trowel. There will always be a need to monitor and collect species from the wild for scientific analysis, but it has to be done with due consideration for their natural populations. Collecting to enhance one's private collection or for commercial gain is not acceptable.

IDENTIFYING SPECIES

Colchicum species are undoubtedly difficult to identify. Studying plants in leaf as well as in flower, if it is possible, will reinforce the accuracy of an identification. However, there are certain morphological characters that are worth paying particular attention to. These are listed below, not in any order of preference, with details about how they should be assessed. They all carry equal importance in interpretation and diagnosis.

1. Corm shape and tunic details, especially for identifying those species with soboliferous corms. Corms can, however, vary much in detail within a species, in size and shape, but also in the presence or absence of a well-developed hypopodium, and the presence or absence and length of a neck.
2. Cataphyll details. Cataphyll length can vary enormously within a given species, this primarily depending on the depth of the corm in the soil. Cataphylls arise from near the base of the corm and reach the soil surface, so they can therefore indicate the underground length of perianth tube or claw. Cataphyll measurements are only given in the species entries when known or where it was possible to obtain measurements from herbarium specimens.
3. Leaf size (at flowering time or after), number and shape, presence of longitudinal ridges or pleats, and margin details (smooth, cartilaginous or ciliate). Leaf measurements are given in the species entries for fully developed leaves of mature plants. However, in species that have leaves partly developed at flowering, leaf measurements as they are at flowering are given where possible.
4. Flower numbers and season.
5. Perianth tube colour, length and thickness, and whether split to the base, as in species formally included under *Merendera* and *Bulbocodium*. The length

Colchicum cupanii subsp. *glossophyllum* has a more restricted distribution than subsp. *cupanii* and has narrower leaves, being found only in Greece and southwest Albania.

of the perianth tube is measured from soil level to the base of the tepals. With species formerly included in *Merendera* and *Bulbocodium* the claw is measured the same way. The perianth tube and claw continue underground to the base of the corm, but this is not measured as it is impractical and destructive to dig up plants in flower.

6. Tepal shape, size (inner and outer whorl), colour, presence or absence of tessellations, ridges or stripes, filament channel details, and presence or absence of basal auricles. Tepals are measured from the base or point that they unite, to the apex. Flower colour can be quite variable in some species, so a range of colour possibilities is presented in the species entries.
7. Stamen details, such as length, colour of filament and of the sometimes swollen base, point of incision of the filament bases, filament surface (glabrous, papillose, ciliate, or ridged with fine lamellae), anther colour, and whether basifixed or versatile. Anther colour can vary a lot in a species, and sometimes between individuals in a single population. It is important not to confuse anther colour and pollen colour, as anther colour may be masked when the pollen emerges, being various shades of yellow, orange-yellow or, in a few instances, green or cream.
8. Style length (shorter, equalling or exceeding the stamens), whether free or fused, colour, and stigma shape (punctiform or decurrent).
9. Fruit capsule position (partly subterranean, at the soil surface, or well above ground on expanded pedicels); fruit surface (glabrous, papillose or part pubescent) and whether adorned with brown or reddish brown dots.
10. Seed details. Published accounts rarely include seed details, but observations can be important.

KEY TO *COLCHICUM* SPECIES

Flowers and mature leaves are required for accurate species identification. There is often a problem matching up the flowers of the species that flower without their leaves, with their later appearing leaves. For these species an attempt should be made to study plants from a given location twice, first when they are in flower in the autumn, and second when they have fully developed leaves and fruits, usually by late spring. However, caution is required, as on occasions two or more species may be growing in close proximity and could be mixed up. If a collecting permit can be obtained, national laws are adhered to, and there are plenty of plants at the site in question, living material could be uprooted in its flowering state. Then it is possible to grow on the corms to leafing maturity and thus link up flowers with leaves. The conservation status of the species is paramount, however, especially with regard to those known only from a few localities. It is sometimes also very difficult to match leaf and flower samples in herbaria, with or without the presence of corms and fruit capsules.

All this makes the construction of keys for identifying species particularly complicated. Species may need to be keyed out more than once to accommodate variation. It is also difficult to combine leaf and flower characteristics into a single key. Some published accounts present two keys, one to flowering material and one to leaves. The keys here are based on the species as they appear in the wild and on authenticated, wild-collected material.

In the keys below, the first one is broken down into five additional keys based on important botanical characters, thus avoiding one very lengthy and cumbersome key. In some instances this means that a variable species might fall into several divisions of a particular key or appear in more than one key.

Keys are only an aid to accurate identification. Other details need to be taken into consideration such as geography, flowering time and habitat.

KEY ONE

1a	Tepals free, separated to the base (species formerly included in *Bulbocodium* and *Merendera*)	**KEY SIX**
1b	Tepals united into a tube (perianth tube)	2
2a	Anthers basifixed, the filaments short	3
2b	Anthers dorsifixed, the filaments long	4
3a	Flowers yellow, tepals striped; leaves (2–)3–4	*C. luteum* (N & NE Afghanistan; NW China – W Xinjiang; Kazakhstan; Kyrgyzstan; NW Pakistan – Chitral, Swat; Tajikistan; Uzbekistan)
3b	Flowers white, the tepals with a broad, violet-black stripe; leaves 2–7	*C. kesselringii* (Kazakhstan; Kyrgyzstan; Tajikistan; Uzbekistan)
4a	Plants flowering in advance of leaf development (hysteranthous species) (*C. macedonicum* flowers in mid-summer, often with last season's leaves still present)	5
4b	Plants flowering with the leaves partly or fully developed (synanthous species)	6
5a	Flowers tessellated mildly to boldly	**KEY TWO**
5b	Flowers not tessellated or obscurely so	**KEY THREE**
6a	Plants flowering in autumn or early winter (September to early January)	**KEY FOUR**
6b	Plants flowering in late winter or spring (late January to June)	**KEY FIVE**

KEY TWO (hysteranthous species with tessellated flowers)

1a	Flowers star-shaped, tepals heavily tessellated; anthers purple-brown or purple-black, pollen yellow or occasionally green	2
1b	Flowers chalice-shaped to campanulate, faintly to heavily tessellated; anthers yellow or greyish yellow (dark purple in *C. feinbruniae*), pollen yellow	3
2a	Tepals pale to mid rose-pink; pollen green or yellowish; leaves large,	

	11–15.5cm wide, strongly pleated, not undulate

       C. macrophyllum (Crete; Rhodes; SW Turkey)

2b Flowers deep red- or violet-purple, rarely paler; pollen yellow; leaves narrow, not more than 2.5cm wide, undulate ***C. variegatum*** (S Greece; Aegean islands (Cyclades); Rhodes; SW Turkey)

3a Flowers deep red-purple, pubescent at base within; leaves glaucous, ciliate at margin ***C. turcicum*** (Bulgaria; N Greece; Republic of North Macedonia; NW Turkey)

3b Flowers pale purple-pink, lilac-purple to rose-purple or whitish, glabrous or pubescent at base within; leaves green, occasionally grey-green, glabrous **4**

4a Filament channels finely pubescent or puberulent; leaves 5–9(–11) **5**

4b Filament channels glabrous, rarely slightly puberulous; leaves 3–5(–6) **6**

5a Flowers 1–4, boldly tessellated; tepals 40–70mm long; stigma decurrent for 3–4mm ***C. bivonae*** (Balkans; Bulgaria; Sardinia; S Italy; Sicily; W Turkey – including Ayiassos and Lesbos)

5b Flowers usually solitary, moderately tessellated; tepals 26–38mm long; stigma decurrent for 1.5–2.5mm ***C. gracile*** (S Italy)

6a Stamens long, two-thirds to four-fifths the length of the tepals; styles usually crimson-purple-tipped, the stigma decurrent for only 1mm, or non-decurrent ***C. cilicicum*** (S & SE Turkey; Syria; Jordan)

6b Stamens short, not more than half the length of the tepals; styles white, seldom purple-tipped, the stigma decurrent for 0.7–5(–7)mm **7**

7a Anthers purple; flowers 3–15, tepals very finely tessellated ***C. feinbruniae*** (Israel; Lebanon; Syria)

7b Anthers yellow, rarely buff; flowers 1–4, tepals coarsely tessellated, sometimes very faintly so **8**

8a Style stout, swollen and curved at the tip; corms irregular, often soboliferous ***C. davisii*** (S Turkey)

8b Style slender, straight or somewhat curved, not swollen at the tip; corms regular, non-soboliferous **9**

9a Anthers 9–15mm long; tepals often longitudinally pleated; leaves usually 3 ***C. euboeum*** (N Greece – Euboea)

9b Anthers 4–9mm long; tepals not longitudinally pleated; leaves 3–6 **10**

10a Flowers moderately tessellated; leaves twisted or undulate **11**

10b Flowers faintly tessellated; leaves not twisted, rarely slightly so, or undulate **13**

11a Flowers deep lilac-purple to rose-purple; filaments white, sometimes flushed purple towards the top; leaves bright shiny green ***C. parnassicum*** (mainland Greece)

11b Flowers pale lilac-purple or whitish with mid-purple tessellations; filaments pale yellow or whitish; leaves grey-green **12**

12a Filament channels glabrous; tepals 3–12mm wide, with a median white stripe for up to three-quarters the length; stigma decurrent for 1.5–2.5mm, rarely more .... *C. sfikasianum* (S Greece – Attica, Peloponnese, Ionian Isles)

12b Filament channels minutely puberulous; tepals 8–11mm wide, without a median white stripe; stigma decurrent for 1.5–5.5mm .... *C. chalcedonicum* (E & NE Greece; NW Turkey)

13a Flowers 4–20; cataphylls appearing above the ground; leaves stiffly ascending .... *C. hierosolymitanum* (Israel; Jordan; Lebanon; Palestine; Syria)

13b Flowers generally 1–5; cataphylls often reaching ground level but not protruding above; leaves ascending to spreading or pressed close to the ground 14

14a Flowers goblet-shaped (infundibular), narrow, mauve-purple; leaves dull green .... *C. autumnale* subsp. *pannonicum* (E Europe)

14b Flowers funnel-shaped, often narrowly so, pinkish purple to lilac-purple; leaves bright green or grey-green 15

15a Stigma decurrent for 2–6(–7)mm; anthers 6–9mm long; leaves not or slightly twisted, ascending to erect .... *C. graecum* (C Albania; S Greece – Peloponnese)

15b Stigma decurrent for just 0.7–6mm; anthers 4–8mm long; leaves not as above 16

16a Stigma decurrent for 1–2mm; tepals less than 8mm wide, without a median white streak; leaves appressed to the ground, subacute .... *C. lingulatum* (S. Greece; NE Turkey; W Anatolia)

16b Stigma decurrent for 1.5–6mm; tepals up to 16mm wide, often with a median white streak; leaves ascending, slightly twisted .... *C. haynaldii* (Adriatic states – Albania, W Bulgaria, SW Romania, N & NE Greece)

KEY THREE (hysteranthous species with non-tessellated flowers)

1a Anthers purple, purple-brown, blackish purple, grey-green, violet-grey or brownish grey, pollen usually yellow 2

1b Anthers yellow or orange yellow (purple-brown in *C. speciosum* Bornmuelleri Group and *C. lingulatum*), pollen yellow 7

2a Flowers deep crimson-purple; stigma decurrent for 3–4mm at style tip; leaves 5–9 .... *C. turcicum* (Bulgaria; N Greece; Republic of North Macedonia; NW Turkey)

2b Flowers pink, pinkish purple to rose- or lilac-purple, or white; stigma punctiform, or decurrent up to 4mm at style tip; leaves 2–5 3

3a Stigma punctiform; tepals not more than 5.5mm wide; leaves 2–8, spreading or appressed to ground 4

3b Stigma decurrent; tepals more than 6mm wide; leaves 2–5, ascending

	to upright	5
4a	Cataphylls protruding above ground for up to 10mm; leaves 2–4, not more than 2mm wide, spreading; anthers grey-green or purple-grey *C. antilibanoticum* (Israel; Lebanon; Syria)	
4b	Cataphylls not protruding above ground; leaves 3–8, 2–3.5mm wide, appressed to, or curving over, the ground; anthers violet-green or brownish green *C. cretense* (Crete)	
5a	Flowers large, tepals 11–24mm wide; filament channels finely pubescent; style swollen at tip; anthers often purple or purple-brown *C. speciosum* Bornmuelleri Group (Turkey; C Anatolia)	
5b	Flowers smaller, tepals 6–14mm wide; filament channels glabrous; style not or scarcely swollen at the tip; anthers blackish or blackish purple or purplish pink	6
6a	Tepals less than 35mm long; leaves linear to linear-lanceolate *C. haynaldii* (Adriatic states – Albania, W Bulgaria, SW Romania, N & NE Greece)	
6b	Tepals 40–60mm long; leaves oblong-lanceolate *C. lusitanum* (C Italy; Portugal; Spain)	
7a	Filament channels finely, often densely, pubescent	8
7b	Filament channels glabrous or practically so	12
8a	Tepals dense and opaque; leaves 3–4(–5)	9
8b	Tepals thin and delicate; leaves 4–12	11
9a	Flowers 3–7, the perianth tube (from ground level) 3–6cm long; filaments white with an orange yellow base *C. kotschyi* (W & NW Iran; Iraq; Turkey – Inner Anatolia)	
9b	Flowers 1–3, the perianth tube (from ground level) 5–12cm long; filaments greenish or yellowish	10
10a	Flowers chalice-shaped; anthers yellow; stigma decurrent for 2–4mm *C. speciosum* (Caucasus; Turkey – N & NE Anatolia; N Iran) & *C. woronowii* (W Caucasus)	
10b	Flowers funnel-shaped; anthers purple-brown; stigma generally decurrent for 0.5–1.5mm *C. speciosum* Bornmuelleri Group (Turkey – NW Anatolia)	
11a	Flowers 1–3; anthers yellow or golden-yellow; leaves yellowish green to mid-green, moderately thick *C. inundatum* (S Turkey)	
11b	Flowers 5–10(–15); anthers straw-coloured; leaves matt green, very thick *C. imperatoris-fridericii*	
12a	Stigma punctiform	13
12b	Stigma decurrent for 0.5–5(–6)mm	28
13a	Flowers often solitary; leaves 3, variable, straight or curling, often hairy *C. guessfeldtianum* (Egypt – Sinai; S Jordan)	
13b	Flowers 2 or more, occasionally solitary; leaves 2–10, not curling, glabrous	14
14a	Cataphyll protruding above ground, red *C. sanguicolle* (Turkey – S Anatolia)	
14b	Cataphyll not or only slightly protruding, white if visible at anthesis	15

15a Leaves 2(–3), the tips often just showing at anthesis or previous season's leaves still present — 16
15b Leaves 3–30, not showing at flowering time — 20
16a Flowers white or pale pink; base of filaments dark yellow — 17
16b Flowers bright lilac-purple, pink or rose-purple; base of filaments not yellow, greenish yellow or orange-yellow — 18
17a Flowers spidery in appearance with even-sized, acute-tipped tepals; corms not soboliferous — *C. antilibanoticum* (Israel; Lebanon; Syria)
17b Flowers not spidery in appearance, with subobtuse tepals, the inner noticeably shorter than the outer; corms soboliferous — *C. rausii* (mainland Greece – S Pindus)
18a Flowers bright lilac-purple, tepals mostly 2.5–4cm long; corms with underground stolons — *C. boissieri* (S Greece; Turkey – W & S Anatolia)
18b Flowers pink to rose-purple or lilac, tepals 1.7–3.5cm long; corms without underground stolons — 19
19a Flowers purplish pink, occasionally white; tepals 1.7–3cm long; flowering August to September — *C. alpinum* (S Alps; south to Corsica, Sardinia & Sicily)
19b Flowers deep purple-rose; tepals 2.5–3.5cm long; flowering June — *C. macedonicum* (Albania; Republic of North Macedonia)
20a Flowers numerous, usually 10–25 per corm; leaves 10–30, spreading horizontally — *C. polyphyllum* (NW Syria)
20b Flowers 1–8(–12) per corm; leaves 3–10, spreading to ascending — 21
21a Tepals mostly 8–30mm long; corm tunic coriaceous — 22
21b Tepals 25–45mm long; corm tunic membranous to subcoriaceous — 24
22a Flowers pink- to mid-rose-purple; leaves 3–4(–5), ascending, channelled, 5–14mm wide, sometimes scabrid along the margin — *C. pulchellum* (Greece – Peloponnese)
22b Flowers lilac- to pink-purple, pale pink or white; leaves 6–8(–10), spreading, not channelled, 1–5mm wide, glabrous — 23
23a Flowers pink-purple or lilac — *C. parlatoris* (Greece – Peloponnese)
23b Flowers white or palest pink — *C. tunicatum* (Israel; Jordan; Palestine; S Syria)
24a Flowers white to pale pink; anthers with or without longitudinal membranous margins; leaves 3–6(–8) — 25
24b Flowers purple-pink to bright pink; anthers without longitudinal membranous margins; leaves 3–5 — 26
25a Flowers white; perianth tube short, equalling the linear-lanceolate tepals; leaves grey-green — *C. maraschicum* (S Turkey)
25b Flowers white or pale purplish pink; perianth tube long, normally 2–3 times longer than tepals; leaves deep green — *C. troodii* (Cyprus; W Syria; Turkey – S Anatolia)
26a Flowers bright pink; leaves usually 3, 24–45mm wide at maturity; corms soboliferous — *C. baytopiorum* (Turkey – SW Anatolia)

26b Flowers purplish pink to lilac purple with a white centre; leaves 3–5, 4–17mm wide at maturity; corms soboliferous or not   27
27a Filaments yellow at the base; flowers 1–4; anthers 4–6mm long; corms regular, without stolons   *C. arenarium* (Bosnia and Herzegovina; Croatia; Czech Republic; Hungary; Kosovo; Montenegro; Republic of North Macedonia; Serbia; Slovakia; Slovenia)
27b Filaments green at the base; flowers 1–2; anthers 5–8mm long; corms soboliferous   *C. chlorobasis* (S Turkey)
28a Stigma long-decurrent, for 1–4(–5)mm   29
28b Stigma short-decurrent, for no more than 0.5–1.5mm   35
29a Anthers 2–3mm long; leaves glaucous   30
29b Anthers 4–10mm long; leaves green, occasionally slightly grey-green   31
30a Flowers rose-pink or rose-purple; tepals 12–13mm long   *C. corsicum* (S Corsica)
30b Flowers pale pink or lilac-pink; tepals 20–31mm long   *C. arenasii* (W Corsica & Sardinia)
31a Anthers 10–17mm long, with longitudinal membranous margins; flowers white to pale purplish pink; corm with a very long fibrous neck to 25cm, reaching ground level, concealing the cataphylls   *C. balansae* (Turkey – S & SW Anatolia; Rhodes – Salakos)
31b Anthers 5–10mm long (rarely to 11mm in *C. parnassicum*), without obvious membranous margins; flowers bright pink to purple-pink or lilac- or rose-purple; corm neck fibrous or not, to 14cm long at most, not concealing the cataphylls   32
32a Flowers large, tepals 40–65mm long; leaves generally (3–)4(–5)   *C. autumnale* (C, S & W Europe; W Russia; Ukraine)
32b Flowers smaller, tepals 10–45mm long; leaves (2–)3–7   33
33a Leaves (2–)3–4, linear-lanceolate, not undulate, generally more than 15cm long at maturity, ascending to erect; anthers yellow   *C. neapolitanum* (European Mediterranean; Portugal)
33b Leaves 4–9, strap-shaped to oblong-elliptic, often undulate, generally less than 15cm long at maturity, flat on the ground; anthers yellow or purplish   34
34a Flowers 4–20 per corm; leaves 5–9, not more than 20mm wide; anthers yellow, 6–10mm long   *C. hierosolymitanum* (Israel; Jordan (rare); Lebanon; Palestine; possibly in Syria)
34b Flowers 1–5 per corm normally; leaves 4–5, 10–33mm wide; anthers yellow or greyish yellow, 4–8mm long   *C. lingulatum* (C Greece – Euboea; SW Turkey)
35a Flowers large, the tepals 50–75 × 12–25mm; leaves 5.5–7(–11)cm wide   *C. cilicicum* (Lebanon; NW Syria; Turkey – S Anatolia)
35b Flowers smaller, the tepals not more than 45 × 12mm; leaves 0.3–4.5cm wide   36

36a Anthers 8–15mm long, longer than the filaments; flowers white or very pale pink; mature leaves large, 5–10cm wide   *C. dolichantherum* (NW Syria; Turkey – S Anatolia)

36b Anthers 3–8mm long (to 10mm in *C. persicum*), shorter than the filaments ; flowers white, pale pink, pinkish purple to deep rose-purple; mature leaves not more than 4.5cm wide   37

37a Tepals distinctly grooved along the median line; leaves 5–many; fruit capsule brown-dotted   *C. persicum* (W Iran)

37b Tepals not as above; leaves 3–6(–8); fruit capsules not brown-dotted   38

38a Tepals 25–45mm long; anthers 5–8mm long   39

38b Tepals 10–31mm long (sometimes to 40mm in *C. umbrosum*); anthers 3–5mm long   43

39a Stigma decurrent for 0.5–1.5(–2)mm; filament channels glabrous or subglabrous   40

39b Stigma decurrent for no more than 0.5mm; filament channels puberulous at least on the margins   42

40a Flowers white, sometimes with a pink flush   *C. paschei* (S Turkey)

40b Flowers lilac to lilac-pink or purple-pink   41

41a Styles greatly exceeding the stamens; leaves 3(–4), ascending to erect   *C. laetum* (SE Russia)

41b Styles equalling or slightly longer than the stamens; leaves 4–7, spreading to pressed close to the ground   *C. lingulatum* (S Greece; Turkey – NE & W Anatolia)

42a Flowers white (sometimes flushed pink) or very pale pink; corm tunics membranous; mature leaves 3–8, dark glossy green, flattish   *C. troodii* (Cyprus)

42b Flowers white to pink or pale purple-pink to purple; corm tunics almost coriaceous; mature leaves 3, mid-green to grey-green, channelled   *C. decaisnei* (N Israel; Lebanon; Syria)

43a Tepals and stamens 4–5; styles 2; ovary and fruit with 2 locules (all other species in the genus have 3 locules)   *C. gonarei* (C Sardinia)

43b Tepals and stamens 6; styles 3; ovary and fruit with 3 locules   44

44a Tepals deep pink to rose-purple, 4–9(–12)mm wide   *C. pulchellum* (N Greece)

44b Tepals pink to lilac-pink or pale purple, 2–6mm wide   45

45a Stigma decurrent for at least 1mm; tepals 10–20(–30)mm long   *C. micranthum* (NW Turkey)

45b Stigma decurrent for no more than 0.5mm; tepals mostly 16–30mm long, occasionally longer   46

46a Flowers (1–)2–6; tepals incurved at apex, 30–40mm long; leaves ligulate, green, 13–20mm wide; corm tunic leathery   *C. umbrosum* (W Caucasus; Romania; S Russia, Crimea; Turkey – N Anatolia)

46b Flowers 1–2(–3); tepals not incurved at apex, 12–23mm long; leaves linear-lanceolate to lanceolate, glaucous, 3–9(–18)mm wide; corm tunic membranous **C. nanum** (S Corsica)

KEY FOUR (synanthous, autumn-flowering species)

1a Styles solitary **C. "ramonensis"** (Israel)
1b Styles 3 2
2a Tepal base within and stamen and style bases all adorned with long hairs; male and female plants present **C. tuviae** (Israel; Palestine)
2b Tepals, stamens and styles glabrous; plants all hermaphrodite 3
3a Anthers purple-black, purple-brown, brownish or greyish green, pollen yellow; stigma punctiform 4
3b Anthers yellow, pollen yellow, sometimes pale or orange-yellow; stigma punctiform or decurrent 11
4a Leaves 6–20, linear to filiform, generally hairy on the dorsal surface and along the margins; flowers white or yellowish, sometimes flushed pink or purple in the upper part **C. crocifolium** (W Iran; N Iraq; Jordan; N Syria; SC Turkey)
4b Leaves 2–8, linear to strap-shaped (ligulate), usually glabrous or sometimes ciliate on the margins (sometimes hairy on the dorsal surface in *C. hungaricum*); flowers white, pink, purple or lilac-purple 5
5a Leaves 2(–3), 5–18mm wide, strap-shaped 6
5b Leaves 2–8, 1–10mm wide, linear or filiform 7
6a Tepals 18–25mm long, not more than 7mm wide; leaves glabrous, occasionally ciliate on the margin at the base; flowering September to December **C. cupanii** (Mediterranean region from SE France eastwards to Greece, including Crete)
6b Tepals 24–30mm long, 5–10mm wide; leaves retrorse-ciliate, also often hairy on the dorsal surface; flowering December onwards **C. hungaricum** (Albania; Bosnia; Bulgaria; N Greece; Hungary; Republic of North Macedonia; Slovenia)
7a Leaves 2(–3); corms stolon-like (soboliferous) **C. psaridis** (S Greece – Peloponnese)
7b Leaves 3–5; corms not stolon-like (not soboliferous) 8
8a Filaments with lanceolate teeth on either side at base; leaves usually 3 **C. ritchii** (Egypt; Israel; S Jordan; Palestine)
8b Filaments without basal teeth; leaves 3–5 9
9a Filaments swollen at dark yellow base; leaves mostly 5–10mm wide **C. schimperi** (Egypt; Israel; Jordan; S Syria)
9b Filaments not as above, greenish or whitish at base; leaves 1–5mm wide 10
10a Cataphylls white and protruding above ground at flowering; flowers 5–7 **C. fasciculare** (Jordan; Syria)

10b Cataphylls not protruding above ground at flowering; flowers 1–4 *C. pusillum* (C & S Greece; Aegean islands)

11a Anthers 8–15mm long, equalling or longer than the corresponding filaments; mature leaves large, 5–10cm wide *C. dolichantherum* (NW Syria; S Turkey)

11b Anthers not more than 6mm long, shorter than the corresponding filaments; mature leaves smaller, not more than 4.5cm wide 12

12a Leaves linear or filiform, 1–4mm wide 13

12b Leaves ligulate, narrow lanceolate or strap-shaped, 5–60(–90)mm wide 14

13a Tepals 4–6mm wide; styles shorter than stamens *C. stevenii* (NE Cyprus; W Syria; Turkey – S & W Anatolia) & *C. osmaniyense* (S Turkey)

13b Tepals 7–10mm wide; styles equalling stamens or slightly longer *C. peloponnesiacum* (Greece – Peloponnese)

14a Leaves 2, not more than 2cm wide at maturity; flowers deep purple-rose *C. macedonicum* (Albania; Republic of North Macedonia)

14b Leaves 3–4–5(–6), 2–6(–9)cm wide at maturity; flowers pinkish purple to rose-purple 15

15a Stigma decurrent for 2–6mm; leaves 3–6; corms not stoloniferous *C. graecum* (Greece – Peloponnese)

15b Stigma punctiform; leaves usually 3; corms stoloniferous *C. baytopiorum* (Turkey – SW Anatolia)

KEY FIVE (synanthous, spring-flowering species; all have punctiform stigmas)

1a Filaments pubescent at least below; leaves with ciliate margins *C. burttii* (W Turkey, including C Anatolia)

1b Filaments glabrous; leaves usually not ciliate, occasionally with a somewhat scabrid margin or lamellae rarely ciliate 2

2a Filaments with lanceolate teeth on both sides at base *C. ritchii* (Egypt; Israel; S Jordan; Palestine)

2b Filaments without teeth at base 3

3a Flowers white, flushed purple (at base or in upper part) 4

3b Flowers pink, pinkish purple, lilac-pink or pure white 5

4a Flowers 1–3; anthers yellow to buff, occasionally greyish purple; leaves 3(–4), to 13mm wide *C. freynii* (W Iran)

4b Flowers 3–6; anthers purple-black to dark grey; leaves 5–6(–8), not more than 6mm wide *C. chimonanthum* (N Greece)

5a Corms soboliferous; leaves often exceeding flowers at anthesis 6

5b Corms not soboliferous; leaves generally shorter than flowers at anthesis 7

6a Filaments and styles white; leaves mid-green; tepals faintly veined *C. minutum* (Turkey – S Anatolia)

6b Filaments and styles yellowish or greenish; leaves dark green; tepals

	distinctly veined	***C. munzurense*** (EC Turkey)

7a Leaves 6–20, not all present at flowering, often hirsute, sometimes glabrous; flowers 4–12; fruit capsule often papillose to hirsute
C. crocifolium (Jordan; Syria)

7b Leaves 2–5, partly developed at flowering, glabrous, occasionally ciliate on the margin; flowers 1–7, rarely more; fruit capsules glabrous 8

8a Anthers purple-black or greyish purple, pollen yellow 9

8b Anthers yellow, greenish black, greenish purple or greyish brown, pollen yellow 11

9a Flowers purple-pink, globular-campanulate, the tepals 6–12mm wide; leaves 3(–4), not more than 10mm wide at anthesis, glabrous or minutely scabrid at margin   ***C. triphyllum*** (NW Africa; Bulgaria; C & S Spain; Greece; Romania; S Russia (including Crimea); Turkey – W, C & S Anatolia)

9b Flowers rose-lilac, purple-carmine, lilac, pinkish or white, chalice-shaped to funnel-shaped, the tepals 6–7mm wide; leaves 2, rarely 3, more than 3cm wide at anthesis, usually ciliate at the margin 10

10a Leaves dorsally pilose, linear to linear-lanceolate; flowers rose-lilac to purplish carmine; corm tunic leathery   ***C. doerfleri*** (Albania)

10b Leaves glabrous except often for a ciliate margin, linear-lanceolate; flowers lilac, pinkish or white; corm tunic membranous
C. hungaricum (Albania; Bosnia; Bulgaria; Republic of North Macedonia; N Greece; Hungary; Slovenia)

11a Base of filaments swollen, dark yellow; leaves pubescent or glabrous
C. schimperi (Egypt; Israel; Jordan; S Syria)

11b Base of filaments scarcely swollen, greenish yellow or whitish; leaves usually glabrous, sometimes ciliate at the margin 12

12a Leaves 2(–3) at maturity, mostly 15–35(–45)mm wide 13

12b Leaves 3–5 or more at maturity, 1–20mm wide 14

13a Tepals auricled at the base; leaves 2–3, convolute and sheathing the perianth tubes; corm tunics subcoriaceous   ***C. szovitsii*** (W & NW Iran; Caucasus; Turkey – N, C & S Anatolia)

13b Tepals not auricled; leaves 2, not convolute; corm tunics membranous or coriaceous   ***C. lagotum*** (NE Turkey)

14a Leaves linear to linear-lanceolate, 1–5(–7)mm wide at maturity; corm tunics membranous   ***C. serpentinum*** (Caucasus; Turkey – C & S Anatolia)

14b Leaves narrow-oblong, straight-sided, 5–20mm wide at maturity; corm tunic leathery (coriaceous) 15

15a Anthers yellow or buff; at least inner tepals with lamellae along the filament grooves   ***C. varians*** (W Iran)

15b Anthers muddy yellow or grey-green; tepals without basal lamellae
C. antepense (SC Turkey)

KEY SIX (species formerly in the genera *Merendera* or *Bulbocodium*; with perianth split to the base, i.e. with separate, clawed tepals)

1a	Styles united above the throat at the height of the anthers	***C. bulbocodium*** (Pyrenees; Alps; N Italy eastwards to S Russia and Ukraine)
1b	Styles free for their entire length	2
2a	Corms soboliferous (horizontal and stolon-like), the individual plants patch-forming and often gregarious; tepals with a pair of basal, tooth-like appendages	***C. soboliferum*** (Turkey; NW Iran)
2b	Corms upright, not soboliferous, the individual plants not patch-forming, although often gregarious; tepals mostly without basal appendages, sometimes auriculate	3
3a	Anthers basifixed, (4–)5.5mm long or more, usually exceeding the filaments; autumn- and spring-flowering	4
3b	Anthers versatile (except for *C. androcymbioides*), not more than 5mm long, not more than half as long as the filaments; spring-flowering (but *C. atticum* is winter-flowering onwards)	6
4a	Lowermost leaves mostly 10–35mm wide; flowers white, sometimes flushed pink or purple; spring-flowering	***C. robustum*** (Afghanistan; N & NE Iran; S Kazakhstan; Kashmir; Pakistan; Tajikistan; Turkmenistan; S Uzbekistan)
4b	Lowermost leaves 1–9mm wide; flowers pinkish purple to rosy purple; autumn-flowering	5
5a	Flowers appearing in advance of the leaves; leaves 3–4, mostly 4–9mm wide at maturity; anthers mostly 8–12mm long	***C. montanum*** (S France; C Pyrenees; Portugal; Spain)
5b	Flowers and leaves appearing together; leaves 5–10, not more than 3mm wide at maturity; anthers mostly 5.5–8mm long	***C. filifolium*** (Balearic Islands; S France; Portugal; Spain)
6a	Leaves 3, lanceolate, mostly 20–30mm wide at maturity; outermost corm tunics membranous, mid-brown	***C. kurdicum*** (N Iraq; SE Turkey)
6b	Leaves 2–6, linear to linear-lanceolate, 2–12mm wide at maturity; outermost corm tunics usually coriaceous, dark brown or blackish brown	7
7a	Anthers basifixed, yellow; leaves with a cartilaginous, scabrid margin; tepals not more than 2mm wide	***C. androcymbioides*** (S Spain)
7b	Anthers dorsifixed, black or greenish black or grey (yellow in *C. wendelboi* and sometimes in *C. raddeanum*); leaves glabrous, occasionally ciliate or scabrid towards the margin base; tepals 2–10mm wide	8
8a	Tepals not more than 5mm wide, usually not auriculate at the base	9
8b	Tepals 6–9mm wide, usually auriculate at the base	11
9a	Leaves usually ciliate or at least scabrid towards the margin base; flowers mostly 3–6 per corm; tepals narrow-elliptical; anthers black	***C. atticum*** (S Bulgaria; S Greece; Turkey)

9b   Leaves glabrous; flowers 1–4 per corm; tepals oblanceolate or elliptic-oblanceolate; anthers yellow to greenish black (sometimes black in C. figlalii) **10**

10a Flowers lilac; anthers black or greenish black; corms with a long tunicated neck that is 4–8cm long ***C. figlalii*** (SW Turkey, rare)

10b Flowers white to pale pink; anthers yellow; corms with a tunicated neck not more than 3cm long ***C. wendelboi*** (W Iran)

11a Leaves 2; flowers white or pale pink; anthers dark yellow or grey ***C. raddeanum*** (N & W Iran)

11b Leaves 2–4; flowers pinkish purple, lilac or violet, rarely white **12**

12a Flowers 1–2; anthers greenish yellow ***C. manissadjianii*** (N Turkey)

12a Flowers 1–5; anthers dark grey to black or olive-brown ***C. trigynum*** (Caucasus; NW Iran; Turkey)

SPECIES DIVISIONS BASED ON GROSS MORPHOLOGICAL CHARACTERS

Juxtaposition of species in the following groupings does not necessarily imply a close genetic affinity.

A. Species with soboliferous (rhizome-like) corms

A1. Autumn-flowering species flowering without their leaves (hysteranthous):
   C. baytopiorum, C. boissieri, C. chlorobasis, C. psaridis, C. rausii, C. sieheanum.

A2. Spring-flowering flowering with leaves part-developed (synanthous):
   C. asteranthum, C. leptanthum, C. minutum, C. munzurense.

B. Autumn- and early winter-flowering species flowering without leaves (hysteranthous) and with a perianth tube and regular corms

B1. Species with a punctiform stigma: *C. alpinum, C. antilibanoticum, C. arenarium, C. asteranthum, C. chimonanthum, C. cilicicum*, C. cretense, C. crocifolium, C. freynii, C. gonarei, C. imperatoris-friderici*, C. inundatum*, C. longifolium, C. maraschicum, C. parlatoris, C. persicum*, C. polyphyllum, C. pulchellum*, C. sanguicolle*, C. troodii*, C. tunicatum.*

B2. Species with a decurrent stigma:

B2(i). Species with untessellated or very faintly tessellated tepals: *C. arenasii, C. autumnale, C. balansae, C. cilicicum*, C. confusum, C. corsicum, C. decaisnei, C. dolichantherum, C. gracile, C. graecum, C. heldreichii, C. hierosolymitanum, C. imperatoris-friderici*, C. inundatum*, C. kotschyi, C. laetum, C. lingulatum, C. lusitanum, C. macedonicum, C. micaceum, C. micranthum, C. multiflorum, C. nanum, C. neapolitanum, C. parnassicum, C. paschei, C. persicum*, C. pulchellum*, C. sanguicolle*, C. speciosum* (including Bornmuelleri Group and Giganteum Group), *C. troodii*, C. turcicum,*

C. umbrosum*, C. woronowii.

B2(ii). Species with well-marked tessellations: C. bivonae, C. chalcedonicum, C. davisii, C. euboeum, C. feinbruniae, C. macrophyllum, C. sfikasianum, C. variegatum.

* Species with punctiform stigmas or very shortly decurrent (not more than 0.5mm).

C. Late winter- and spring-flowering species with leaves at flowering (synanthous) partly developed and with a perianth tube and regular corms

C1. Species with punctiform stigmas: C. antepense, C. burtii, C. cupanii, C. doerfleri, C. fasciculare, C. hirsutum, C. hungaricum, C. kesselringii, C. lagotum, C. luteum, C. osmaniyense, C. peloponnesiacum, C. pusillum, C. serpentinum, C. stevenii, C. szovitsii, C. triphyllum, C. tuviae, C. varians.

C2. Species with decurrent stigmas: C. guessfeldtianum, C. haynaldii, C. "ramonensis", C. ritchii, C. schimperi.

D. Species with free, clawed tepals (without a perianth tube) and regular corms

D1. Species flowering without their leaves (hysteranthous): C. filifolium, C. montanum.

D2. Species flowering with leaves partly developed (synanthous): C. androcymbioides, C. atticum, C. bulbocodium, C. figlalii, C. ignescens, C. kurdicum, C. manissadjianii, C. raddeanum, C. robustum, C. soboliferum, C. trigynum, C. wendelboi.

HYBRIDS

Four hybrids with binomial names are recognised in the genus. These are C. × agrippinum, C. × alberti, C. × ambiguum and C. × byzantinum are discussed at the end of this chapter, after the species. Out of these, only one, C. × alberti, is naturally occurring and known from the wild, and is derived from spring- to summer-flowering species. The others are hybrids of autumn-flowering species. It is likely that other autumn-flowering species have crossed in cultivation, giving rise to plants simply recognised as cultivars. Among the winter- and spring-flowering species, only a handful of hybrids have so far arisen in cultivation, although their number is likely to rise.

Even where several members of the genus co-occur in the wild, such as C. confusum growing with C. haynaldii and C. autumnale in the northern Pindos in Greece, hybrids have not been found. This is because flowering periods do not necessarily overlap and many are markedly habitat-specific, occupying different niches and forming discrete populations. In addition, chromosome counts may be mutually incompatible.

Identification characteristics of *Colchicum* species

Species	Spring-flowering	Summer-flowering	Autumn-flowering	Winter-flowering	Leaves present at flowering time	Leaves absent at flowering time	Flowers with free, clawed tepals	Corm soboliferous	Tepals untesselated or faintly tessellated	Tepals with well-marked tessellation	Stigma punctiform or shortly decurrent	Stigma decurrent
C. alpinum		■				■				■	■	
C. androcymbioides	■				■		■		■		■	
C. antepense	■				■					■	■	
C. antilibanoticum				■	■				■		■	
C. arenarium			■			■				■		■
C. arenasii			■			■				■		■
C. asteranthum			■			■	■			■	■	
C. atticum	■					■				■	■	
C. autumnale			■			■				■		■
C. balansae			■			■				■		■
C. baytopiorum			■			■			■		■	
C. bivonae			■			■				■		■
C. boissieri			■			■		■	■			■
C. bulbocodium	■				■				■		■	
C. burtii	■					■				■	■	
C. chalcedonicum			■			■			■			■
C. chimonanthum				■	■				■		■	
C. chlorobasis			■			■	■		■		■	
C. cilicicum			■			■				■		■
C. confusum			■			■			■		■	
C. corsicum			■			■			■			■
C. cretense			■		■				■		■	
C. crocifolium	■				■				■		■	
C. cupanii			■		■				■		■	
C. davisii			■			■				■	■	
C. decaisnei			■			■			■			■
C. doerfleri	■					■			■		■	
C. dolichantherum			■			■			■		■	
C. erdalii	■				■				■		■	
C. euboeum			■			■				■		■
C. fasciculare				■	■				■		■	
C. feinbruniae			■			■			■		■	
C. figlalii	■					■	■		■		■	
C. filifolium			■		■				■		■	
C. freynii	■				■				■		■	
C. gonarei			■			■			■		■	
C. gracile			■			■				■		■
C. graecum			■			■				■		■
C. guessfeldtianum				■	■				■		■	

Species	Spring-flowering	Summer-flowering	Autumn-flowering	Winter-flowering	Leaves present at flowering time	Leaves absent at flowering time	Corm soboliferous	Flowers with free clawed tepals	Tepals untesselated or faintly tessellated	Tepals with well-marked tessellation	Stigma punctiform or shortly decurrent	Stigma decurrent
C. haynaldii			●		●				●			●
C. heldreichii		●			●				●			●
C. hierosolymitanum			●	●		●			●		●	
C. hirsutum	●				●				●		●	
C. hungaricum	●			●					●		●	
C. ignescens			●			●		●	●		●	●
C. imperatoris-friderici			●		●					●		●
C. inundatum			●		●					●		●
C. kesselringii	●					●			●			●
C. kotschyi			●			●			●		●	
C. kurdicum	●				●					●		●
C. laetum			●			●			●			●
C. lagotum				●	●				●			●
C. leptanthum	●				●			●	●			●
C. lingulatum			●			●			●		●	
C. longifolium			●		●				●			●
C. lusitanum			●		●				●			●
C. luteum	●				●				●			●
C. macedonicum			●		●				●			●
C. macrophyllum			●			●			●		●	
C. manissadjianii			●		●				●			●
C. maraschicum	●				●				●			●
C. micaceum			●		●				●			●
C. micranthum			●		●				●			●
C. minutum	●				●	●			●			●
C. montanum		●			●				●			●
C. multiflorum			●		●				●			●
C. munzurense	●				●				●			●
C. nanum			●		●				●			●
C. neapolitanum			●			●				●		●
C. osmaniyense				●	●				●		●	
C. parlatoris			●		●				●			●
C. parnassicum		●			●				●		●	
C. paschei			●		●				●			●
C. peloponnesiacum			●			●	●		●			●
C. persicum			●		●				●			●
C. polyphyllum			●		●				●			●
C. psaridis				●		●			●			●
C. pulchellum		●			●				●			●

Identification characteristics of Colchicum species

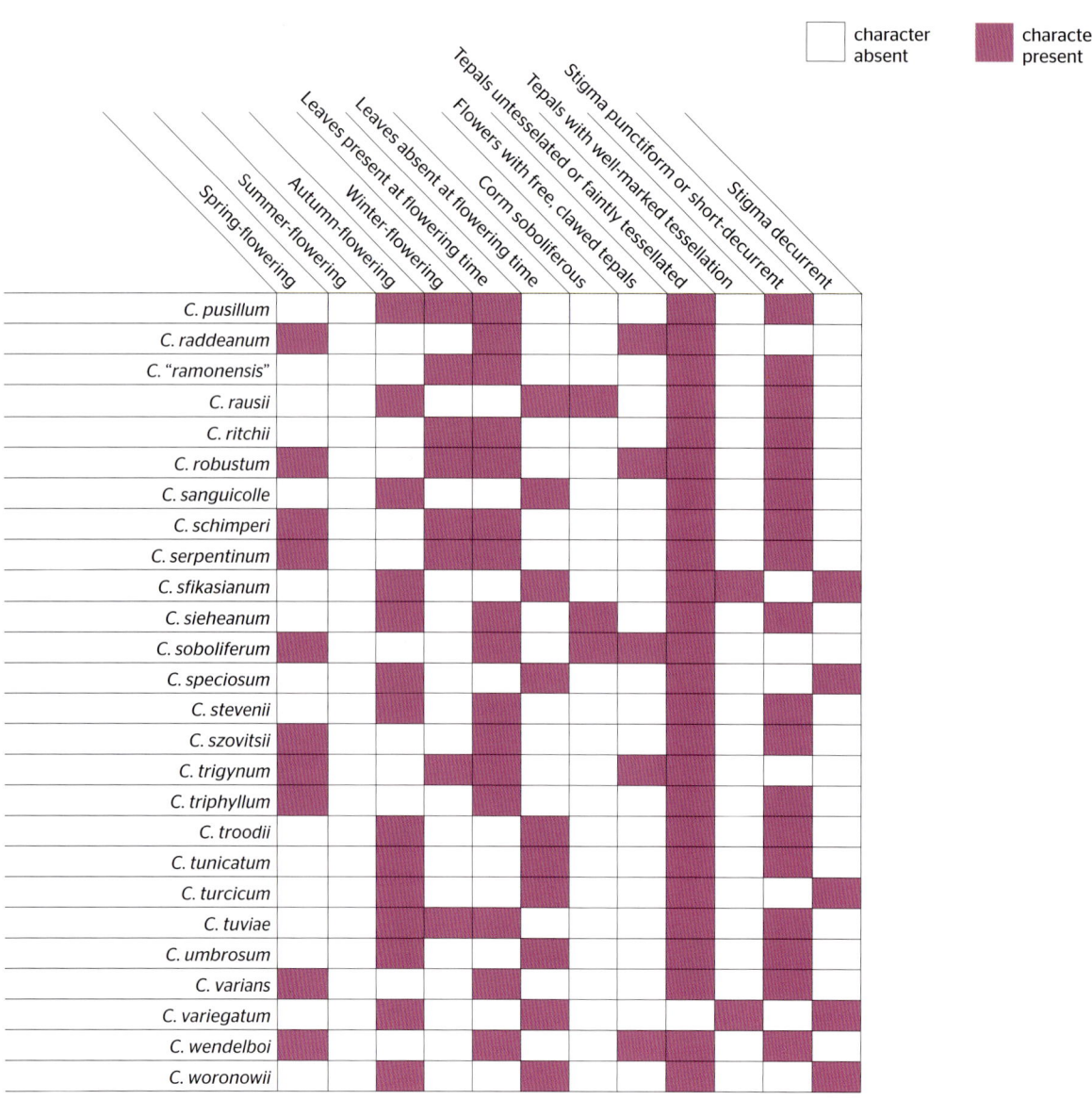

TOP RIGHT A colour variant of *Colchicum arenarium* in Moldova.

RIGHT *Colchicum laetum* thriving on the steppes of the Caucasus.

Colchicum alpinum

Colchicum alpinum DC., *Fl. Franc.* [de Candolle & Lamarck], 3rd edn. 3: 195 (1805)
Type: W Alps, Chamouny to le Col du Bon-Homme, De Candolle (holotype G)
Syns *C. alpinum* var. *parvulum* (Ten.) Baker, *C. merenderoides* E.P. Perrier & Songeon, *C. parvulum* Ten., *C. pseudoparvulum* Lojac.

Corm: Subglobose to ovoid, 1–2.5 × 1–1.5cm, with a membranous to sub-leathery, reddish brown tunic with a short foot and a long, thin neck
Cataphylls: White, slender, 2.4–6cm long
Leaves: 2(3), appearing after flowering, ascending, strap-shaped to linear-lanceolate, 8–12(–15) × 0.2–1(–1.4)cm, glabrous, apex subacute to subobtuse
Flowers: 1(–3), funnel-shaped, purplish pink, sometimes very pale or white, finely parallel veined; perianth tube white or greenish white, up to twice the length of the tepals
Tepals: Narrow-oblong to narrow-elliptic, 1.7–3(–3.5) × 0.4–0.8(–1)cm, apex obtuse
Stamens: About one third the length of the tepals; filaments white; anthers mid-yellow, 3–4mm long
Styles: White, occasionally tipped pink, straight to curved, sometimes rather contorted, overtopping the stamens; stigmas punctiform
Fruit capsules: Oblong-elliptic, 1.5–2cm long, glabrous
Habitat: Meadows and stony places, generally on acid substrates
Flowering period: (July–)August–September
Distribution: SW Alps (SE France, NW Italy (Brescia westwards), W and S Switzerland (Ticino and Valais)), Apennines, Sicily; 900–2,000m
Cytology: 2n=54, 56

A small and rather demure species sometimes referred to as alpine meadow saffron, and flowering in the wild in early autumn. It resembles *C. arenarium*, found further north in central Europe, from which it differs in its fewer, generally narrower leaves and smaller, often paler flowers. Hybrids between *C. alpinum* and *C. autumnale* are recorded from the French Alps, reportedly with intermediate chromosome numbers (2n=46–47) and characters (Brickell 1980).

Closely related to *C. alpinum* are two relatively little-known species, *C. gonarei* and *C. nanum*.

The small nature of *C. alpinum* does not endear it to gardeners. However it is quite delightful and has been maintained in several private gardens in England over a number of years, lending itself to large troughs and raised beds.

Flowering in early autumn, *Colchicum alpinum* performs well in the garden in a raised bed.

Colchicum androcymbioides

Colchicum androcymbioides (Valdés) K. Perss., *Bot. Jahrb. Syst.* 127(2): 169 (2007)
Type: Spain, Andalucia, Ronda, S. Pedro de Alcantara, 900m, *Talavera & Valdes 28484* (holotype SEV)
Syn. *Merendera androcymbioides* Valdés

Corm: Oblong-ovoid, 1–3.5cm long, with a subcoriaceous, brown tunic, extended above into a short neck
Cataphylls: Whitish, greatly exceeding the neck of the corm, up to 4cm long
Leaves: 3–6, linear to linear-lanceolate, 13–18(–25) × 0.4–1.3cm at maturity, with a wide cartilaginous and scabrid margin
Flowers: 1–4(–5), white to pale purplish pink or pale lilac; tepals split to the base, claw very slender, 2–3 times longer than the blade
Tepals: Linear-lanceolate to narrow-elliptic, 14–40 × (1.5–) 2.5–4.5mm
Stamens: About one third the length of the tepal limbs; filaments white; anthers yellow or lilac, 3–5.5mm long
Styles: Filiform, somewhat shorter to, or slightly exceeeding, the stamens
Fruit capsules: Ellipsoid, 1.2–2.5cm long, abruptly pointed, located at ground level or slightly above
Habitat: Rocky meadows
Flowering period: February–April
Distribution: S Spain (Serranía de Ronda), endemic; 5,00–1,200(–1,750)m
Cytology: 2n=54

This is a Spanish endemic restricted to the Serranía de Ronda in Andalucia. Its closest ally is *C. atticum*, a native of south-eastern Europe and Turkey. *Colchicum androcymbioides* is readily distinguished on account of its broader leaves, narrower tepals and larger anthers that are yellow rather than violet.

With such a restricted distribution *C. androcymbioides* must be considered to be a threatened species.

A spring-flowering species, *Colchicum androcymbioides* has narrow tepals that are split to the base.

Colchicum antepense

Colchicum antepense K. Perss., *Bot. Jahrb. Syst.* 127(2): 169 (2007)
Type: Turkey, Gaziantep, 600–900m, *Hausknecht 924* (holotype G-BOISS; isotypes FI, M, W, WU)

Corm: Narrow-ovoid, with finely striated tunics, with a short but broad foot, with or without a neck that is up to 4 cm long when present
Leaves: 3–4(–5), linear to linear-lanceolate, erect to ascending and equalling or shortly overtopping the flowers at anthesis, spreading at maturity, 10–15 × 0.5–1.5cm, somewhat twisted, scabrid or ciliate at the margin, sometimes hirsute on dorsal surface
Flowers: 2–6(–9), funnel-shaped, whitish to pinkish purple, the tube generally 1–4cm long above ground
Tepals: Narrow-oblanceolate to narrow-oblong, to 2.5 × 0.6 cm, pronouncedly channelled in the lower half and with basal ridges or lamellae, subacute to somewhat hooded at the tip
Stamens: Filaments greenish white, swollen and orange or brownish yellow at the base; anthers grey-green to muddy yellow
Styles: Stigma punctiform
Fruit capsules: Not recorded
Habitat: Rocky and grassy places on calcareous soils
Flowering period: March–April
Distribution: SC Turkey (Aintab, Gaziantep, Marvin Dağ); 600–900m
Cytology: 2n=28

This recently described species is restricted, as far as it is known, to the Gaziantep region of south-central Turkey. It finds its closest ally in *C. serpentinum*, differing primarily in its more membranous, several-layered, matt-brown corm tunics and in its deep green, somewhat glaucous leaves that often equal or exceed the flowers at anthesis. At maturity the leaves are strongly twisted and undulate. In addition, the tepals are clearly channelled, almost folded in the lower part. It is one of the earliest spring-flowering species.

Although it has small and not particularly brightly coloured flowers, *C. antepense* has some appeal to enthusiasts because it is one of the first to flower in spring. It is grown at Gothenburg Botanical Garden.

Colchicum antepense is valued by enthusiasts because it is one of the first spring-flowering species to appear.

Colchicum antilibanoticum

Colchicum antilibanoticum Gomb., *Bull. Soc. Bot. France* 104: 286 **(1957)**
Type: Syria, Damascus, Antiliban, Dec. 1932, *Gombault 3477* (holotype P; syntype S)
Syn. *C. tauri* sensu Mouterde ex Feinbrun, non Siehe ex Stef.

Corm: Ovoid-oblong, 3-4cm long, with a reddish brown, membranous to subcoriaceous tunic, extending above into a short neck
Cataphylls: White, protruding above ground by about 10mm, greatly exceeding the corm neck
Leaves: 2-4(-5), appearing after flowering, in spring as the snow begins to melt (occasionally showing at flowering in populations on Mount Hermon), linear, 2-10mm wide, glabrous, channelled
Flowers: 3-7, occasionally more, funnel-shaped, often with a spidery appearance, white or very pale pink; perianth tube white, equalling or slightly longer than the tepals
Tepals: Linear-oblong to linear-elliptic, 2.3-2.8 × 0.3-0.4cm, often incurving to some degree, apex subobtuse to acute
Stamens: About half the length of the tepals; filaments slender, white, swollen and egg yolk-yellow at the base; anthers 4-4.5mm long, grey-green, purplish brown or purple-grey
Styles: White, very slender, generally exceeding the stamens; stigmas punctiform
Fruit capsules: Ovoid, 1.2-2.5cm, glabrous, located at ground level
Habitat: Rocky slopes, especially of northerly exposure, among hawthorn, oak or spiny shrubs
Flowering period: Late September-November(-early December), before snowfall usually
Distribution: Israel, Lebanon and W Syria, possibly also Jordan; 1,500-2,800m
Cytology: 2n=14

This is a pretty little species with pale, often white, rather spidery flowers with contrasting dark grey-green anthers. It is often found growing in association with *Crocus hermoneus* subsp. *hermoneus* with which it overlaps in its flowering time.

Found in rocky habitats in the wild, *Colchicum antilibanoticum* has pale, spidery flowers with contrasting dark anthers.

Colchicum arenarium

Colchicum arenarium Waldst. & Kit., *Descr. Icon. Pl. Hung.* 2: 195, t. 179 (1803–1805)
Type: Hungary, Comitatus Pesthiensis, Waldstein (BP, PR), Ukraine, Odesska obl.: Grebeniki (Hrebenyky), near the Moldavian border town of Tiraspol, 19 Sep. 1999, *Dyatlov s.n.* (neotype KW)
Syn. *C. fominii* Bordz.

Corm: Ovoid, 2–3 × 1–1.5cm, with a membranous to somewhat leathery, dark reddish or blackish brown tunic and a long neck, 5–8cm long
Cataphylls: White, slender, slightly longer than the corm neck
Leaves: 3–5, appearing after flowering, strap-shaped to linear-lanceolate or narrowly lanceolate, 0.8–2.5 × 0.4–1.7cm, glabrous, channelled, apex obtuse, margin flat, somewhat scabrid or ciliate
Flowers: 1 or 2(–4), funnel-shaped, purplish pink to lilac-rose, with a whitish centre; perianth tube white, often flushed pink towards the top, up to twice as long as the tepals, 4.5–6.5cm long
Tepals: Narrow-oblong to narrow-elliptic, 2.5–4.2 × 0.3–1.2cm, 7–9-veined, obtuse
Stamens: One third the length of the tepals; filaments white, 12–14mm long, slightly swollen at the base; anthers mid-yellow, 5–7mm long
Styles: White, straight, about as long as the stamens; stigmas punctiform to very slightly decurrent for 0.5–0.75mm
Fruit capsules: Oblong-ovoid to oblong, 1–2(–3)cm long, glabrous, located at ground level
Habitat: Sandy and stony habitats and fields
Flowering period: September–October
Distribution: Hungary, Romania, Ukraine, Croatia, Serbia; 350–1,250m
Cytology: 2n = 36, 54

Colchicum arenarium is a small-flowered species restricted to eastern Europe, eastwards as far as Crimea, but little studied and seldom seen in cultivation. It is probably most closely allied to *C. umbrosum* which is found further east. The differences are slight, but *C. arenarium* has larger, deeper coloured flowers with narrower leaves, while the stigma is usually punctiform.

Grintescu *et al.* (1966) in *Flora of Romania* treat *C. arenarium* and *C. fominii* as separate species, mainly on corm and leaf characters. The apparent differences overlap in their ranges and it is difficult to uphold them as distinct species.

Colchicum arenarium growing in steppe habitat near Budzhak in Moldova.
INSET **Leaves of *Colchicum arenarium* in cultivation.**

Colchicum arenasii

Colchicum arenasii Fridl., *Acta Bot. Gallica* 146(2): 158, fig. 1 (1999)
Type: France, Corsica, l'Anse de Chevanu, Nov. 1994, *Fridlender 335* (holotype P)
Syn. *C. corsicum* auct., non Baker

Corm: Ovoid, 2.3–2.6cm long, with a membranous brown tunic, extended above into a short neck, to 3.5cm long
Cataphylls: White, slender, 5–6.5cm long, not appearing above ground
Leaves: 3–4, appearing after flowering, ascending to spreading, narrow oblong-elliptic to strap-shaped, 12–20 × 1.3–1.8cm, glaucous green, glabrous, apex obtuse to subobtuse
Flowers: Solitary, occasionally paired, funnel-shaped, opening broadly in sunshine, white to rose-pink, often moderately tessellated, or bicoloured with a white base and central line; perianth tube white, sometimes flushed pink, 2.5–4cm long
Tepals: Elliptic-oblong to elliptic-oblanceolate, 2.1–3.1 × 0.5–1.2cm, the outer somewhat longer than the inner, all obtuse or subobtuse at the apex
Stamens: One third to half the length of the tepals; filaments white, sometimes flushed pink in the upper part, somewhat swollen at the base; anthers yellow or orange-yellow, 2.5–4mm long
Styles: Equalling the stamens, slender, slightly expanded towards the top, white or pinkish; stigmas decurrent for 3–5mm
Fruit capsules: Oblong, 2.5–3cm long, with a short beak, glabrous, located at ground level
Habitat: Coastal littoral in rocky and sandy places
Flowering period: September–October
Distribution: SW Corsica and Sardinia; low altitudes close to sea level
Cytology: 2n=162

Closely allied to *C. corsicum*, this is a recently named species. Plants formerly included in *C. corsicum* from the south-western side of Corsica were recognised as *C. arenasii* in 1999. Quite a few records of *C. corsicum* from Sardinia are also referable to *C. arenasii*. The differences between the two species are slight. While the flowers of *C. corsicum* are mid-rose-pink to lilac-pink, those of *C. arenasii* are white to rose-pink, often with a hint of tessellation. *Colchicum corsicum* also has rather smaller flowers. Differences are also apparent in the stamens and stigmas.

Drawing of *Colchicum arenasii* from Fridlender (1999).

Colchicum asteranthum

Colchicum asteranthum Vassiliad. & K. Perss., *Preslia* 74(1): 57 (2002)
Type: Greece, Peloponnese, Arkadia, Mt Lirkio, 950–1,450m, Jan. 1999, *Vassiliades* (holotype GB)

Corm: Soboliferous, the central part oblong, small, about 1cm diameter, with grub-like, horizontal, shoot-bearing lobes to 4.5 × 0.5cm, the tunic membranous and evanescent, yellowish or reddish brown, with or without a very short neck
Cataphylls: Slender and sinuous, white, greenish at the top, mostly 6–10cm long
Leaves: (3–)4(–5), partly developed at flowering, then not more than 4cm long, at maturity ascending to arcuate, spreading to recurved, linear, 8–15 × 0.2–0.5cm, deep green, glabrous to somewhat scaberulous, deeply channelled, apex acute to subacute, the margins finely cartilaginous
Flowers: 1–2, small, funnel-shaped to starry, near white to pinkish lilac, faintly honey-scented; perianth tube white, slender, 1–2.5cm long, sometimes longer
Tepals: Oblanceolate to oblong-oblanceolate, 1.8–2.8 × 0.4–0.8cm, apex subacute to subobtuse, the inner with tiny, basal lamellae along the filament channels
Stamens: 7–11mm long; filaments white, flushed with pinkish lilac, base slightly swollen and yellowish; anthers dark brown to purplish grey, 1.7–2.5mm long
Styles: More or less equalling the stamens, whitish, straight or somewhat curved, the stigmas punctiform
Fruit capsules: Ellipsoid to almost globose, 1–1.5cm long, brown when mature, apex shortly apiculate, located at or just below ground level
Habitat: Stony terra rossa over limestone in dry areas normally with some winter snow cover
Flowering period: December–January
Distribution: Greece (Mount Lyrkeio in C Peloponnese only); 950–1,450m
Cytology: 2n=36

This is a very localised Greek endemic which probably finds its closest ally in the Turkish endemics, *C. leptanthum* and *C. minutum*. All three are small-flowered with slender leaves, but *C. asteranthum* usually has four leaves while the Turkish species usually have three. *Colchicum leptanthum* has erect leaves and white flowers with very narrow tepals that lack basal lamellae. *Colchicum minutum* has thicker, shallowly channelled leaves, keeled beneath, tapered tepals, and larger anthers.

Colchicum asteranthum has a chromosome number shared by just four other species, *C. autumnale*, *C. bivonae*, *C. cretense* and *C. szovitsii*.

Colchicum asteranthum is endemic to a small area of southern Greece.
INSET **Leaves of *Colchicum asteranthum* in cultivation.**

Colchicum atticum

Colchicum atticum Spruner ex Tommas., *Flora* 23: 730 (1840)
Type: Greece, Sterea Ellada, Marathon to Lycabettus, Jan. 1840, *Spruner s.n.* (lectotype G-BOISS)
Syns *Bulbocodium atticum* (Spruner ex Tommas.) Nyman, *Merendera attica* (Spruner ex Tommas.) Boiss. & Spruner

Corm: Regular, oblong-ovoid, 1.5–2.5cm long, occasionally larger, with a dark brown to blackish brown coriaceous tunic, this extending above into a short neck, rarely much more than 1cm long
Cataphylls: Whitish or green, generally 2–4cm long
Leaves: (2–)4(–6), present at flowering, then up to 5cm long, at maturity linear to linear-lanceolate, 18 × 0.3–0.5(–0.8)cm, green, channelled, apex acute, with a thin cartilaginous and somewhat scabrid or ciliate margin
Flowers: 3–5, wide funnel-shaped to starry, 3–4cm diameter, white to pale pinkish purple; tepals split to the base, narrowed into a whitish claw equal in length to the tepals or somewhat longer
Tepals: Linear-elliptic or linear, 1.5–2.5cm × 3–4(–5)mm, apex obtuse or subobtuse, the inner tepals rather narrower than the outer
Stamens: Filaments white; anthers versatile, black, purplish or greenish black, 2–3mm long, shorter than the filaments
Styles: Filiform, straight, slightly exceeding the stamens
Fruit capsules: Ovoid to ellipsoid, 1.5–2.5cm long, glabrous, located at ground level or partly below
Habitat: Limestone hillsides
Flowering period: November–April
Distribution: S Bulgaria, SE and NE Greece, East Aegean Islands (Lesbos, Kalimnos, Samos), W and C Turkey; 100–2,000m
Cytology: 2n=54

Colchicum atticum is readily confused with other small species, particularly *C. pusillum* and *C. stevenii*. However, it has the typical split perianth tube of species formally included in the genus *Merendera*, although plants often have to be examined closely to observe this important distinction. All three of these species have rather starry, small flowers with dark anthers. While *C. atticum* and *C. stevenii* overlap in the wild, *C. pusillum* has a more western and southerly distribution concentrated on southern Greece, the Greek islands and Cyprus.

Although they have split to reveal the pollen, the dark anthers of *Colchicum atticum* are still visible.

Colchicum autumnale

Colchicum autumnale L., *Sp. Pl.* 1: 341 (**1753**)
Type: Southern Europe, Herb. Burser III: 70 (lectotype UPS)
Syns *C. borisii* Stef., *C. bulgaricum* Velen., *C. drenowskii* Degen & Rech. f. ex Kitan. in part, *C. orientale* Friv. ex Kunth, *C. rhodopaeum* Kov., *C. transsilvanicum* Schur, *C. vranjanum* Adamović ex Stef.

Corm: Large, ovoid, 3–6 × 1.5–4.5(–5.5)cm, the tunic brown to reddish or blackish brown, leathery to membranous, extending upwards into a 2–7cm long neck
Cataphylls: White, 5–12cm long, extending just to the soil surface
Leaves: 3–4(–5), appearing long after flowering, erect to ascending or somewhat spreading, linear-lanceolate to lanceolate or lanceolate-oblong, 15–35(–40) × 2–5(–7)cm, dull green, and often somewhat twisted, apex obtuse to subacute
Flowers: 2–5(–8), goblet-shaped, narrow at first but soon opening more widely, lilac-pink to lilac-purple, occasionally pale pink or rose-purple, the tepals often with a medium, white or pale stripe towards the base within, occasionally faintly tessellated; perianth tube white, often flushed with pink or purple, especially towards the top, two to three times as long as the tepals
Tepals: Elliptic to oblong, oblanceolate or oblong-obovate, mostly 3–6 × 1–2cm, occasionally larger, apex obtuse to subobtuse
Stamens: Filaments white or cream, somewhat swollen and yellow or golden at the base, the three inner longer than the outer; anthers yellow, 5–8mm long
Styles: White, sometimes purplish towards the top, overtopping the stamens; stigmas decurrent for 1.5–5mm
Fruit capsules: Ovoid to ellipsoid, 3–6cm long, one or several clustered together, glabrous, located well above ground level; seeds 2–4mm
Habitat: Damp meadows in the lowlands or mountains, grassy woodland clearings and margins, humusy pockets in rocky terrain, dolines and poljes, over limestone, schistose or serpentinite, often gregarious
Flowering period: August–early October
Distribution: W, C and S Europe (including Britain) away from the Mediterranean (naturalised in Denmark and Sweden); sea level to 2,200m
Cytology: 2n=36, 38

A species of meadows in the wild, *Colchicum autumnale* is suitable for growing in grassy places in the garden.

Along with *C. triphyllum*, *C. autumnale* is one of the most widespread and variable species in the genus. It is closely related to *C. confusum*. *Colchicum autumnale* is familiar in many non-Mediterranean European countries, including Britain. Its range covers much of temperate Europe, Ukraine and southern Russia.

It is primarily a species of moist grassy meadows and waysides. In some areas it has been eradicated in many of its former sites because of its toxicity to cattle in particular. In Britain the main populations of *C. autumnale* are in the West Midlands and Welsh borders, although it has been introduced at other sites. In the Alps it is found in every département and canton.

An important plant in the medieval herb gardens of Europe and elsewhere, it has been used as a medicinal plant for centuries, particularly in the treatment of gout. It has also been an important source of colchicine.

In cultivation plants are highly varied in appearance, depending on their provenance. In leaf, at least, the more robust forms can be up to 50cm tall. This species has undoubtedly played a significant part in the development of the large-flowered, autumn-blooming cultivars of gardens.

Colchicum autumnale in some of its forms can be mistaken for *C. speciosum* in cultivation. The two do not grow together in the wild so the confusion does not occur there. Both species are extensively cultivated and numerous fine garden hybrids have one or other, or indeed both, species in their genetic constitution. In flower the two can look very similar. However, the flowers are smaller in *C. autumnale*, the tepals mostly in the range 3–6cm long, while in *C. speciosum* they are 4.5–8cm long. The former has glabrous filament channels, but in the latter the ridges are puberulous to finely pubescent. In cultivation the leaves of most types of *C. autumnale* develop very late, often not appearing much above ground before the end of winter and continuing to develop through the spring. They are more upright and narrower, the lowermost mostly 2–5cm wide, linear-lanceolate or lanceolate. In contrast, those of *C. speciosum* are well-developed by the end of winter. Upright at first, at maturity they are spreading, oblong-lanceolate to elliptic, the lowermost mostly 5–9.5cm wide.

Colchicum rhodopaeum, described from near Kritschim in the Rhodope Mountains of Bulgaria, is sometimes treated as a separate, although closely allied, species. However, it fits well within the present concept of *C. autumnale* and is treated as a synonym here.

Botanists are not generally agreed on the best way to treat *C. autumnale* and its variants. For instance, Grintescu *et al.* (1966) in *Flora of Romania* recognise two varieties, var. *autumnale* and var. *pannonicum*, along with six formas. While it would be difficult to uphold all these in a botanical sense, these divisions do have some merit horticulturally. The most prominent are the spring-flowering variants of *C. autumnale*, best treated as *C. autumnale* subsp. *vernum* (Reichard) K. Richt. (syns. *C. autumnale* var. *vernum* L., *C. praecox* Spenn., *C. vernale* Hoffm., *C. vernum* (Reichard) Georgi). Spring-flowering in colchicums that normally flower without their leaves does occur from time to time, but seems to be especially prevalent in *C. autumnale*.

Cultivars of *C. autumnale* include 'Annecy', 'Dorothee Kersen', 'Drama Bunch', 'Elizabeth', 'Karin Persson', 'Lausanne' and 'Pleniflorum', and white-flowered

RIGHT **In *Colchicum autumnale* subsp. *pannonicum* the flowers are a deeper colour than those of subsp. *autumnale* and the tepals scarcely overlap.**

FAR RIGHT **The dull green, upright leaves of *Colchicum autumnale* mature in sprng, long after the flowers have disappeared**

BOTTOM RIGHT ***Colchicum autumnale* thriving in the recent trial at RHS Garden Hyde Hall.**

variants such as 'Alboplenum' and 'Album'.

We do recognise *C. autumnale* subsp. *pannonicum* which was described from Transylvania in Romania. In cultivation, at least, it is an easily recognised and popular plant.

Colchicum autumnale subsp. pannonicum (Griseb. & Schenk) Nyman, *Consp. Fl. Eur.*: 743 (1882)
Type: Romania, Sibiu: 'In pratis montanis pr. Gr. Scheuern Transsylv.', 12 Sep. 1852, *Grisebach 295* (lectotype GOET)
Syns *C. autumnale* var. *pannonicum* (Griseb. & Schenk.) Baker, *C. pannonicum* Griseb. & Schenk.

> **Corm:** Ovoid to narrowly ovoid, 3.8–7.4 × 3.5–4.2cm, foot short and blunt 0.7–1.2cm long, neck slender 2.5–4.8cm long
> **Leaves:** 5–6(–7), ascending to spreading, recurved in the upper half, slightly yellow-green, moderately glossy, with several ridges lengthways, the lowermost leaf elliptical 29–32 × 6.2–6.9cm, subobtuse
> **Flowers:** 2–7, narrowly chalice-shaped, opening eventually to wide bowls with tepals scarcely overlapping, deep lilac-pink to deep pink, with very faint tessellation in the uppr half and a faint white stripe extending to about half way from base along each tepal within, perianth tube pink-flushed, paling towards base, 8.8–10.6cm long
> **Tepals:** Elliptic-oblanceolate, 5.6–6 × 2–2.2cm, the inner somewhat shorter than the outer
> **Fruit capsules:** Occasionally produced
> **Habitat:** Meadows and woodland margins
> **Flowering period:** September–October
> **Distribution:** Romania (Cluj)

The main features distinguishing subsp. *pannonicum* from subsp. *autumnale* are detailed in the descriptive details above. In cultivation it is represented primarily by the long-established cultivar 'Nancy Lindsay', which matches the type description of this subspecies and reportedly came from Romania in the first instance. Several wild-collected accessions from the Cluj region grown at Royal Botanic Gardens, Kew, also closely match 'Nancy Lindsay'.

Colchicum atropurpureum Stapf ex Stearn, *J. Bot* 72: 341 (1934)
Although described more than 80 years ago, this plant presents somewhat of a mystery. William Stearn based his description on a name suggested by Otto Stapf, in turn based on plants in cultivation at Cambridge University Botanic Garden. The type specimen located at Royal Botanic Gardens, Kew, clearly indicates its Cambridge origin and the plants had originally gone to Cambridge from the Dutch

RIGHT **A pale-flowered variant of *Colchicum autumnale* growing near Bologa in Romania.**

BOTTOM RIGHT **At the same site in Romania there is considerable variation in the flower colour and tepal shape of *Colchicum autumnale*.**

firm of Van Tubergen. Plants were also in more general cultivation in the 1930s, often under the name *C. autumnale* var. *atropurpureum*.

As *Colchicum autumnale* var. *atropurpureum* the plant first appeared in Weston's *Universal Botanist* (1771) and later in Loudon's *Hortus Britannicus* (1839). It is impossible to be in any way certain that the plants described by these early authors were the same plant recognised as a species by Stearn. Edward A. Bowles (1924) equates the plant in question with that described in Parkinson's *Paradisi in Sole Paradisus Terrestris* (1629), stressing that it might have been of Balkan origin, although apparently only known from cultivated plants. Charles Grey (1937) states that Van Tubergen introduced it into cultivation and that they obtained it from the valley of the Meuse in France. He goes onto say that it must have been an escape from cultivation, as the area had been well known to field botanists for years.

Karin Persson (2007) equates it with *C. autumnale*, which is why we are discussing it here. Measurement of the type specimen, even allowing for some possible shrinkage, gives a tepal size of 3.7–4.1 × 1–1.1cm. The very variable and widespread *C. autumnale* has tepals primarily 3–6cm long and 1–2cm wide. Thus *C. atropurpureum* has flowers at the lower range of those of *C. autumnale*. There are other differences, the two most notable being the undulate leaves and the compartively short perianth tube.

In the general features of the flowers, the descriptions of *C. atropurpureum* come close to those of *C. turcicum*. However, again there are differences, as the flowers of *C. atropurpureum* are untessellated and fewer in number, as are the leaves, while the tepals are generally smaller. *Colchicum turcicum* was certainly in cultivation early in the 20th century, along with other species primarily of Turkish origin. It is possible that the plant grown as *C. atropurpureum* is of hybrid origin, involving *C. turcicum* as one of its parents. Brian Mathew (1987) saw a similar connection when he suggested that 'The garden plant known as *C. atropurpureum* is not unlike *C. turcicum*'. We regard it as best treated as a cultivar, *Colchicum* 'Atropurpureum'.

The mystery is what has become of the plant today. It is not listed in the *RHS Plant Finder* and the name has not appeared in any recent catalogues. This requires further investigation.

RIGHT **The type specimen of *Colchicum autumnale* var. *atropurpureum*.**

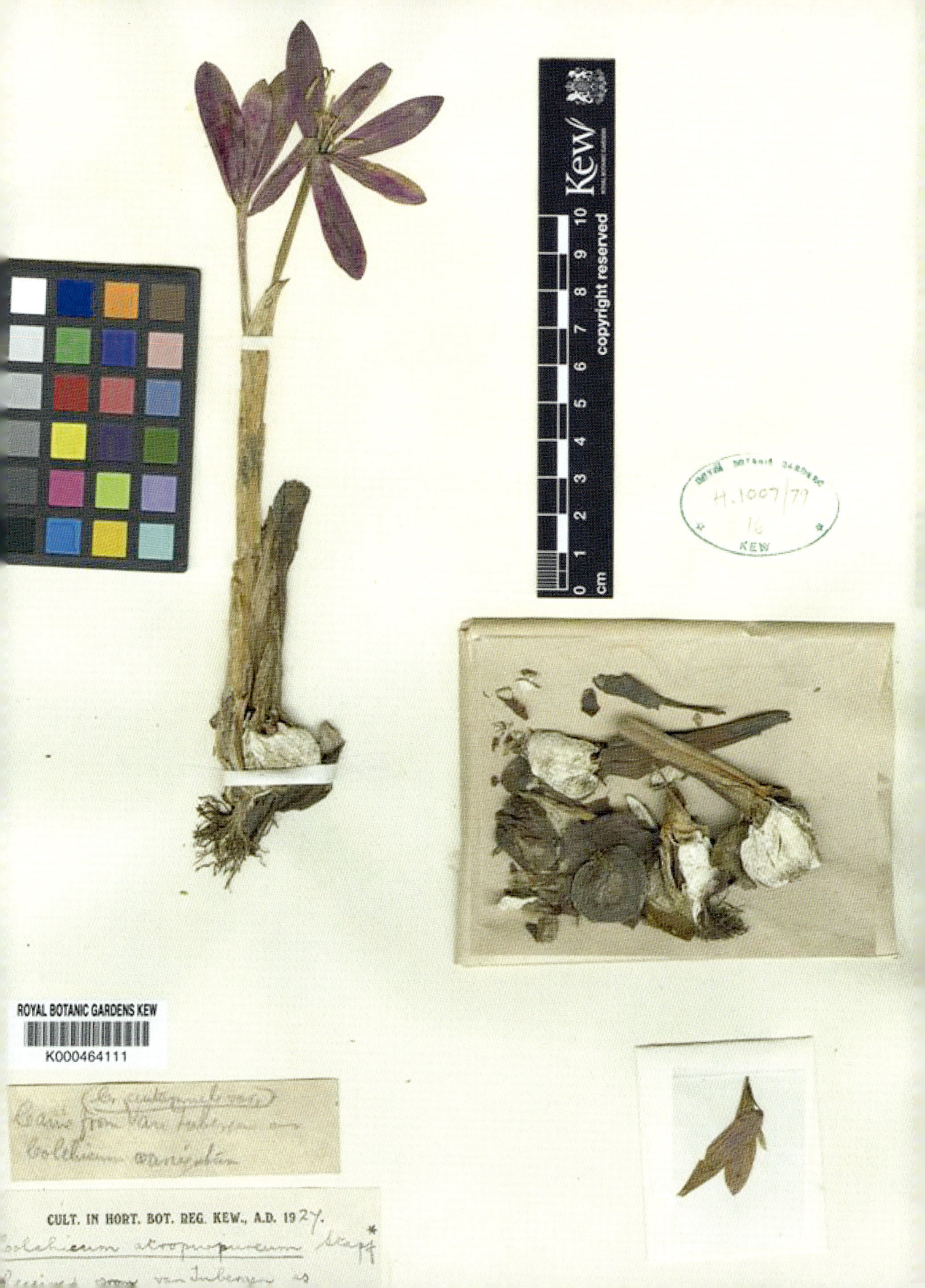

CULT. IN HORT. BOT. REG. KEW., A.D. 1927.

Colchicum atropurpureum Stapf

Colchicum balansae

Colchicum balansae Planch., *Ann. Sci. Nat., Bot.* sér. 4, 4: 145 (1855)
Type: Turkey, Içel, Taurus, Cicilian Gates, *Balansa s.n.* (lectotype P; isotypes G, K)
Syns *C. candidum* Schott & Kotschy ex Baker, *C. laetum* sensu Baker pro parte, non Steven, *C. speciosum* sensu Stef., non Steven

Corm: Ovoid to almost globose, 4–5cm long, often developing a long foot, with a prominent dark brown or almost blackish brown, fibrous, coriaceous tunic, extended above into a narrow, extremely long neck, to 30(50)cm long
Cataphylls: Yellowish white, sometimes somewhat greenish at the top, equalling the tunic neck
Leaves: 4–5, appearing after flowering, in the late winter and early spring, oblong to oblong-elliptic or oblong-lanceolate (linear-lanceolate in young plants), 15–30 × 3–9cm, dull glaucous green, flattish, sometimes somewhat twisted, with a sunken midvein, apex obtuse to subobtuse, margin thinly cartilaginous
Flowers: 3–6, funnel-shaped, white to pale lilac-purple or purplish violet, occasionally faintly tessellated; perianth tube whitish, 2–9cm long
Tepals: Linear to elliptic-oblong, 4–7.5 × 0.4–1.3cm, apex obtuse to subacute, filament channels glabrous
Stamens: A third to half as long as the tepals; filaments, whitish, glabrous, swollen and yellow at the base; anthers yellow, 8–16mm long
Styles: Overtopping the stamens, white, usually purple-flushed towards the top, curved and thickened at the tip; stigmas decurrent for 1.5–4mm
Fruit capsules: Ovoid to ellipsoid, 2.5–4cm long, sometimes faintly brown-dotted, shortly beaked, located at ground level
Habitat: Rocky places, clay slopes, meadows, juniper scrub, oak scrub and pine forests, generally over limestone or schist
Flowering period: September–October
Distribution: Greece (mainland Attica and Poros, Samos and Rhodes islands), S and SW Turkey; near sea level to 2,000m
Cytology: 2n=108

An interesting little species, *C. balansae* is remarkable for the depth at which the corms can be found. This affects the length of the tunic neck as well as the cataphylls and perianth tubes. While 20–30cm is perhaps normal, depths of up to 50cm long have been recorded. This has helped protect the species from predatory plant collectors who find extracting whole specimens extremely difficult and

In cultivation, Colchicum balansae takes a while to bulk up.

time-consuming. In the absence of corm details the large, glaucous leaves and particularly long anthers are useful diagnostic characters.

Colchicum balansae is slow to reproduce, both in the wild and in cultivation and large colonies are rare. The species is a high polyploid, actually a dodecaploid, and such plants tend to prove less fertile, with a tendency to favour vegetative reproduction. But in this species such reproduction is probably hindered by the great depth at which the corms are located.

One collection (J. & J. Archibald 6134), made in Turkey in 1985, is especially intriguing. While it matches the prime diagnostic features of the species its chromosome number is just 2n=54, half that recorded for the species. Persson (1999b) has suggested that this plant, which is completely sterile, is 'an example of the unusual phenomenon of haploidy', meaning it has just one set of chromosomes, half the full complement.

RIGHT **A pink variant of *Colchicum balansae*.**

FAR RIGHT **This species usually grows in small colonies as it is slow to reproduce.**

BOTTOM RIGHT **In white-flowered variants the yellow base of the filaments is usually visible.**

Colchicum baytopiorum

Colchicum baytopiorum C.D. Brickell, *Notes Roy. Bot. Gard. Edinburgh* 41(1): 49 (1983)
Type: Turkey, Antalya, Termessos, 550m, *T. Baytop ISTE 36255* (holotype ISTE)

Corm: Ovoid to subglobose, 2.5–3.5cm long, adorned with a reddish brown, membranous and evanescent tunic, producing short, stout, horizontal or inclined soboles, up to 6 × 2cm, corm neck absent
Leaves: 3, present at flowering, then rarely more than 5cm long, or appearing shortly afterwards, at maturity ascending to spreading and recurving, linear-lanceolate to lanceolate, 22–32 × 2.4–4.5cm, glabrous, apex acute to obtuse
Flowers: 1–3, occasionally more, narrow funnel- to chalice-shaped, deep rosy purple, concolorous, rarely with a whitsh throat; perianth tube white, flushed purple in the upper part often, equalling or up to twice the length of the tepals
Tepals: Linear-elliptic to lanceolate-elliptic or oblanceolate, 2.5–3.5 × 0.5–0.8cm, occasionally somewhat larger, apex obtuse to acute
Stamens: Filaments white, twice as long as the anthers, swollen and yellow or yellow-orange at the base; anthers 5–7mm long, lemon-yellow
Styles: White, sometimes suffused with purple in the upper part; stigmas punctiform
Fruit capsules: Narrow ellipsoid, 2–2.5cm long, glabrous, brown with darker dots at maturity, apex apiculate, located just above ground level
Habitat: Moist and shady places, rocky places, stony meadows, light woodland (pine or oak), maquis
Flowering period: October–November
Distribution: SW Turkey (Antalya and Isparta); 50–1,500m
Cytology: Not recorded

This Turkish endemic has obvious affinities with the Greek *C. boissieri*. Both are autumn-flowering species with bright purplish pink flowers, yellow anthers and punctiform stigmas. While the corms of *C. boissieri* are soboliferous with slender soboles, those of *C. baytopiorum* are usually more conventional but they do, at the same time, produce thick soboles. Perhaps the most obvious difference is seen in the leaves. In *C. baytopiorum* they are broader and part developed at flowering, while in *C. boissieri* they are not developed at flowering and are much narrower.

Brickell (1983) comments on the development of soboles as observed in cultivated specimens: 'Initial development of the soboles of *C. baytopiorum* appears to take place during the flowering period as active growth begins in the

The partly developed leaves of Colchicum baytopiorum are usually visible when it flowers in autumn,

autumn. The soboles then extend during leaf development and at the end of the growing season usually form a new 'conventional' corm into which the food material from the sobole is absorbed. Some corms may still retain this soboliferous character at the end of the growing season.'

The species is named in honour of Prof. Asuman Baytop and Prof. Turhan Baytop whose collections in Turkey and work on the Turkish flora have greatly enhanced the botanical knowledge of the region.

RIGHT **A cultivated specimen of** *Colchicum baytopiorum* **with narrow tepals originally collected from near Termessos in Turkey.**

FAR RIGHT **The broad, spreading leaves of** *Colchicum baytopiorum*, **here growing in south-west Turkey.**

BOTTOM RIGHT **Corm development of** *Colchicum baytopiorum* **has been elucidated using cultivated specimens.**

BOTTOM FAR RIGHT **The leaves of** *Colchicum baytopiorum* **are not always visible at flowering time.**

Colchicum bivonae

Colchicum bivonae Guss., *Cat. Pl. Hort. Boccadifalco* 1821: 4 (1821)
Type: Italy, Sicily, Palermo, S. Maria di Gesu, Monte Petruso, la Nivera, Gibilmessa, Monte Cofani, *Gussone* (type material not located at G (Persson 2007); NAP)
Syns *C. amabile* Heldr., *C. bowlesianum* B.L. Burtt, *C. illyricum* Friv. ex Kunth, *C. latifolium* sensu Griseb., non Sm., *C. latifolium* Sm., *C. sibthorpii* Baker

Corm: Ovoid to subglobose, 2.5–5cm long, with a papery to subcoriaceous, mid to reddish brown tunic, extended above into a short neck, not more than 2cm long, sometimes rather obscure
Cataphylls: White, relatively stout, greatly exceeding the neck of the corm
Leaves: 5–9(–11), appearing after flowering, in winter, suberect to spreading, linear-lanceolate to ligulate, mainly 20–30 × 2–4cm, oblong, mid to deep green and somewhat shiny, glabrous, apex rounded to obtuse
Flowers: 1–4, occasionally up to 6, large goblet-shaped, pink to pinkish or rose-purple, paler or whitish towards the base, strongly tessellated, sometimes weakly so; perianth tube whitish, sometimes purple-flushed towards the top
Tepals: Narrow- to broad-elliptic, occasionally elliptic-obovate, 4–7(–8.5) × 2–3(–3.5) cm, often with one or several longitudinal ridges, apex obtuse to subacute, the ridges of the filament channels pubescent
Stamens: Two-thirds to half the length of the tepals; filaments white, often flushed pinkish purple towards the top; anthers dark brown or purple-brown, 7–9(–12)mm long
Styles: White, often pinkish purple towards the curved and slightly swollen tip; stigmas decurrent on the style for 3–4mm
Fruit capsules: Not recorded
Habitat: Rocky and grassy places, olive groves, *Quercus* and *Fagus* woodlands and woodland margins, *Quercus coccifera* scrub
Flowering period: September–mid-November
Distribution: Balkans from Croatia southwards, Greece (incl. Rhodes, Samos and Symi), S Italy (incl. Sicily), Sardinia, S and SW Turkey; sea level to 1,400m
Cytology: 2n=36, 48, 54, 90

This is a distinctive but quite variable species, found in the wild in the central Mediterranean region and relatively common in Greece. With its sizeable, well-tessellated flowers it is not easily mistaken for any other species, although it has acquired a number of synonyms over the years, much to the irritation of gardeners.

Tessellation is a prominent feature of the flowers of *Colchicum bivonae*.

Colchicum bivonae as a parent has had a great influence on the large autumn-flowering cultivars seen in gardens today. Cultivars such as 'Autumn Queen', 'Disraeli', 'Glory of Heemstede' and 'Oktoberfest' clearly have *C. bivonae* in their parentage. Other cultivars also appear to have at least some *C. bivonae* in their ancestry, perhaps most notably 'Rosy Dawn'.

Strangely, the species itself is not as common in gardens as might be supposed. However, in recent years several fine cultivars have been selected, such as 'Apollo', 'Mount Etna', 'Papa Rema' and 'Vesta'. Most of these are lightly scented and occasionally set seed in the open garden, flowering primarily in September and early October.

RIGHT *Colchicum bivonae* flowering in the Peloponnese.

FAR RIGHT The large and tessellated flowers of *Colchicum bivonae* have been inherited by many garden cultivars.

BOTTOM RIGHT A clone of *Colchicum bivonae* once cultivated by artist Cedric Morris.

Colchicum boissieri

Colchicum boissieri Orph., *Atti Congr. Int. Bot. Firenze* 1874. 1876: 31 **(1876)**
Type: Greece, Peloponnese, Messinia, Taigetos, *Orphanides 4016* (lectotype G-BOISS)
Syns *C. pinatziorum* Rech. f., *C. procurrens* Baker

Corm: Soboliferous, uneven, horizontal or inclined, ovoid-oblong, to 6.5cm long, with horizontal lobes and a papery, reddish or yellowish brown tunic, neck absent or very short
Cataphylls: White, sometimes greenish at the top, 9-11cm long, reaching the soil surface or slightly above
Leaves: (2)3, appearing shortly after flowering, erect to ascending, linear, mostly 8-22 × 0.4-1.2cm, usually glabrous, channelled and with a distinct midvein, sometimes with a few hairs below towards the base, margin sometimes ciliate
Flowers: 1(-3), funnel-shaped to starry, bright rosy lilac to deep pink, greenish at the base, with or without a whitish throat within, with a whitish midvein in the lower half of each tepal; perianth tube white or greenish white, generally equalling or shorter than the tepals
Tepals: Narrow oblong to oblanceolate, 2-4(-5) × 0.5-1.6cm
Stamens: Filaments white, base greenish and slightly swollen; anthers pale yellow
Styles: White, curved, sometimes straight, equalling or longer than the stamens; stigmas punctiform
Fruit capsules: Oblong-ellipsoid, 1.5-2cm long, subterranean or at soil surface
Habitat: Rocky and grassy places, stony ground, open scrub, woodland margins and clearings (*Abies*, *Cedrus*, *Pinus* and *Quercus*), juniper scrub
Flowering period: September-November
Distribution: Greece (C mainland southwards, and Euboea (Kantili), Chios and Samos islands), extreme W Turkey; 400-1,800m
Cytology: $2n=54$

This is an attractive species with wide, funnel-shaped flowers that are starry when fully open. The curious corms are grub-like, reminiscent of those of wood anemone, *Anemone nemorosa*, but thicker and rather more uneven. In cultivation it is a fine plant for pan culture. Planted in a bulb frame, its soboliferous habit means that it will wander and may invade other plantings.

Colchicum pinatziorum, included in synonymy here, was described from the island of Euboea. It is said to be a smaller and slenderer version of *C. boissieri*, but the latter can be very variable, even within a single population.

Colchicum boissieri flowering on Mount Parnassus in Greece.

Colchicum bulbocodium

Colchicum bulbocodium Ker Gawl, *Bot. Mag.* tab. 1028 **(1807)**
Type: 'Habitat in Hispania', Herb. Linn. No. 417.1 (Mathew & Lopez Gonzalez in Jarvis *et al.* 1993: 27) (lectotype LINN)
Syns *Bulbocodium vernum* L., *Colchicum vernum* Ker Gawl. ex Stef., *Merendera verna* (L.) Bubani

Corm: Ovoid, 1.5–3cm long, adorned with a mid to dark brown tunic, this extended above into a short neck or neck absent
Cataphylls: Whitish, slender, 2–6cm long, not appearing above ground
Leaves: 3–4, present at flowering, then shorter than the flowers, at maturity ascending, linear-lanceolate to lanceolate, 11–15 × 0.8–1.5cm, deep green, sometimes flushed with purple, glabrous, channelled, sheathing at the base, apex obtuse to subacute
Flowers: Solitary, occasionally 2–3, funnel-shaped to starry, mid to deep pink, often rather vibrant, rarely white; tepals split to the base, claws whitish, often flushed pink in the upper part, equalling or slightly longer than the tepal limbs, knitted together by teeth-like auricles at the base of the limbs
Tepals: Linear-lanceolate to linear-oblanceolate, 2.2–4.2 × 0.3–0.6cm, glabrous, apex acute to subacute
Stamens: About half the length of the tepal limbs; filaments whitish, slender, attached to the base of the tepal limb; anthers purplish to purplish green or yellow, 3–4.5mm long
Style: White or pinkish, especially in the upper part, solitary, 3-branched and curved towards the top, extending beyond the stamens; stigmas punctiform
Fruit capsules: Oblong, 1.5–2cm, glabrous, located at ground level or half buried
Habitat: Alpine meadows and other grassy places, often near melting snow
Flowering period: (February–)March–May
Distribution: S Austria (Carinthia), S and SE France (SW Alps, Pyrenees), N Italy (Aosta, Belluno, Cuneo, Turin), N Spain (Pyrenees) and Switzerland (Valais); 800–2,000m
Cytology: 2n=22

An attractive little plant, this species is sometimes referred to as spring meadow saffron. It is readily distinguished from all other species of *Colchicum* by its solitary, three-branched style. In all other species there are three separate styles, although in *C. luteum* and *C. kesselringii* the styles are fused inside perianth tube and appear to be free otherwise.

A three-branched style distinguishes *Colchicum bulbocodium* from all other species.

In its darker flowered forms, *C. bulbocodium* is a very desirable species. In cultivation it is well suited to pan culture, and best kept moist at all times and given a cool summer rest. It is not suitable for sunny bulb frame cultivation and will quickly diminish if grown in this way.

This species has for many years been known as *Bulbocoium vernum*, and as such it appears widely in European literature. The position of *Bulbocodium* is discussed in the Classification chapter, and current research indicates that it is well embedded in the genus *Colchicum*. It is also often mistaken for species formerly included in *Merendera*, particularly the autumn-flowering *Colchicum montanum* which flowers in advance of its leaves. Molecular phylogenetic studies (Persson *et al.* 2011, Chacòn *et al.* 2014) have shown a very close relationship of *C. bulbocodium* to *C. trigynum* (formerly *Merendera trigyna*) and to other taxa previously included in *Merendera*.

RIGHT **Darker flowered variants of *Colchicum bulbocodium* are favoured in cultivation.**

FAR RIGHT **The illustration in *Botanical Register* (1821) chosen as the lectotype for *Colchicum bulbocodium* subsp. *versicolor*.**

BOTTOM RIGHT **Also known as spring meadow saffron, *Colchicum bulbocodium* inhabits grassy places.**

Colchicum bulbocodium* subsp. *versicolor (Ker Gawl.) K. Perss., *Bot. Jahrb. Syst.* 127(2): 178 (2007)
Type: South Russia, Ker Gawler (lectotype l.c., illustration p. 571 in *Bot. Reg.* (1821))
Syns *Bulbocodium dioszegianum* Repaics, *B. montanum* Fisch., *B. ruthenicum* Bunge, *B. vernum* f. *dioszegianum* (Rapaics) Soó, *B. versicolor* (Ker Gawl.) Spreng., *Colchicum versicolor* Ker Gawl.

> **Distribution:** Croatia, Hungary, C Italy, Macedonia, Romania, S Russia (European) and Ukraine; 100–2,400m
> **Habitat:** Similar to *Colchicum bulbocodium* subsp. *bulbocodium*, but often rather drier
> **Flowering period:** March–May

This subspecies is very similar to *C. bulbocodium* subsp. *bulbocodium* but is a smaller plant with corms that are 1–1.5cm long, and leaves that are 0.5–0.8cm wide, occasionally wider. The flowers are smaller, the tepal limbs 1.2–1.8cm long, the basal auricles obtuse.

A variety lacking the basal tepal limb auricles, and confined to the Transylvanian Alps in Romania, has been identified as *C. bulbocodium* subsp. *versicolor* var. *edentatum* (Schur) K. Perss. (syn. *Bulbocodium edentatum* Schur).

Colchicum burttii

Colchicum burttii Meikle, Bot. Mag. 181(3): 134 (1977)
Type: Turkey, Çanakkale, Eceabat to Abide, Mar. 1975, *T. Baytop ISTE 31322* (holotype K)

Corm: Subglobose to narrow-ovoid, 3–5cm long, with blackish brown, coriaceous, longitudinally corrugated, tunics, the inner tunics thinner and paler, extended above into a fibrous, laciniated neck, to 5cm long
Cataphylls: White, slender, 7–9.5cm long
Leaves: 2–4, often 3, present at flowering, then 2–4cm, at maturity ascending to spreading, linear to linear-lanceolate, 10–15 × 0.8–1cm, the apex obtuse and hooded, ciliate on the margins, otherwise glabrous or shortly pilose
Flowers: 1–4, funnel-shaped at first, becoming starry, white to pale purplish pink; perianth tube white, equalling or rather longer than the tepals
Tepals: Narrow-oblanceolate, 2–3.5 × 0.4–0.6cm, apex subacute to obtuse, glabrous
Stamens: Filaments white, hairy at least in the lower part; anthers black or purplish black anthers, 2–3mm long
Styles: White, straight, shorter than or equalling the stamens; stigmas punctiform
Fruit capsules: Narrow-ellipsoid, 10–20 × 3–8mm, pilose at the apex
Habitat: Stony and rocky slopes, grazed meadows, often on terra rossa
Flowering period: January–March
Distribution: Greek island of Chios, W Turkey (Çanakkale, Konya, Kütahya and Muğla); sea level to 1,450m
Cytology: 2n=54

This is a localised species, restricted primarily to south-western Turkey. In the past it has been likened to *C. triphyllum* but the corm tunics and the flowers are very different. It is also sometimes confused with another Turkish endemic, *C. serpentinum*, which is found further to the east and north-east. Both species have rather starry flowers at maturity, both white to purplish pink, with the leaves shortly developed at anthesis. There is a difference in the tough, corrugated corm tunics of *C. burtii*. But without resorting to digging up a plant, *C. burtii* can at once be distinguished by its partly or wholly hairy filaments.

Colchicum burttii is named in honour of Bill Burtt (1913–2008), a botanist based at Royal Botanic Garden Edinburgh, whose extensive research covered many different plant groups including *Colchicum* and *Crocus*.

A mainly Turkish species, *Colchicum burttii* has hairy filaments.

Colchicum chalcedonicum

Colchicum chalcedonicum Azn., *Bull. Soc. Bot. France* 44: 174 (1897)
Type: Turkey, Istanbul, Caïiche Dağ, 250m, Aug. 1893, *Aznavour s.n.* (lectotype LY)

Corm: Ovoid to somewhat ellipsoid, 2–3.5cm long, with a many-layered, rather coriaceous, reddish brown to deep brown tunic, extended above into a relatively short neck, 1–7.5cm long
Cataphylls: White or yellowish white, sometimes flushed purple towards the top, 5–12cm long, somewhat exceeding the neck of the corm
Leaves: 3–4(–6), appearing after flowering, in late winter and early spring, spreading and procumbent, linear-lanceolate to oblong or elliptic-oblong, 4–10 × 1–2(–2.5)cm, glabrous or rarely somewhat scaberulous, apex obtuse to rounded or somewhat cucullate, margin undulate and distinctly cartilaginous
Flowers: 1–2 generally, funnel-shaped to somewhat chalice-shaped, deep pink, paler at the base, strongly tessellated with deep purple or violet-purple, generally with a median white stripe on the inside extending up to three quarters the length of the tepals; perianth tube white to yellowish white, often flushed lilac-purple towards the top, to 7cm long
Tepals: Oblong to oblanceolate, 3–4.5 × 0.5–1.3cm, occasionally a little larger, filament channels sparsely papillose
Stamens: To two thirds the length of the tepals, the outer appearing shorter than the inner and inserted in the top of perianth tube; filaments white, slightly expanded and yellow at the base; anthers yellow or brown, 4–9mm long
Styles: White, flushed purple towards the top, equalling or exceeding the stamens; stigmas decurrent for 1.5–5mm
Fruit capsules: Ovoid to somewhat ellipsoid, 1.2–2.5cm long, often flushed with purplish carmine, located at ground level
Habitat: Stony and sandy places, maquis, on siliceous and ultramafic substrates, occasionally on limestone
Flowering period: late August–October
Distribution: E mainland Greece (Stereá Elláda, W Macedonia and Thessaly), NW Turkey (Istanbul area); near sea level to 300m
Cytology: 2n = 54

Clear tessellation is a feature of the flowers of *Colchicum chalcedonicum*.

A relatively little-known species, this has beautifully tessellated flowers It is native to a relatively small area that encompasses both sides of the Bosphorus in Greece and Turkey.

It is occasionally seen in cultivation, especially as a plant for pan culture, and is frequently misidentified as *C. cilicicum*. Its affinity, however, lies with *C. lingulatum*.

Colchicum chalcedonicum subsp. punctatum K. Perss., *Candollea* 53: 405 (1998)
Type: Turkey, Aydin, 15km from Nazilli to Beydağ, near Samailli, *Quercus-Styrax* thickets, under shrubs in deep, stony soil (micaschist), 700–800m, 23 Apr 1991, flowered in cultivation 21 Sep 1994, *K Persson 521* (holotype BG; isotype K)

> **Flowering period:** September–October
> **Habitat:** Among deciduous shrubs over mica schist
> **Distribution:** W Turkey (near Labraunda archaeological site in Muğla province); 700–1,000m
> **Cytology:** 2n = 50

This subspecies is similar to *C. chalcedonicum* subsp. *chalcedonicum* but usually has five leaves that are spreading to erect, larger, mostly 15–22 × 2–3cm, and scarcely undulated at the margin. The flowers are tessellated or dotted, the tepals rather larger at 4–5.5 × 0.7–1.5cm, and the anthers are often yellow.

RIGHT & BOTTOM RIGHT
Larger tepals than the typical species are a feature of *Colchicum chalcedonicum* subsp. *punctatum*.

Colchicum chimonanthum

Colchicum chimonanthum K. Perss., *Pl. Syst. Evol.* 217: 56 (1999)
Type: Greece, Macedonia, SE of Nea Zichni, near Mesorrachi, 20m, Mar. 1976
K. Persson 315 (holotype GB)

Corm: Ovoid to more or less subglobose, with membranous yellow-brown tunics and a short neck, not more than 1.5cm long
Cataphylls: Yellowish white, much exceeding the tunic neck
Leaves: (5–)6(–8), often 6, partly developed at flowering, then 1–4cm, at maturity ascending, linear to linear-lanceolate, up to 14 × 5.5cm, dark green or grey-green, parsely to densely scabrid, channelled
Flowers: 3–6, narrow goblet-shaped, pinkish lilac or white flushed purplish lilac towards the tips; perianth tube slender, up to 3.5cm long
Tepals: Oblong to oblanceolate, to 2.2 × 0.5cm
Stamens: Filaments golden at the slightly swollen base; anthers grey to purplish black
Styles: White, filiform, straight or slightly curved at the top, equalling or somewhat exceeding the stamens; stigmas punctiform
Fruit capsules: Not recorded
Habitat: Short grassy and stony places, generally on gritty, clay soils
Flowering period: January–February
Distribution: N Greece (C and E Macedonia); *c.*200m
Cytology: 2n=32

Currently known from just two localities in eastern Greek Macedonia, both not far from Thessaloniki, this is a particularly ornamental species. As well as the occasional multiplication of the corms during the annual replacement cycle, *C. chimonanthum* also often produces axillary or reserve buds in the axils of the second leaf. These tiny cormlets become detached and can develop into new individuals, a useful means of propagation for gardeners.

This species was described in 1999 by Karin Persson. It bears comparison with the eastern Mediterranean *C. crocifolium* which has a greater number of hairier leaves per corm and rather smaller, more numerous flowers. The chromosome numbers of the two species are very different.

Colchicum chimonanthum growing wild in the Thessaloniki region of north-eastern Greece.

Colchicum chlorobasis

> **Colchicum chlorobasis** K. Perss., *Edinburgh J. Bot.* 62: 182 (2006)
> **Type:** Turkey, Konya, SW of Sorkun, 1,730m, Oct. 1996, *Kerndorff & Pasche 96–09* (holotype GB)
>
> **Corm:** Rounded, although irregular in shape and often with one or two short projections, with a brown or reddish brown tunic and a neck, projecting along the cataphylls for up to 6cm
> **Leaves:** 4–5(6), appearing above ground as the flowers begin to fade, not exceeding 1.2cm wide
> **Flowers:** 1–2, rose-lilac or rose-purple, the tepals fading to white towards the base
> **Tepals:** Narrow-oblanceolate, 3.5–5cm long, distinctly veined, subobtuse; filament channels ridged but glabrous
> **Stamens:** Filaments white, greenish at slightly expanded base; anthers yellow
> **Styles:** White, equalling or slightly exceeding the stamens; stigma punctiform
> **Fruit capsules:** Located below or just at ground level
> **Habitat:** In spiny, open scrub and among junipers
> **Flowering period:** September–October
> **Distribution:** SE Turkey (Bozkir region); 1,000–1,750m
> **Cytology:** 2n=54

This fairly recently described species is confined, as far as is known, to a small area well to the east in southern Turkey. It is to the east of the known localities of *C. baytopiorum*, to which it bears some similarity, but *C. chlorobasis* has more substantial and wider leaves.

Colchicum chlorobasis comes closest to the more widespread *C. boissieri*. Both share the same chromosome number, and have similar leaf and flower shapes and colours. Differences between the two species can be seen in the corms and in the leaf number, *C. boissieri* regularly having three leaves per corm.

The only other species that might be confused with *C. chlorobasis* is *C. sieheanum*, but the latter is found well isolated from the other species discussed here. *Colchicum sieheanum* bears narrower, funnel-shaped, thinner-textured flowers of a rich violet-purple with very slender perianth tubes and less stout filaments that are yellow rather than green at the base.

Colchicum chlorobasis was first collected in 1960 by Guichard. However, the type description was later based on specimens collected by Kerndorff and Pasche in 1996, quite close to Guichard's site. Corms collected in 1996 were cultivated at Gothenburg Botanical Garden in Sweden.

Colchicum chlorobasis grows in open sites protected by spiny scrub in south-eastern Turkey.

Colchicum cilicicum

Colchicum cilicicum (Boiss.) Dammer, *Gard. Chron. ser. 3*, 23: 34 **(1898)**
Type: Turkey, Içel, Güllek, *Kotschy 84d* (lectotype G-BOISS)
Syns *C. autumnale* var. *tenorei* (Parl.) Fiori, *C. balansae* var. *macrophyllum* Siehe ex Hayek, *C. byzantinum* var. *cilicicum* Boiss., *C. decaisnei* sensu Lynch, non Boiss., *C. tenorei* Parl.

Corm: Ovoid to subglobose, 4–6cm long, with a dark brown, dull, sub-leathery tunic, the inner tunic paler and more papery, extended above into a neck, 5–12(–17)cm long
Cataphylls: White, sometimes purple-flushed in the upper part, up to 17cm long
Leaves: 3–4(5), appearing long after flowering, in early spring, suberect to half-spreading, narrow-elliptic to elliptic-lanceolate, 30–40 × 5.5–9(–11.5)cm, deep bright green and moderately glossy, glabrous, often slightly undulate, apex obtuse
Flowers: 3–15, occasionally as many as 25 per corm, funnel to broad chalice-shaped, pale lilac-purple to vivid pinkish purple or rose-purple, generally with a deeper tip, often with a whitish central stripe in the lower two thirds of each tepal, occasionally lightly to moderately tessellated, especially in the upper part; perianth tube whitish, sometimes flushed purple towards the top, generally one and a half to twice as long as the tepals, 8.5–11cm long
Tepals: Elliptic to oblanceolate, 5–7.5 × 1.2–3.5cm, apex obtuse to subacute, pubescent along the filament channel ridges, keeled outside in the lower half
Stamens: Two thirds to four fifths the length of the tepals; filaments white, pinkish or purple with a swollen yellow base; anthers bright yellow, 6–10mm long, shorter than the filaments
Styles: White, pinkish or purple, straight or slightly curved, often very long and equalling the tepals and flopping sideways, often with a purple or dull crimson, curved or hooked tip; stigmas very shortly decurrent (no more than 0.5mm) or punctiform
Fruit capsules: Ellipsoid to obovoid, 3–4cm long, glabrous, pale brown with dark red-brown spots at maturity, with a short beak, located just above the ground on pedicels up to 5cm long
Habitat: Rocky slopes, stream banks, margins or forest or open forest (primarily *Cedrus*, *Pinus* and *Quercus*), on limestone
Flowering period: (Late August–)September–early November, occasionally later
Distribution: Lebanon, NW Syria, S and SC Turkey (S Anatolia: Adana, Mersin, Isparta, Muğla); 350–2,300m
Cytology: 2n=54

Colchicum cilicicum thriving in the trial at RHS Garden Hyde Hall.

A highly attractive and floriferous species, it is restricted in the wild to the eastern Mediterranean where it can be locally common. *Colchicum cilicicum* is one of the more commonly seen species in the Taurus Mountains of Turkey but it can also be found as far afield as northern Syria and Lebanon. The long stamens are a distinctive and useful diagnostic character of the species, whether or not the flowers are lightly tessellated.

As a species it is not seen as much in gardens as it deserves. However, it has been cultivated since 1896 and is a parent of several hybrid cultivars, 'Felbrigg' being perhaps the finest. The selection *C. cilicicum* 'Purpureum' with reddish purple flowers is probably the most widely available. Another selection, with rose-purple flowers, is *C. cilicicum* 'Cilician Gates'.

In cultivation, both *C. dolichantherum* and *C. imperatoris-friderici* have been mistaken for *C. cilicicum*. However, the deeper coloured flowers with long stamens (at least two-thirds the length of the tepals) and the thinner, more mobile, long anthers are particularly useful diagnostic characters for *C. cilicicum*. In addition, the styles of *C. cilicicum* are stouter, more curved, and swollen at the apex. In leaf, *C. cilicicum* has less shiny laminae, more tapered to the apex and less clearly pleated.

RIGHT **Colchicum cilicicum has been in cultivation for more than 100 years.**

BOTTOM RIGHT **This broad-tepalled variant of Colchicum cilicicum shows the long styles typical of the species.**

Colchicum confusum

Colchicum confusum K. Perss., *Pl. Syst. Evol.* 217: 60 (1999)
Type: Greece, Thessalia, Trikala, Pertouli to Elati, 20 Sep. 1989, *K. Persson 486* (holotype GB; isotypes B, K, UPA)

Corm: Ovoid, sometimes narrowly so, 2.5–5cm long, occasionally larger, with a dark brown to blackish brown, many layered tunic, extended above into a 3–8(–13)cm long neck
Cataphylls: Yellowish white, 5–14cm long, often flushed green or purple at the top
Leaves: 3–5 normally, appearing after flowering, often rather spreading or curved, lanceolate to linear-lanceolate, to 20 × 3.5cm, occasionally broader, grey-green to mid-green, with a distinct keel and with a narrow, cartilaginous margin
Flowers: 2–5, narrow chalices, purple or reddish purple; perianth tube whitish, flushed purple towards the top, up to 12cm long but often only half that length
Tepals: Narrow oblong to more or less oblanceolate, to 4.5 × 1cm, occasionally larger, with a median white streak for up to half way from the base, the outer tepals noticeably longer than the inner
Stamens: Filaments white, slightly swollen and yellow at the base and purple-flushed towards the top; anthers yellow
Styles: White, often flushed purple towards the top, overtopping the stamens; stigmas decurrent and slightly hooked, 1.5–4.5mm long
Fruit capsules: Oblong-ellipsoid to ovoid, 2–5cm long, located at ground level
Habitat: Moist meadows on deep serpentinite soils, dolines, margins and clearings in pine forest, primarily over schistose and ophiolitic substrates
Flowering period: September–October, occasionally into early November
Distribution: C and N Greece (Corfu and Paxi, Mount Olympus, Mount Parnassus, Pindus mountains); 300–1,400m
Cytology: 2n=40

Endemic to Greece, *C. confusum* is found down the length of the Pindus mountains southwards as far as Mount Parnassus, with outliers on Mount Olympus in Thessaly, and on the islands of Corfu and Paxi. Plants tend to be gregarious, sometimes found growing in close proximity to both *C. autumnale* and *C. haynaldii*.

An ally of the widespread and variable *C. autumnale*, *C. confusum* was described from Trikala in Thessaly, Greece, based upon a 1989 Karin Persson collection. To the casual observer this appropriately named species might be confused with *C. autumnale*, and the two look very similar in flower.

Colchicum confusum growing near Argyratika on the Ionian island of Paxi near Corfu.

This similarity has meant that *C. confusum* has not made a great impression on gardeners and is rare in cultivation. It is present in a few collections, such as those at Gothenburg Botanical Garden, Sweden, and Royal Botanic Gardens, Kew. It is reportedly very slow to increase.

The differences between this species and *C. autumnale* are best summed up by quoting directly from the Persson (1999a): 'The tunics of *C. confusum* are mostly darker in colour and often also thicker than in *C. autumnale* (tunics reddish brown to mid-brown, mostly submembranous to subcoriaceous). Furthermore, the flowers of *C. autumnale* are frequently larger and paler, with tepal segments tending more towards oblong-elliptic or oblanceolate than linear or narrow oblong, and more often rather distinctly tessellate. Leaf characters, above all shape, are more clear-cut: *C. autumnale* has mostly suberect, ± oblong leaves, widest at, or just below, the middle and distally short-tapering (versus erecto-patent to patent, ± lanceolate, widest in basal part, narrowly long-tapering towards apex), and they have rather obscurely cartilaginous margins (versus distinctly cartilaginous). They are, morover, quite often larger, less twisted, and of a somewhat duller and darker green hue.'

RIGHT **Colchicum confusum flowering in a meadow on Corfu.**

FAR RIGHT **Colchicum confusum grown from a collection (K. Persson 456) made in central Greece.**

BOTTOM RIGHT **The tepal tessellation of Colchicum confusum is barely discernible.**

Colchicum corsicum

Colchicum corsicum Baker, *J. Linn. Soc., Bot.* 17: 431 (**1879**)
Type: Italy, Corsica, Bonifacio, *Serafino s.n.* (lectotype K)
Syns *C. autumnale* var. *corsicum* (Baker) Firoi, *C. neapolitanum* var. *corsicum* (Baker) Fiori, *C. verlaqueae* Fridl.

Corm: Ovoid to subglobose, 2–3cm long, with a membranous brown tunic, extended above into a slender neck, at least as long as the corm
Cataphylls: White, slender, 4–6.5cm long
Leaves: 3–4, appearing after flowering, ascending to spreading, linear-lanceolate, 12–20 × 1–1.8cm, green, glabrous, tapering towards the obtuse to subobtuse apex
Flowers: 1–2(–3), funnel-shaped, pale to mid-rose-pink or lilac-pink; perianth tube white, greatly exceeding the tepals, 2–3 times as long, to 7.5cm
Tepals: Oblong- to linear-elliptic, 16–27 × 2.3–5.5mm, glabrous, apex obtuse
Stamens: About a third of the length of the tepals; filaments white; anthers yellow, 2–3mm long
Styles: White, straight, curved towards the apex; stigmas decurrent for 1.7–3mm
Fruit capsules: Oblong, 2–3cm long, glabrous, located at ground level
Habitat: Rocky and grassy places, maquis, open woodland
Flowering period: September–early October
Distribution: S Corsica (Bitalza, Cagna and Zerubia), Sardinia (Maddalena archipelago: La Maddalena, Caprera, Segnalata, Spargi); at low altitudes
Cytology: 2n=216

Confined to the southern part of Corsica and the Maddalena archipelago off north Sardinia, *C. corsicum* is a small-flowered species. It is very localised and is included on the IUCN's list of threatened species where its status is considered to be vulnerable.

Colchicum verlaqueae is very closely related to *C. corsicum* and comes within the circumscription of the latter. In 1999 the western Corsican population of *C. corsicum* and some populations in Sardinia were recognised as a closely related yet distinct species, *C. arenasii*.

Although small, *C. corsicum* is nonetheless attractive and has found its way into a number of private collections. It is probably best suited to pan culture in an alpine house or frame, but it will also succeed outdoors in a raised bed or large trough. In John Morley's garden near Beccles in Suffolk it has self-sown successfully, and flowers annually in paving cracks.

Colchicum corsicum thriving between paving cracks in a Suffolk garden.

Colchicum cretense

Colchicum cretense Greuter, *Candollea* 22: 246 (1967)
Type: Crete, Chania, Kidonia, 1,350m, Oct. 1966, *Greuter* 7737 (holotype Herb. Greuter; isotypes ATH, B, E, FI, G, GB, K, LD, M, P, S, UPA, W)
Syn. *C. creticum* sensu Rech. f. & P.H. Davis., non Turrill

Corm: Ovoid to almost globose, 1.5–3 × 1–1.5(–2)cm, with a membranous brown tunic and a slender neck, to 5cm long, occasionally longer
Cataphylls: White, very slender, to 7.5cm long, occasionally longer
Leaves: 4–8, usually appearing after flowering, occasionally very short then, at maturity arching and pressed to the ground, linear, mostly 7–15 × 0.2–0.35cm, occasionally larger, dull dark green, sometimes flushed purple at the apex, glabrous but occasionally scabrid at the margin
Flowers: 1–3, occasionally as many as 7, funnel- to star-shaped, pale lilac to pinkish lilac or white; perianth tube very slender, two to three times longer than the tepals
Tepals: Narrow oblong to elliptic-oblong, 1.5–2.5 × 0.3–0.55cm, with some narrow lamellae present at the base of the inner tepals
Stamens: Filaments white, yellow and swollen at the base; anthers violet-, yellowish or brownish grey, 2.5–4.5mm long
Styles: White, equalling the stamens; stigmas punctiform
Fruit capsules: Oblong-ovoid, 0.7–1.5cm long, glabrous, located at ground level
Habitat: Open stony and rocky slopes and screes, short grassy places
Flowering period: Late September–November, occasionally in late winter
Distribution: Crete; 1,200–2,300m, occasionally lower
Cytology: $2n = 36$

This small-flowered endemic species is similar to the more widely-distributed *C. pusillum*. Both are found on Crete, with *C. pusillum* up to 1,400m and *C. cretense* above 1,200m, on the principal high mountains of the island such as Lefká Óri, Psilorítis and Díkti. Some botanists consider them to be forms of a single species. For instance, Greuter (1971) suggested that *C. cretense* might be a higher altitude form of *C. pusillum*.

The dark anthers of *C. cretense* are a useful diagnostic feature, whereas those of *C. pusillum* are more usually yellow, especially on Crete. The anthers are 2.5–4.5mm long in *C. cretense*, but only 1.5–3mm in *C. pusillum*. *Colchicum pusillum* also differs in its chromosome number of $2n=54$. Forms of *C. cretense* with double the number of tepals have been recorded from the wild.

The starry flowers and dark anthers of *Colchicum cretense*.

Colchicum crocifolium

Colchicum crocifolium Boiss., *Diagn. Pl. Orient.* ser. 1, 5: 67 (**1844**)
Type: Iran, Fars, Shiraz, *Aucher-Éloy 5365* (holotye G-BOISS; isotypes BM, FI, G, LE, W)
Syns *C. crocifolium* var. *lasiophyllum* Bornm., *C. montanum* sensu Baker, non L., *C. stenanthum* Bornm.

Corm: Ovoid to somewhat oblong, 2.5–4.5cm long, with a several-layered membranous to subcoriaceous chestnut-brown tunic, extended at the top into a slender neck, 3–7.5cm long
Cataphylls: Slender, whitish, 7–17cm long
Leaves: 6 or more, occasionally as many as 20 but only a few present at flowering, then partly developed and 4–6cm long, at maturity spreading to arcuate, filiform to linear, 8–17 × 0.3–1.3cm, dull grey-green, often finely pubescent or hairs confined to the margins, backs or sheaths, sometimes glabrous, grooved, often twisted, apex acute, sometimes somewhat hooded
Flowers: 4–12, narrow funnel-shaped, rarely more than 2cm across when fully open, white or yellowish flushed lilac-pink to purple-pink in the upper part; perianth tube slender, 1–2.5cm long, white or straw-coloured, sometimes greenish towards the top, shorter than or equalling the tepals
Tepals: Narrow-elliptic to linear, 1.2–2.4 × 0.1–0.45cm, spreading and often recurving, apex acute or subacute
Stamens: Filaments slender, white with a slightly swollen orange-brown or brown base; anthers dark grey-green to purplish grey, 1.5–3.5mm long, shorter than the filaments, the outer stamens shorter than the inner ones
Styles: White, slender, overtopping the stamens, straight or angled to one side of the flower; stigmas punctiform
Fruit capsules: Globose to ovoid, 1–1.8cm long, apex pointed, finely papillose or short-hirsute overall, located at ground level
Habitat: Sandy and stony deserts, rocky slopes, open oak scrub, fallow fields
Flowering period: Late December–March
Distribution: W Iran, N Iraq, Jordan (rare), N Syria (rare) and SC Turkey; 200–2,100m, occasionally at lower altitudes in Iran
Cytology: 2n=14

Fine pubescence is often visible on the leaves of *Colchicum crocifolium*.

This is a very small and relatively little-known, semi-desert species, closely related to *C. chimonanthum*. It is nonetheless charming, and very crocus-like in its small flowers and filiform leaves.

Colchicum cupanii

Colchicum cupanii Guss., *Fl. Sicul. Prodr.* 1: 452 (1827), as '*cupani*'
Type: Italy, Sicily, Piana de' Greci, Trapani, Mazzara, *Gussone* (NAP)
Syns *C. bertolonii* Steven, *C. cousturieri* Greuter, *C. creticum* Turrill, *C. cupanii* var. *bertolonii* (Steven) Rouy, *C. cupanii* var. *pulverulentum* Batt. ex Maire & Weiller, *C. gussonei* Lojac., *C. parviflorum* Biv., *C. valeryi* Tineo

Corm: Ovoid, 1.8–2 × 1.3–1.5cm, with a dark brown or reddish brown leathery tunic and a short or no neck
Cataphylls: White, 2–7cm long, occasionally longer
Leaves: 2(–3), present at flowering, then short, ascending to erect at first, later spreading to arcuate, at maturity linear to linear-lanceolate, up to 16 × 1.8cm, glabrous, channelled and with a distinct midvein, apex obtuse to subacute, rarely slightly ciliate at the base
Flowers: 1–4(–10), occasionally more, chalice-shaped to starry, pink to purplish pink, with a whitish throat, rarely entirely white, distinctly veined, with or without deeper parallel striping, almond-scented; perianth tube whitish or yellowish white, often flushed purple, 2–3 times as long as the tepals
Tepals: Elliptic, often rather narrow, 1.8–2.5 × 0.3–0.7cm, apex acute to subobtuse
Stamens: Filaments white, somewhat swollen and greenish yellow at the base, glabrous; anthers purplish black, 2–3mm long
Styles: White, straight, equalling or longer than the stamens; stigmas punctiform
Fruit capsules: Ovoid-oblong, 1.5–1.8m, located at ground level or subterranean; seeds with a whitish appendage
Habitat: Rocky and short grassy places, generally over limestone, archaeological sites, hill slopes, olive groves
Flowering period: late September–December
Distribution: NW Africa (Algeria and Tunisia), Mediterranean region (SE France, Sardinia and Sicily from Malta and S Italy to Crete (rare, confined to the north), Aegean islands including the Cyclades); sea level to 1,100m
Cytology: 2n=54

The distinctive veins are clearly visible on this plant of *Colchicum cupanii* subsp. *glossophyllum* growing near Monemvasia in southern Greece.

A familiar sight in many parts of the western and central Mediterranean including North Africa, this is one of the most delightful little species. It sometimes grows in substantial colonies with a peak flowering time in September and October. In the southern Peloponnese, as subsp. *glossophyllum*, it often forms attractive small posies close to the ground

While the leaves can be partially developed at flowering time, they are often very short, clasping the perianth tubes. The parallel veins on the tepals can be well marked and rather prominent. Variants from islands off the south coast of Crete (Chrysí, Koufonísi and Mikronísi) with purple-striped tepals have been separated as *C. cousturieri*. However, such plants can be found in other parts of the range of *C. cupanii* and it is difficult to uphold *C. cousturieri* as a species, despite its rather isolated distribution in the wild, so it is regarded as a synonym here.

Colchicum cupanii makes a delightful subject for pans in an alpine house or bulb frame. It will also succeed very well outdoors in a large trough or on a raised bed, where it will multiply slowly, generally flowering in cultivation in September and October. In some years it may come into flower towards the end of August. It requires space as the mature leaves spread out close to the ground.

This species looks superficially similar to *C. psaridis*, but the latter differs in its soboliferous, 'grub-like' corms and narower leaves, rather smaller flowers and short filaments.

RIGHT **Forming a small posy of flowers near the ground is how *Colchicum cupanii* subsp. *glossophyllum* is sometimes encountered.**

FAR RIGHT **The smaller and narrower leaves of *Colchicum cupanii* subsp. *glossophyllum*.**

BOTTOM RIGHT ***Colchicum cupanii* growing near Amfissa in Greece.**

Colchicum cupanii subsp. glossophyllum (Heldr.) Rouy, *Bull. Soc. Bot. France* 52: 646 (1905)
Type: Greece, Peloponnese, Messinia, Kalamata, Heldreich Herb. Graec. Norm. no. 1496 (lectotype WU; isotypes FI, G, GB, JE, LD, P, S). 2n=54.
Syn. *C. glossophyllum* Heldr.

Distribution: SW Albania and Greece (mainland and Ionian islands)

This subspecies has rather smaller and narrower leaves than *C. cupanii* subsp. *cupanii* and generally has paler flowers, but is found in similar habitats and altitudes.

Colchicum davisii

Colchicum davisii C.D. Brickell, *The New Plantsman* 5(1): 15 **(1998)**
Type: Turkey, Adana, Bahce Dumnali Dağ, 1,300m, Apr. 1957, *Davis 26938* (holotype E; isotypes GB, WSY)

Corm: Irregularly ovoid, often asymmetric, 4–5 × 3.5–5cm, often with horizontal or subvertical, fleshy soboles, these somewhat flattened and channelled, to 12cm long, tunic pale to mid-brown, papery, flaking, the inner tunics darker brown and more membranous, extended above into a substantial neck, 12–17cm long, sometimes much shorter or apparently scarcely present
Cataphylls: Yellowish white, pale green at the split tip, extending to the soil surface or somewhat protruding
Leaves: 3, appearing well after flowering, in late winter and early spring, erect at first but then spreading in the upper part, oblong-lanceolate to oblong elliptic, 19–23(–31) × 3.5–6cm, glossy mid-green, not plicate but with well-defined central veins as seen from above, apex obtuse to subacute, narrowing abruptly into the sheath of the pseudostem towards the base, with a blunt keel beneath in the basal part
Flowers: 2–5(6), funnel-shaped to more or less goblet-shaped, whitish with pale to deep purple or violet-purple tessellations in the upper two-thirds or half, white towards the base of the tepals, particularly in the throat, parallel-veined and with a median white stripe extending up the tepals for up to two-thirds the distance; perianth tube white, sometimes flushed purple, 4–6cm long, occasionally longer
Tepals: Elliptic-oblong to oblanceolate, 4–6 × (0.8–)1–2.1cm, obtuse to subobtuse at the apex, the inner tepals noticeably rather narrower and shorter than the outer
Stamens: Filaments white, 1.8–2.5cm long, slightly swollen and pale yellow at the base; anthers pale yellow to golden-yellow, 10–15mm long
Styles: White, occasionally flushed purple, somewhat shorter or overtopping the stamens; stigmas strongly curved, decurrent at the end of the style for 3–5mm
Fruit capsules: Narrow oblong-ellipsoid to ellipsoid, 2.5–3cm long, glabrous, shortly beaked
Habitat: Margins of beech (*Fagus*) forests
Flowering period: (August–)September–October
Distribution: S Turkey, with a limited distribution, confined to the Nur Mountains, notably Dumanli Dağ and Kartál Dağ; *c.*1,300m
Cytology: 2n=46

The tessellation of *Colchicum davisii* is just visible at the apex of the tepals.

A Turkish endemic, *C. davisii* finds its closest allies in two other Turkish species, *C. baytopiorum* and *C. cilicicum*. *Colchicum baytopiorum* produces soboles like

C. davisii and it usually has three leaves but these are partly developed at flowering time, and darker green, less glossy than those of *C. davisii*, and markedly recurving at maturity. The flowers of *C. baytopiorum* are untessellated. *Colchicum cilicicum* usually has four to five, broader, clearly plicate leaves. *Colchicum davisii* is very distinct from these species in its strongly curved, often hooked stigmas. *Colchicum speciosum* also bears long-decurrent, curved stigmas, but it differs in many other respects, not least being the lack of soboles, the fewer, untessellated flowers with far longer perianth tubes, and the more numerous, markedly plicate, leaves.

This charming species is named in honour of Dr Peter Davis who contributed enormously to the *Flora of Turkey* project at Royal Botanic Garden Edinburgh. At the same time as material was first collected from Dumnali Dağ in 1957, live corms were also introduced to Royal Botanic Garden Edinburgh where they have continued to thrive until the present day.

The species is established in cultivation and occasionally offered for sale, but is not always correctly identified. Despite this, it has proved to be a good and reliable, hardy plant in the open garden, flowering and multiplying freely, especially in the more northerly parts of Britain. The fact that the leaves do not develop until late winter or early spring means that they remain undamaged through the winter period, unlike those of the related *C. baytopiorum* whose leaves develop soon after flowering in the autumn and can be damaged by hard frosts.

RIGHT *Colchicum davisii* is hardy in the open garden and flowers well.

FAR RIGHT The type collection of *Colchicum davisii* is still growing at Royal Botanic Garden Edinburgh.

BOTTOM RIGHT The prominent tessellations of this cultivated selection of *Colchicum davisii* give a deeper coloured flower.

Colchicum decaisnei

Colchicum decaisnei Boiss., *Fl. Orient. [Boissier]* 5(1): 157 (1882)
Type: Lebanon, Mont Libani, Ghazir, Oct. 1861, *Gaillardot 2804* (lectotype G-BOISS; isotypes JE, LE, P, S)
Syns *C. brevistylum* Feinbrun, *C. troodi* sensu C.D. Brickell, non Kotschy

Corm: Ovoid to almost globose, 2–4cm long, adorned with a pale brown or reddish brown membranous tunic, extended above into a neck, to 9cm long
Cataphylls: Whitish, 8–12cm long, mostly concealed within the corm neck
Leaves: 3–5, occasionally 6, appearing after flowering, spreading to arcuate, linear-lanceolate to lanceolate, occasionally narrow ovate, mainly 12–20 × 1.2–4cm, occasionally up to 7cm wide, mid-green, channelled and somewhat often twisted, sometimes pubescent at the base dorsally, tapering towards the subacute apex, the margin glabrous to ciliate
Flowers: 3–10, occasionally more, funnel-shaped, white, often pink-flushed, to pinkish lilac to pinkish purple, not tessellated; perianth tube white, sometimes pink-flushed towards the top, equalling or up to twice the length of the tepals
Tepals: Linear-elliptic to narrow oblanceolate, 2.6–4.5 × 0.3–1.2cm, apex obtuse to subobtuse, often channelled at the base, filament channels indistinct, often puberulous
Stamens: One-third to half the length of the tepals; filaments whitish, yellowish and slightly swollen at the base; anthers yellow, often pale, 4–8mm long
Styles: Slender, white, equalling or slightly overtopping the stamens, the tip curved, the stigmas decurrent for 0.5–1mm or more or less punctiform
Fruit capsules: Ellipsoid-oblong to ellipsoid or ovoid, 1.5–3.5cm long, obscurely to finely brown-dotted, apex acute to slightly acuminate
Habitat: Maquis, open pine woodland and oak scrub, generally in part shade, locally common
Flowering period: Late September–November(–early December)
Distribution: N Israel, Lebanon, N Palestine, Syria and S Turkey; sea level to 1,000m
Cytology: 2n=54

Corms of this Middle Eastern species tend to clump, with several clustered tightly together, resulting in posies of 30 flowers or more. Young plants produce droppers, long vertical extensions from the corm base, sinking them deeper into the earth. These are, in effect, an extension of the foot or hypopodium. The leaves generally appear shortly after flowering although they are not fully mature until early spring.

Turkish plants of *C. decaisnei* tend to be smaller than those from other parts

The numerous flowers indicate that this is a clump of corms of Colchicum decaisnei, *here growing in Upper Galilee in Israel.*

of the range, both in the size of the corms as well as in the relative sizes of the leaves and flowers.

This species has long been associated with C. troodi from Cyprus and the two were considered to be synonymous in the *Flora of Cyprus* (Meikle 1985). In characters such as the leaves, the stamens and anthers the two have much in common. However, Persson (1999b) argues that the differences between *C. decaisnei* and *C. troodi* are real and outlines the characters that separate them. The leaves of *C. troodi* are flattish, scarcely channelled and dark glossy green, versus narrow-lanceolate, mid or glaucescent green, less glossy and channelled in *C. decaisnei*. The flowers are very pale in *C. troodi*, and white or purple-flushed, while in *C. decaisnei* they are white to pink or deeper pinkish purple. The corms of *C. troodi* have membranaceous tunics, whereas the tunics are subcoriaceous in *C. decaisnei*.

Colchicum decaisnei can also be likened to *C. hierosolymitanum* but there are differences which can be readily observed at various times of the year. *Colchicum decaisnei* produces fewer, wider leaves, often 3–4, occasionally 5, 2.5–4cm wide, versus 5–9, 1–2cm wide in *C. hierosolymitanum*. The flowers of *C. decaisnei* are generally paler, white to pale pink and often starrier in appearance, versus pale to deep rose or purple-pink and funnel-shaped in *C. hierosolymitanum*. Differences can also be observed in the stamens, which in *C. decaisnei* have shorter anthers that are 5–6mm long, but the anthers are 6–10mm long in *C. hierosolymitanum*.

RIGHT & FAR RIGHT **The flowers of *Colchicum decaisnei* can be various shades of pink or white.**

BOTTOM RIGHT **This specimen of *Colchicum decaisnei* from Slinfah in western Syria is clumping up well in a pan.**

Colchicum doerfleri

***Colchicum doerfleri* Halácsy**, *Denkschr. Kaiserl. Akad. Wiss., Wien. Math.-Naturwiss. Kl.* 64: 739 **(1897)**
Type: Republic of Macedonia, Nerisi to Üsküb (Skopje), Apr. 1893, Doerfler, It. Turc. Sec. 564 (lectotype W; isotypes B, W)
Syn. *C. drenowskii* Degen & Rech. f. ex Kitan. pro parte

Corm: Narrow-ovoid, 2.5–4cm long, with a deep brown, subcariaceous tunic
Cataphylls: White, 5–10cm long
Leaves: 2(–3), partly developed at flowering and then as long as, or somewhat shorter than, the flowers, lanceolate to linear-lanceolate, to 24 x 2.5cm at maturity, deep green or grey-green, often slightly hooded at the tip, pilose on the ventral surface, the margins retrorse-ciliate
Flowers: 1–5(–8), narrowly chalice-shaped at first, soon opening more widely, rose-lilac to purple-carmine
Tepals: Narrow-elliptic to lanceolate-elliptic, 2.8–4cm long, subacute
Stamens: Anthers purplish black; pollen deep yellow or orange-yellow
Styles: White or pink, filiform, exceeding the stamens, straight or curved towards the top; stigma punctiform
Fruit capsules: Not recorded
Habitat: Dry slopes over serpentinite or limestone, short grassy places, and open scrub of *Juniperus* or *Quercus*
Flowering period: February–April
Distribution: E Albania, SW Bulgaria, N and NE Greece (on various mountains such as Athos, Olympus, Orvilos, Ossa, Vellia and Vermio) and Republic of North Macedonia; 700–2,000m
Cytology: $2n=54$

This species is very closely related to *C. hungaricum* and considered to be conspecific by some. However, *C. hungaricum* tends to have a more northerly distribution, descending to sea level along the Dalmatian coast, whereas *C. doerfleri* is primarily a Balkan species. *Colchicum hungaricum* is also not found in Greece, and plants recorded as such are referable to *C. doerfleri*.

Colchicum drenowskii, regarded in part as a synonym above, is based on a type from material collected in Bulgaria. However, it is in fact a mixed gathering, the leaves belonging to *C. doerfleri* and the flowers to *C. autumnale*.

The cultivar *C. doerfleri* 'Valentine' has pale rose-pink flowers and usually flowers in mid-February.

Colchicum doerfleri forming a spectacular clump on Mount Ossa in Greece.

Colchicum dolichantherum

Colchicum dolichantherum K. Perss., *Edinburgh J. Bot.* 56(1): 126 (1999)
Type: Turkey, Adana-Gaziantep, Nur Dağ, 1,150m, Oct. 1988 (holotype GB; isotype E)

Corm: Ovoid to almost globose or oblong-ovoid, 3.5–6cm long, with a tough, leathery, mid to reddish brown tunic, extended above into a slender, fibrous neck, to 15cm long
Cataphylls: Substantial, yellowish white, often flushed carmine-purple towards the top, to 14cm long
Leaves: 4–5(–6), appearing long after flowering, spreading from a short pseudostem, narrow-lanceolate to broad oblong or elliptic-oblong, 25–50 × 5–10cm, bright glossy green, the tips often flushed crimson-purple when young, pleated longitudinally, margin thinly cartilaginous, sometimes obscurely so, glabrous overall or sometimes short-ciliate or slightly scabrid towards the base
Flowers: 5–15, funnel-shaped to more or less goblet-shaped, whitish to pale pink or rose-lilac, sometimes obscurely tessellated, tepals sometimes with a median white stripe on the exterior; perianth tube creamy white, 3–6.5cm long
Tepals: Linear to narrow-oblong or oblong-oblanceolate, 2.8–5.5 × 0.5–1.5cm, the inner noticeably shorter than the outer, keeled on the back, apex obtuse or subobtuse, margin and filament channels glabrous or finely ciliate
Stamens: To half the length of the tepals; filaments white with a swollen orange-yellow base; anthers dark yellow, 8–15mm long, equalling or longer than filaments
Styles: White, equalling or slightly longer than the stamens, thickened and somewhat hooked at the tip; stigmas usually decurrent for 0.7–2mm
Fruit capsules: Ellipsoid to ellipsoid-oblong, 2.5–4cm long, sometimes with irregular brown dots, short-beaked, located at ground level or slightly above on short pedicels to 5cm long
Habitat: Meadows, oak scrub, maquis, dolines, rocky woodland, occasionally in vineyards, generally on limestone
Flowering period: Late August–early November
Distribution: NW Syria, S Turkey (Taurus and Nur mountains); sea level to 1,500m
Cytology: 2n=54

The large number of flowers per corm is clearly visible in this plant of *Colchicum dolichantherum* seen west of Göynük in southern Turkey.

This is a charming species producing a large number of flowers per corm. It can also be distinguished by its long anthers and relatively short filaments. The large leaves are also indicative, being probably the longest in the genus, excluding some garden hybrids.

Colchicum erdalii

Colchicum erdalii Özhatay, *Istanbul Üniv. Eczac. Fak. Mecm.* 46(2): 126 (2016)
Type: E Turkey, Erzincan, Kemaliye, Sırakonak village, 1,739m, 27 Jun. 2008, *E. Kaya 591* (holotype ISTE 96117)

Corm: Irregularly oblong-ovoid, to 3.7 cm long, with 1-2 small lobes, with or wiithout a short neck, outer tunic dark reddish brown and coriaceous
Cataphylls: Yellowish white, with a brownish membranous top at soil level
Leaves: 3-5, partly developed at flowering time, then shorter or slightly longer than the flowers, linear-lanceolate, shiny deep green, often with a purplish tip, glabrous, ascending, deeply channelled, spreading at maturity, then 20–21 x 2.4–2.5cm
Flowers: 2-3, funnel-shaped, white or pink-flushed, the tube stout, 20–22mm long, shorter than, or equalling the tepals, white, pale green towards the top
Tepals: Narrow-lanceolate to triangular-lanceolate, 2–2.8cm long, acute to subacute or shortly acuminate, the inner three somewhat shorter than the outer
Stamens: About half the length of the tepals; filaments white, greenish at the slightly swollen base; anthers dark purplish, about a third of the length of the filaments; pollen yellow
Styles: Filiform, equalling or slightly exceeding the stamens, purplish; stigma punctiform
Fruit capsules: Oblong, 8–9mm long, probably immature when measured
Habitat: Rocky slopes and screes
Flowering period: February–March
Distribution: E Turkey (Kemaliye, Erzincan Provinces), endemic; 1,700–1,800m
Cytology: $2n=14$

A recently described species, this is known at present only from a very small area near Kemaliye in Turkey. It was named in honour of Erdal Kaya who has contributed much to our understanding of *Colchicum* in that country in recent years.

Colchicum erdalii is closely related to the more widespread Turkish *C. serpentinum*, differing primarily in the stout perianth tube of the former which widens towards the top, and the pointed, often acuminate, tepals. Also, the chromosome numbers differ slightly.

Colchicum erdalii in cultivation, grown from the type collection.

Colchicum euboeum

Colchicum euboeum (Boiss.) K. Perss., *Candollea* 53: 400 (1998)
Type: Greece, Sterea Ellada, Evvia, Chalkida, Aug. 1871, *Orphanides 4027* (lectotype G-BOISS)
Syns *C. bivonae* subsp. *euboeum* (Boiss.) Nyman, *C. latifolium* var. *euboeum* Boiss.

Corm: Large, ovoid to almost globose, 3.6–6.5cm long, covered in a subcoriaceous or more or less coriaceous, dark brown to chestnut-brown tunic, extended above into a stout neck, 6–5(–15)cm long, occasionally longer
Cataphylls: Yellowish white, equalling or slightly longer than the corm neck
Leaves: 3(–4), appearing after flowering, congested on a short stem up to 8cm above the ground, ascending to spreading, oblong-elliptic to lanceolate-elliptic or elliptic, mainly 10–19 × 2.5–6cm, sometimes larger, deep green, glabrous, slightly keeled on the back and somewhat pleated, otherwise rather flat, the margin rather obscurely cartilaginous
Flowers: 1–3, funnel-shaped, pale lilac-purple, moderately to strongly tessellated with lilac- or violet-purple, with a median white stripe within for most of the tepal length; perianth tube yellowish white to white, 4–10cm long
Tepals: Narrow-oblong to oblong-oblanceolate, generally 4.5–6.5 × 0.6–1.6cm, occasionally larger, the inner tepals generally rather shorter than the outer, often with slight longitudinal pleating and somewhat twisted, keeled on the outside, filament channels glabrous or somewhat puberulous along the margins
Stamens: A third to half the length of the tepals, the outer rather shorter than the inner; filaments yellowish white, thickened and yellow and somewhat swollen at the base; anthers yellow to buff, 9–15mm long
Styles: Yellowish white, extending beyond the stamens; stigmas decurrent for 3–4mm, occasionally more
Fruit capsules: Oblong-ellipsoid to ellipsoid, 2–5cm long, glabrous, short-pointed, located above ground on a short pseudostem
Habitat: Rocky slopes and cliffs, on limestone
Flowering period: July–August
Distribution: Greece (Macedonia, Stereá Elláda, Euboea island and Halkidiki); 900–1,350m
Cytology: 2n=54

Colchicum euboeum on Mount Dirphē on the Greek island of Euboea.

This interesting and geographically restricted species was first recognised by Greek botanist Theodoros Orphanides as long ago as 1876 but, unfortunately, the name was never properly published. Boissier (1882) took up the name but as a variety of *C. latifolium* and it was later upgraded to a subspecies but under *C. bivonae*. Persson (1998a) recognised the true status of this taxon, upgrading it to species level, thus confirming the original concept of Orphanides.

Colchicum euboeum is often compared to the more widespread and better known *C. bivonae*. The latter species has more membranous reddish brown tunics, the corms with scarcely a neck, while the flowers are darker and more pronouncedly tessellated, with generally wider tepals and brown anthers. Comparisons can also be made with another Greek endemic, *C. parnassicum*, which has smaller corms and flowers, generally 3–4(–5) leaves and paler, faintly tessellated flowers with yellow, smaller anthers 5–10mm long. *Colchicum graecum* also has affinities with these species. It also has yellow anthers and generally 4–5(–6) leaves, while the flowers can be faintly tessellated or not. *Colchicum parnassicum* has the same chromosome count of 2n=54 as *C. euboeum*, while *C. graecum* has 2n=42–44 on plants studied to date.

RIGHT & FAR RIGHT *Colchicum euboeum* flowering in August on Mount Dirphē on the Greek island of Euboea.

BOTTOM RIGHT A cultivated plant of *Colchicum euboeum*, revealing some tessellation.

Colchicum fasciculare

Colchicum fasciculare (L.) R. Br., *Narr. Travels Africa [Denham & Clapperton]* 1826: 243 (1826)
Type: Syria, Haleb, Aleppo, (lectotype Russell, illustration in *Nat. Hist. Aleppo*, t. 2 facing p. 34 (1756))
Syns *C. halepense* Freyn, *C. illyricum* Stokes

Corm: Narrow ellipsoid-ovoid, 1.8–4cm long
Cataphylls: White, long and slender, 4–12.8cm long, extending above ground by up to 10mm
Leaves: 3–5, appearing at flowering, then equalling or slightly exceeding the flowers, up to 11cm long, at maturity curving and spreading to recurved, narrow-lanceolate, 14–22 × 0.5–2.2cm, bluish green, glabrous or occasionally sparsely hairy, channelled, apex acute to subacute
Flowers: 4–7, occasionally more, funnel-shaped, white or palest pink to pinkish purple; perianth tube white, shorter than, or equalling the tepals
Tepals: Narrow-elliptic to narrow elliptic-oblong, 1.8–3.5(–4) × 0.3–0.9(–1.2)cm, apex acute to subacute
Stamens: About two-thirds the length of the tepals; filaments, slender, white, slightly swollen and greenish yellow at the base; anthers shorter than the filaments, buff, grey-green
Styles: White, straight, shorter than or equalling the stamens; stigmas punctiform
Fruit capsules: Not recorded
Habitat: Rocky and stony habitats in desert and semi-desert regions, decomposed shales, earthy slopes, locally common
Flowering period: October–January
Distribution: W Iran, Jordan and Syria; (700–)1,550–2,850m
Cytology: 2n=14

Colchicum fasciculare can be readily mistaken for *C. schimperi*. While the two overlap in distribution in Syria and Jordan, the latter species extends down into Israel and Egypt and throughout its distribution it is said to be a rare species. Various differences between the two can be observed. In *C. schimperi* the cataphylls do not appear above ground level and the base of the filaments are egg-yolk yellow and there are usually three leaves only. While the flowers of *C. schimperi* are often pale pink they can on occasion be deep pink.

The narrow leaves of *Colchicum fasciculare* seen on a plant growing south-west of Madaba in Jordan.

Colchicum feinbruniae

Colchicum feinbruniae K. Perss., *Israel J. Bot.* 41(2): 75 (1993)
Type: Israel, Golan, Har Taba, NW of Har Shipon, Oct. 1979, *Heyman s.n.* (holotype HUJ 9437)
Syn. *C. bowlesianum* sensu Feinbrun, non B.L. Burtt.

Corm: Ovoid to more or less ellipsoid, 3.5–5.5cm long, adorned with a dark maroon-brown or orange-brown, membranous tunic, glossy or not, extended above into a split neck, 2.5–12cm long
Cataphylls: Yellowish white, sometimes purple-flushed at the top, greatly exceeding the corm neck, to 18cm long, more often only half that length
Leaves: 6–10, appearing after flowering, in mid-winter, crowded at ground level, ascending to slightly spreading, linear-lanceolate to linear-oblanceolate, 15–22 × 1.5–2.5cm, bluish to mid-green, young leaves often flushed with purplish crimson at the top, glabrous to short-ciliate, especially towards the base, shallowly channelled and with a blunt keel beneath, apex obtuse to subacute or sometimes slightly retuse, margins thinly cartilaginous
Flowers: 2–10, occasionally more, goblet-shaped, pale to deep pinkish purple, finely but distinctly tessellated overall, sometimes with a paler central stripe; perianth tube white flushed pinkish purple, especially towards the top, equalling or longer than the tepals, 3–7cm long
Tepals: Elliptic-oval to oval-obovate, 3.5–6(–7) × 0.7–2(–2.3)cm, the inner rather shorter than the outer, apex obtuse to subobtuse, often keeled at the base externally and usually with a fine groove above, especially in the lower half, the filament channels glabrous to densely puberulous
Stamens: About half to two-thirds the length of the tepals; filaments whitish, flushed with pinkish purple, pale yellow at the slightly swollen base, the outer stamens somewhat shorter than the inner; anthers dark grey-green or purple, 8–12mm long
Styles: White, sometimes purplish towards the tip, usually overtopping the stamens, slightly swollen and hooked at the tip; stigmas shortly decurrent for 0.7–2mm on the slightly curved tip
Fruit capsules: Ellipsoid, 2.5–3cm long, brown-dotted when dry, shortly pointed, located at ground level
Habitat: Rocky habitats on heavy terra rossa or basaltic soils
Flowering period: October–January
Distribution: Israel, Lebanon and NW Syria
Cytology: 2n=22

Fruit capsules from the previous year can be seen on this plant of *Colchicum feinbruniae* growing on the Golan Heights.

A highly attractive, showy species, this has perhaps the largest flowers of any of the Middle Eastern species outside Turkey. However, it has a rather limited distribution in the wild.

Its finely tessellated flowers meant it was at one time associated with *C. bivonae* but the two species are not closely related. *Colchicum feinbruniae* probably finds its closest allies in non-tessellated species, particularly *C. hierosolymitanum* and *C. polyphyllum*. It is to be found in Post & Dinsmore (1933) erroneously as *C. bowlesianum*.

A rare plant in cultivation, it is certainly in the collections at Gothenburg Botanic Garden, Sweden, but seldom, if ever, seen in private gardens.

RIGHT **For a Middle Eastern species, *Colchicum feinbruniae* has particulalrly large flowers.**

FAR RIGHT **The styles of *Colchicum feinbruniae* usually exceed the stamens in length.**

BOTTOM RIGHT **Distinct tessellation over the entire tepal surface is a feature of *Colchicum feinbruniae*.**

Colchicum figlalii

Colchicum figlalii (Varol) Parolly & Eren, *Willdenowia* 37: 267 (2007)
Type: Turkey, Muğla, Sandras Dağ, Kartal Gölü, 1,900–2,100m, May 2001, *Varol 4016* (holotype Muğla Univ. Dept. Biology)
Syn. *Merendera figlalii* Varol

Corm: More or less ovoid, 1–2.5cm long, adorned with dark blackish brown tunics which are extended above into a long neck, 4–8cm long
Cataphylls: Whitish to pale or deep maroon, extending beyond the corm neck by 1–2cm
Leaves: 3(–4), present at flowering, then as long as, or somewhat longer than the flowers, at maturity spreading widely apart, strap-shaped to linear-lanceolate, 8–12 × 0.3–0.6(–0.9)cm, glabrous, apex acute to subacute
Flowers: 1(–2), narrow funnel-shaped, bright lilac; tepals split to the base, narrowed rather abruptly into a filiform claw, two or three times longer than the limb
Tepals: Linear to narrow-elliptic, mostly 1.4–1.9 × 0.2–0.5mm, without auricles
Stamens: Filaments whitish lilac; anthers greenish black or black, 2–3.5mm long, about half the length of the filaments
Styles: Slender, straight or curved, about as long as the tepals
Fruit capsules: Oblong-ovoid, about 10mm long, located just above ground level
Habitat: Serpentinite rock clearings, associated species include *Corydalis rutifolia* subsp. *erdelii*, *Gagea glacialis* and *Muscari bourgaei*
Flowering period: April–early June
Distribution: SW Turkey (Sandras Dağ), endemic; 1,900–2,100m
Cytology: $2n=54$

Discovered at the very beginning of the present century, this species of restricted distribution finds its closest ally in the more widespread and better known *C. trigynum* (often found in literature as *Merendera trigyna*). The prime differences are to be found in the long tunicated neck of the corm of *C. figlalii*, which also has rather smaller leaves and flowers, linear to elliptic rather than variously oblanceolate tepals, and longer filaments that are 5–7mm, versus 1.5–4mm in *C. trigynum*. In addition, the flowers of *C. figlalii* tend to be brighter lilac rather than purplish pink or whitish.

Colchicum figlalii, which is classed as endangered in the wild, was named in honour of the constituent Rector of Muğla University, Prof. Dr Ethem Ruhi Fığlalı. It has proved amenable to cultivation at Gothenburg Botanical Garden, Sweden, but is little known elsewhere.

Endemic to a small region of Turkey, *Colchicum figlalii* has striking lilac-coloured flowers.

Colchicum filifolium

Colchicum filifolium (Cambess.) Stef., *Sborn. B'lghar. Akad. Nauk* 22: 58 (1926)
Type: Spain, Balearic (Majorca), *Trias s.n.* (holotype MPU)
Syns *Bulbocodium balearicum* Nyman, *Merendera filifolia* Cambess.

Corm: Ovoid to ovoid-oblong, 1.5–2cm long, adorned with a membranous, mid-brown tunic
Cataphylls: White, 4–8cm long, just reaching the soil surface
Leaves: 5–10, appearing at about the same time as the flowers, at maturity linear, to 15cm long but only 1–3mm broad, flat
Flowers: Solitary, pinkish purple
Tepals: Elliptic to elliptic-obovate, 2–4 × 0.2–0.5(–0.8)cm
Stamens: Anthers basifixed, yellow, 5.5–8mm long, equal to or longer than the corresponding filaments, curving outwards
Styles: White, straight and erect, shorter than the stamens
Fruit capsules: Oblong, 8–12mm long, glabrous, held shortly above ground on slender pedicels
Habitat: Pine woodland, scrub, generally on light sandy and stony ground
Flowering period: August–November
Distribution: NW Africa (Algeria and Morocco), SW Europe (S France, Spain, including Balearic Islands); sea level to 200m
Cytology: 2n=54

Only known from low altitudes, this species is closely related to the better known and more widespread *C. montanum*. Both species bear basifixed anthers that are generally equal to or longer than the corresponding filaments. In all other species formerly included in *Merendera*, with the exception of *C. robustum*, the anthers are clearly shorter than the filaments.

Colchicum filifolium differs from *C. montanum* primarily in its more numerous, very narrow leaves and rather smaller flowers.

A finely grown pan of *Colchicum filifolium* exhibited by Royal Botanic Gardens, Kew.

Colchicum freynii

Colchicum freynii Bornm., *Notizbl. Bot. Gart. Berlin-Dahlem* 7: 172 (1917)
Type: Iran, Zanjan, 500–600m, Jan. 1892, *Bornmüller 4725* (lectotype JE; isotypes BM, LD, LE, P, W, WU)
Syns *C. falcifolium* sensu C.D. Brickell pro parte, non Stapf., *C. zangezurum* Grossh.

Corm: Ovoid to oblong, mostly 2–3.5cm long, adorned with dark brown, many-layered, rather leathery tunics, these extended above into a rather ragged neck, often rather short, not more than 4.5cm long
Cataphylls: Whitish, 5–9cm long, generally extended well beyond the corm neck
Leaves: 3, occasionally 4, partly developed at flowering, then shorter than the flowers, rarely more than 3.5cm long, at maturity spreading to somewhat recurved, linear to linear lanceolate, 8–16 × 0.2–0.6cm, occasionally broader, green or grey-green, margins and apex often flushed crimson-purple, channelled, base and margins sometimes finely papillose or short-ciliate, otherwise glabrous
Flowers: 1–3, funnel-shaped, white, generally flushed pinkish purple or rose-purple especially in the lower part; perianth tube whitish or yellowish white, 0.5–2cm long
Tepals: Linear to more or less oblanceolate, mostly 1.5–2.5 × 0.1–0.4cm, apex acute to subobtuse, the filament channels glabrous
Stamens: About two-thirds to half the length of the tepals; filaments, pale yellow or greenish, yellowish green or orange-yellow at the swollen base; anthers yellow to purplish grey, small, 2–3.5mm long
Styles: Equalling or slightly exceeding the stamens, pale yellow to greenish; stigmas punctiform
Fruit capsules: Oblong to more or less ovoid, 0.8–1.5cm long, often puberulous or papillose in the lower part, apex short-apiculate, located at ground level
Habitat: Dry, stony and sandy slopes
Flowering period: January–early April
Distribution: Armenia, Azerbaijan and NW Iran; 400–1,600m
Cytology: $2n=54$

A cultivated pan of *Colchicum freynii* grown from a collection made south of Julfa in Azerbaijan.

A relatively little-known species, this is often compared to *C. crocifolium*. However, the latter has a more westerly distribution, although the two just overlap in north-western Iran. In contrast to this species, *C. crocifolium* has more membranous corm tunics, more leaves per corm, longer cataphylls, a greater number of flowers per corm and $2n=14$. *Colchicum chimonathum* is related to *C. freynii* but that has a chromosome number of $2n=32$.

Colchicum gonarei

Colchicum gonarei Camarda, *Boll. Soc. Sarda. Sci. Nat.* 17: 227 **(1978)**
Type: Italy, Sardinia, Orani, Mount Gonare, 1,070m, Sep. 1976, *Comarda s.n.* (holotype FI; isotypes FI, SS)

Corm: Ovoid to pear-shaped, often with a well-developed oblong foot, the tunic brown, extended above into a short neck, to 2cm long
Cataphylls: White, greatly exceeding the neck of the corm
Leaves: 1-2, 5-10cm long
Flowers: Solitary, occasionally paired, rose-lilac with deeper veins
Tepals: 4-5, linear-lanceolate, 10-20mm long
Stamens: 4-5, with white filaments and yellow anthers, 2-3mm long
Styles: 2, equalling the stamens
Flowering period: September–October
Fruit capsules: Oblong, 1.8-3.3cm long, 2-loculi
Habitat: Rocky mountain slopes, open herb communities
Flowering period: September–October
Distribution: C Sardinia (Monte Gonare); *c.*1,070m
Cytology: 2n=180, 182

This small and rather insignificant species is confined, as far as is known, to Monte Gonare in central Sardinia. The mountain is well known as being the site of the 17th-century sanctuary of Nostra Signora di Gonare, from where there are commanding views of the Sardinian mountains.

This species has clear affinities with *C. nanum*, but *C. gonarei* has several odd features, the most obvious being the reduction of the floral parts. There are only four or five tepals instead of the usual six, and a similarly reduced number of stamens. In addition, there only two styles and a bilocular ovary and fruit capsule.

Growing here on Mount Gonare in Sardinia, its only known location, *Colchicum gonarei* has fewer tepals than most other species.
INSET **Leaves of** *Colchicum gonarei* **in cultivation.**

Colchicum gracile

Colchicum gracile K. Perss., *Bot. Jahrb.* 127(4): 284 (2008)
Type: Italy, Basilicata, Potenza, south-east of Lauria between Pestiere and motorway, 650m, 28 Sep. 1990, *K. Persson 492* (holotype GB)

Corm: Narrow-ovoid, to 3.3cm long, with a subcoriaceous chestnut-brown tunic, the inner more membranous and reddish brown, extended into a neck 3–4cm long
Cataphylls: White, slender, extending about 10mm beyond the corm neck
Leaves: Generally 5, absent at flowering, linear to linear-lanceolate, to 17 x 1.5cm at maturity, plain green, shallowly channelled and with a slight twist, glabrous
Flowers: 1(–2), funnel-shaped, lilac-purple, generally rather faintly tessellated; perianth tube yellowish white, 2–4cm long from the top of the cataphyll
Tepals: narrowly oblanceolate to suboblong, 26–38 x 4–11mm, subacute, the filament channels puberulous
Stamens: About half the length of the tepals, filaments white with a swollen yellow base; anthers purplish black to purplish or occasionally yellow, pollen yellow
Styles: White flushed with lilac or purple towards the top, distinctly overtopping the stamens, hooked at the apex; stigma decurrent for 1.5–2.5mm
Fruit capsules: Narrow-ellipsoid, located just above ground level
Habitat: Dry grassy places, maquis
Flowering period: September–mid October
Distribution: S Italy (Basilicata and Calabria); 325–1,175m
Cytolgy: 2n=80, rarely 82

This species appears to be endemic to a small area in southern Italy and is found in similar habitats to those of *C. bivonae*. Plants of the latter species in the area are generally quite small and can be distinguished by their more campanulate, more distinctly tessellated flowers, the shape and tunics of the corm, the more obovate tepals and in the thicker, stiffer, more rigidly fixed oblong anthers.

Cochicum neapolitanum and *C. lusitanum* are also recorded from the area. The former is a plant of woodland while the latter is a plant of grassy places and open scrub, often in rather moist settings. Persson, when describing *C. gracile*, said that some records of *C. lusitanum* might be of *C. gracile*, but the possibility remains (pers. obs.) that *C. gracile* is of hybrid origin, *C. bivonae* × *C. lusitanum*. At 2n=80 *C. gracile* comes midway between *C. bivonae* (2n=48) and *C. lusitanum* (2n=108).

The holotype herbarium specimen of *Colchicum gracile* collected from southern Itay.

Colchicum graecum

Colchicum graecum K. Perss., *Willdenowia* 18(1): 34 **(1988)**
Type: Greece, Peloponnese, Kalamata, Taigetos, Xerovouna, 1,700–1,480m, Sep. 1970, *J. Persson 4035*, (holotype GB; isotypes C, K)
Syn. *C. taygeteum* Heldr., in sched.

Corm: Globose to ovoid, 3–6 × 2.5–4.5cm, occasionally somewhat larger, with or without a basal foot up to 3cm long, with a dark brown or blackish brown tunic in several layers, extended upwards into a thick neck, to 13cm long, occasionally longer
Leaves: 4–5(–6), appearing after flowering, generally in late winter and early spring, ascending to suberect, oblong-lanceolate to elliptic-oblong, mostly 15–25 × 2–7cm, dull dark green, glabrous, occasionally slightly twisted, apex obtuse to subobtuse
Flowers: 2–5(–7), rarely solitary, funnel-shaped, pinkish purple to rose-purple, paler outside towards the base, not or obscurely tessellated, the tepals with a median white stripe in the lower half within; perianth tube cream, equalling or twice as long as the tepals
Tepals: Narrow-oblong to oblanceolate, 2.5–6(–6.5) × 0.5–1.5cm, apex obtuse to subobtuse, filament channels glabrous or puberulent along the ridges
Stamens: Filaments white to cream or purplish, those of the inner stamens noticeably longer than the outer; anthers yellow, 6–10mm long
Styles: White, flushed purple in the upper part, overtopping the stamens, curved at the top; stigmas decurrent for 2–6mm
Fruit capsules: Oblong-ellipsoid, 2–4.5cm long, glabrous, located above ground level by 6–10cm; seeds 2.5–5mm
Habitat: Rocky and stony habitats, ravines, screes, woodland margins, often on limestone but also on schistose rocks
Flowering period: July–early September
Distribution: Mainland Greece (Aroania, Giona, Killini, Mainalo, N and S Pindus, Taygetus, Tymfristos, and other mountains on both sides of the Corinth Sea); (900–)1,300–2,400m
Cytology: 2n=42, 44

A localised species, this is found on various mountains in Greece, generally at greater altitudes than most other native *Colchicum* species. *Colchicum graecum* has affinities with *C. parnassicum* and was confused with it in the past, mainly through misidentification. Certainly since the 1870s it was generally treated as

The white tepal stripe is visible in this specimen of *Colchicum graecum*.

C. parnassicum. Karl Hermann Zahn, who collected plants in the Taygetus Mountains in 1897, realised its distinctiveness by naming it provisionally as *C. taygeteum*. Despite this, it was not formally recognised at species level until 1988 when Karin Persson's meticulous research recognised its value and she made a formal description under the name *C. graecum*.

Colchicum parnassicum is actually restricted to Mount Parnassus and Mount Helicon, and is a slighter plant with fewer, brighter green leaves, the flowers distinctly tessellated. In addition, the tunics of *C. parnassicum* are thinner and paler. The chromosome numbers are also different, *C. parnassicum* being 2n=54 and *C. graecum* being generally 2n=44.

Colchicum graecum has been cultivated at various botanic gardens for a number of years, notably at Gothenburg Botanical Garden, Sweden, and Royal Botanic Gardens, Kew, and is also represented in several private collections. It is reportedly quite easy to grow given a sunny position and a well-drained soil. However, the rather pale, medium-sized flowers win it little favour among gardeners who, on the whole, prefer the larger, more brightly coloured species and cultivars.

RIGHT **The ascending leaves and fruit of *Colchicum graecum* in northern Greece, held above ground by up to 10cm.**

BOTTOM RIGHT **_Colchicum graecum_ thrives in cultivation in a well-drained, sunny position.**

Colchicum guessfeldtianum

Colchicum guessfeldtianum Asch. & Schweinf., *Ill. Fl. Égypte, Suppl.* 2: 774 (1889)
Type: Egypt, Galalah du Sud, 1887, *Schwinfurth*, (holotype B – destroyed)
Syn. *C. ritchii* var. *guessfeldtianum* (Asch. & Schweinf.) Stef.

Corm: Ovoid to ovoid-oblong, 2–3.5cm long, adorned with a deep brown papery tunic, extended above into a narrow neck, to 5cm long, occasionally longer
Cataphylls: White, slender, extending well beyond the neck of the corm, appearing shortly above ground
Leaves: 3, appearing after flowering, straight to strongly curled, ligulate, 12–28 × 0.5–3cm, grey-green, densely hairy to occasionally glabrous, channelled
Flowers: 1–3, funnel-shaped, pale to mid-pink; perianth tube white or pinkish, up to twice the length of the tepals
Tepals: Elliptic-oblanceolate, 1.8–3.2 × 0.3–0.8cm, apex subobtuse to subacute, the filament channels glabrous, without teeth
Stamens: About two-thirds the length of the tepals; the filaments slender, white or with a flush of pink, yellowish and slightly swollen at the base; anthers pale yellow
Styles: Slender, overtopping the stamens
Fruit capsules: Not recorded
Habitat: Dry rocky slopes over limestone or granite
Flowering period: November–December
Distribution: Sinai Mountains in Egypt (Sinai Peninsula) and S Jordan, rare; 1,100–2,000m
Cytology: 2n=14

Found only in the Sinai Mountains straddling the Egypt-Jordan border area, this is a little known, montane-desert species. The leaves are particularly variable, even on the same plant, ranging from straight to markedly curved or coiled, and glabrous to quite densely pubescent. Plants often make small, congested clumps of a number of close-set corms.

Colchicum guessfeldtianum is similar to, and easily mistaken for, *C. schimperi*, which is included in *C. guessfeldtianum* by some authorities, and sometimes also with *C. ritchii* and *C. tuviae*. From *C. schimperi* it differs in its often very hairy leaves and in the pale yellow anthers, while the styles overtop the stamens, whereas they are shorter or equal them in *C. schimperi*. Both *C. ritchii* and *C. tuviae* bear characteristic teeth at the base of the filaments, but these are lacking in the other two species. In *C. guessfeldtianum* there are characteristically three leaves which appear after flowering.

In the wild *Colchicum guessfeldtianum* is found on dry, rocky slopes.

Colchicum haynaldii

Colchicum haynaldii Heuff., *Verh. K.K. Zool.-Bot. Ges. Wien* 8: 213 (1858), as '*haynaldi*'
Type: Romania, 'Banatus in Danubii', Sep., *Heuffel* (holotype BP)
Syns *C. callicymbium* Stearn & Stef., *C. jankae* Freyn, *C. pantocratoris* Spreitz. ex Stef., *C. visianii* Parl.

Corm: Ovoid to almost globose, 2–4.5cm long, occasionally larger, with a deep reddish or dark brown subcoriaceous or somewhat membranous tunic, extended above into a neck, up to 11cm long
Cataphylls: Yellowish white, often flushed with green or purple at the apex, exceeding the neck of the corms by up to 5cm, occasionally more
Leaves: (3–)4–6(–8), appearing after flowering, erect-patent to patent, linear to linear-lanceolate, 10–24 × 0.5–4cm, pale or greyish green, glabrous or minutely puberulous beneath, flattish or shallowly channelled, twisted, with a narrow keel beneath, apex attenuated, obtuse to subobtuse, the margin narrowly cartilaginous, often ciliate or scabrid towards the base
Flowers: 1–4(–7) or sometimes more, narrow funnel-shaped to more or less goblet-shaped, purple to rosy-purple or –lilac, usually very faintly tessellated, especially in the upper part, with a white streak for at least the lower third, sometimes extending the full tepal length; perianth tube white to somewhat yellowish, generally flushed pink or purple towards the top, 2.5–15cm long
Tepals: Linear-elliptic to oblanceolate or subobovate, mostly 4.5–10 × 0.3–1.6cm, the outer tepals noticeably longer than the inner, all usually keeled on the back, 10–15-veined, filament channels glabrous to puberulous or ciliate on the margins
Stamens: One third to half the length of the tepals; filaments, whitish or yellowish white, sometimes flushed purple in the upper part, with a greenish or yellowish, somewhat swollen base; anthers yellow, rarely purple, often rather pale
Styles: Whitish, sometimes flushed purple in the upper part, equalling to slightly exceeding the stamens; stigmas decurrent for 1.5–5mm on the hooked tip
Fruit capsules: Oblong to oblong-ellipsoid, generally 2–4.5cm long, brown with darker dots at maturity, glabrous, located at ground level
Habitat: Meadows, rocky scrub, dolines and poljes, forest glades (beech, oak and pine), primarily on limestone, occasionally on siliceous substrates
Flowering period: September–October
Distribution: Albania, W Bosnia and Herzegovina, W Bulgaria, W Croatia, N and NE Greece, North Macedonia, SW Romania and W Serbia; sea level to 1,500m
Cytology: $2n=96$

Faint tessellation can be seen on this specimen of *Colchicum haynaldii* growing on the Ionian island of Paxos.

This species was described in 1858. However, plants grown at Cambridge University Botanic Garden from the 1900s were described in 1934 under the name *C. callicymbium*. It was stated in the description of that species that the original corms had been received from German horticulturist Max Leichtlin from a consignment collected 'near the monastery of Hagios Dionysias next to Kraterino [modern day Katerini], Gulf of Salonica'. This is a known locality for *C. haynaldii* and details from the description of *C. callicymbium* indicate that the two are the same species, the former having precedence as the earliest epithet.

The flowers of *C. haynaldii* perhaps come closest to those of *C. autumnale* and *C. confusum*, all three flowering in autumn in advance of leaf production. In its tessellated forms the flowers look quite distinct. Persson (1999a) observes that 'As commonly observed in this group of autumn-flowering colchicums with hysteranthous leaves appearing in the spring, flowering *C. haynaldii* at least for the botanist has a facies which is fairly characteristic but nonetheless difficult to render into words. Features of the tunics are mostly among the most useful diagnostic characters, and leaf characters are conclusive. Flowers of *C. haynaldii* are perhaps most similar to *C. confusum* and *C. autumnale*, both of which may grow together with the species'.

RIGHT **A less tessellated specimen of *Colchicum haynaldii* on Paxos.**

FAR RIGHT ***Colchicum haynaldii* has been in cultivation since 1902.**

BOTTOM RIGHT **The limestone and woodland habitat of *Colchicum haynaldii* on Paxos.**

Colchicum heldreichii

Colchicum heldreichii K. Perss., *Edinburgh J. Bot.* 56: 98 (1999)
Type: Turkey, Konya, 3 km ENE of Derbent, 1,600m, Aug. 1999, *K. Persson 555* (holotype GB)

Corm: Ellipsoid to ovoid, 3–4.5cm long, with a membranous tunic extended above into a neck, up to 9cm long
Cataphylls: 6.5–11cm long
Leaves: 3–4, spreading to rather curved, linear to narrow-oblong, 7–15 × 1–2.5cm, with a somewhat undulate, thinly cartilaginous margin
Flowers: 1–4, funnel-shaped, white to pale lilac-pink
Tepals: Linear to narrow-oblanceolate, mostly 2.5–4 × 0.3–0.8cm, the filament channels glabrous or pubescent at the very base
Stamens: about half the length of the tepals, with white filaments egg-yolk coloured and slightly swollen at the base; anthers yellow, 4–7mm long
Styles: White, curved at the tip, the stigmas decurrent for 1–4.5mm
Fruit capsules: Not recorded
Habitat: Grassy meadows and snow hollows
Flowering period: August–September
Distribution: WC Turkey (Antalya and Konya); 1,600–2,350m
Cytology: $2n=54$

A relatively little-known species, this is restricted to a limited area in southern Anatolia in Turkey. Described in 1999, it had previously been regarded as *C. kotschyi*.

Colchicum heldreichii tends to be smaller than that species in all its parts, the leaves being crowded at the soil surface and not projected on a short pseudostem. Also, the filament channels of *C. heldreichii* are glabrous or only finely pubescent at the very base, but densely pubescent overall in *C. kotschyi*. There are also differences in the relative proportions of perianth tube and the stamens. The two species have different chromosome numbers, with *C. kotschyi* having $2n=20$.

A restricted species in the wild, Colchicum heldreichii is occasionally seen in cultivation.

Colchicum hierosolymitanum

Colchicum hierosolymitanum Feinbrun, *Palestine J. Bot., Jerusalem ser.* 6: 84 **(1953)**

Type: Israel, Jerusalem, Oct. 1927, Zohary & Feinbrun Fl. Exsic. Pal. 33 (lectotype HUJ)

Corm: Ovoid, 4–6cm long, with a membranous, chestnut-brown tunic, extended above into a short neck, up to 6cm long
Cataphylls: White, greatly exceeding the neck of the corms, 12–25cm long, the tip just poking through the ground
Leaves: 5–9, appearing after flowering, in late winter and early spring, stiffly ascending to patent, narrow-elliptic to narrow-lanceolate, 15–30 × 0.8–2cm, dark green, channelled, apex obtuse to subacute
Flowers: 4–20, funnel-shaped, pale to deep rose or purple-pink, faintly tessellated towards the top; perianth tube pink to purple, to three times the tepal length
Tepals: Elliptic to elliptic-obovate or almost oblong, 3.6–4.5 × 0.7–1.2cm, obtuse to subobtuse, often slightly ridged longitudinally
Stamens: Filaments whitish, pink or deep purple, 12–15mm long; anthers yellow, 6–10mm long
Styles: White, sometimes purple-tinged, overtopping the stamens, thickened towards the top; stigmas decurrent for 1.5–3mm on the slightly curved tip
Fruit capsules: Oblong-ellipsoid, 1.5–3cm long, covered in golden-brown dots, located at ground level.
Habitat: Fields, olive groves, cultivated land, open forest, mainly on deep terra-rossa and basaltic soils, sometimes in large numbers
Flowering period: October–early January
Distribution: Israel, Jordan (rare), Lebanon, Palestine, possibly Syria
Cytology: 2n=18

Closely related to *C. decaisnei*, both species produce large numbers of flowers from single corms, forming small posies. In many respects *C. hierosolymitanum* looks like some of the smaller flowered variants of *C. autumnale*, which, although widespread in Europe, is absent from the Middle East, preferring cooler, moister habitats. The corms of *C. hierosolymitanum* grow deep in the soil, often 40–50cm down, so are not disturbed by conventional ploughs on cultivated land.

Colchicum hierosolymitanum often produces numerous flowers from a single corm.

Colchicum hirsutum

Colchicum hirsutum Stef., *Sborn. B'lghar. Akad. Nauk* 22: 34 (1926)
Type: Turkey, Erzinjan, Hodschadurdagh, May 1890, *Sintenis Iter Orientale 1890 s.n.* (lectotype LD)
Syn. *C. falcifolium* sensu C.D. Brickell pro parte, non Stapf.

Corm: Ovoid, 3-4cm long, adorned with a reddish brown, papery or somewhat coriaceous tunic, extended into a very short neck, or neck absent
Cataphylls: Whitish, slender, 6.5-10cm long
Leaves: 3-4(-6), present at flowering, then 1-6cm long, at maturity linear, 9-15 × 0.1-0.3cm, occasionally wider, glabrous or more often hispid below and on the margins, sometimes densely so, grooved, apex acute to obtuse
Flowers: 2-5(-8), narrow funnel-shaped, opening fully to broad stars, white to purplish pink, often with darker veining; perianth tube whitish, about as long as the tepals
Tepals: Narrow-elliptic to narrow-oblanceolate, 1.3-2.5 × 0.2-0.4(-0.6)cm, apex acute to subobtuse, the filament channels glabrous
Stamens: Filaments whitish, slender, glabrous, somewhat swollen at the base; anthers black, greenish black or brownish black, 2-5-3.5mm long
Styles: White, filiform, straight; stigmas punctiform
Fruit capsules: Subglobose to narrow-ovoid, 1-1.5cm long, glabrous or hispid, apex apiculate, located at ground level or part buried
Habitat: Bare rocky limestone areas often by melting snow, steppe
Flowering period: February–April, occasionally later
Distribution: SE Turkey (Euphrates-Tigris region); 250-1,600m
Cytology: 2n=14

Relatively little known in the wild, this species belongs to a complex of little, small-flowered species centred upon eastern Turkey, Iran, Iraq and the Caucasus. These include *C. fasciculare*, *C. serpentinum* and *C. varians*.

In cultivation it often flowers very early, typically around the end of December. The dying blooms are prone to botrytis infection at that time of the year.

A cultivated specimen of *Colchicum hirsutum* derived from a collection made in Turkey.

Colchicum hungaricum

Colchicum hungaricum Janka, *Természetrajzi Füz.* 10: 75 **(1886)**
Type: Hungary, Baranya, Harsány, Feb. 1867, *Janka s.n.* (lectotype CI; isotypes BM, FI, G, GOET, JE, K, P, UPS, W, WU)
Syns *C. croaticum* Dykes, *C. montanum* var. *croaticum* A. Grove

Corm: Oblong to ovoid, to 1.3–3 × 1–2cm, with a dark brown, papery to sub-leathery tunic extended above into a short neck
Cataphylls: White or cream, fairly stout, 5.5–9cm long
Leaves: 2, occasionally 3, present at flowering, then somewhat shorter than, or as long as the flowers, at maturity ascending, linear-lanceolate, 14–20 × 1–2cm, channelled, apex acute, surface glabrous but margin with downward-directed hairs
Flowers: Usually 2–6, occasionally more, broad chalices, pink, purplish pink, pinkish violet or white, with a greenish throat, parallel-veined; perianth tube lime-green or greenish white, about twice the length of the tepals
Tepals: Oblong-elliptic to lanceolate-elliptic, 2.4–3 × 0.5–0.7(–1)cm, apex subacute to subobtuse
Stamens: Filaments greenish white, swollen at the base; anthers purplish black or violet-brown, 2–3.5mm long
Styles: White, about as long as the stamens, straight or slightly curved; stigmas punctiform
Fruit capsules: Subglobose, 1–2cm long, apex shortly acuminate, located at ground level
Habitat: Dry slopes and stony meadows, steppes, often on clay soils
Flowering period: December–April
Distribution: Albania, Bosnia and Herzegovina, Bulgaria, Croatia, Hungary, Kosovo, S and North Macedonia; sea level to 900m. Reports of the species from France are erroneous
Cytology: 2n=54

This is one of the most attractive of the spring-flowering species. The darker flowered forms are extremely desirable, as are the pure white ones (f. *albiflorum*) with their contrasting dark anthers highlighting the centre of each bloom.

Colchicum doerfleri has been long confused with *C. hungaricum*. *Colchicum doerfleri* has a more southerly distribution on the whole, centred on the Greek mainland, excluding the Peloponnese. The differences are discussed under that species.

Being a particularly floriferous species, this is well-established in cultivation

An attractive, spring-flowering species, *Colchicum hungaricum* is native to eastern Europe.

It is most often seen as a subject for pan culture under glass but occasionally on a raised bed in the open garden.

Colchicum hungaricum f. albiflorum (K. Maly) Hayek & Markgraf, *Fedde Rep. Sp. Nov.*, Beihefte 30: iii (1932)

Flowers: White, with pencil-veined tepals

White-flowered forms are sometimes seen among populations of the pink or purplish forms in the wild. There are several white or near white selections in cultivation, including seed-raised 'Velebit Star', white forms of which can be assigned to this forma.

RIGHT **In cultivation, *Colchicum hungaricum* is floriferous and most commonly grown in a pan under glass.**

FAR RIGHT ***Colchicum hungaricum* f. *albiflorum* is distinguished by its white flowers.**

BOTTOM RIGHT **The hairs on the leaf margins are just visible on this specimen of *Colchicum hungaricum* f. *albiflorum*.**

Colchicum ignescens

Colchicum ignescens K. Perss., *Bot. Jahrb. Syst.* 127 (2): 191 (2007)
Type: Turkey, Sunliurfa, Karacadağ, 1,650m, Apr. 1990, *Kammerlander et al. 90–149* (holotype GB)

Corm: Ovoid to almost globose, 1.8–3cm long, adorned with a brownish black, coriaceous tunic, not extended above or with a short neck not more than 1cm long
Cataphylls: White, 3–5.5cm long, soon withering
Leaves: 3–4, partly developed at flowering, then somewhat shorter than the flowers, ascending, oblong to elliptic-oblong, 10–17 x 2–4cm at maturity, glossy mid-green, often suffused with purple when young, glabrous
Flowers: 1–4, dark rose-purple, carmine outside in the lower part, funnel-shaped; tube split to the base, the claws white
Tepals: Oblong to oblanceolate, 13–23mm long, with short lamellae along the base of the filaments; limbs interlocked at the base by small triangular or filiform auricles
Stamens: Filaments straw-coloured or yellowish green, slightly swollen and yellow at the base; anthers purplish brown, purplish brown or dark grey
Styles: Filiform, pale yellow or greenish yellow; stigma punctiform
Fruit capsules: Ellipsoid, located at ground level
Habitat: Rocky ground on deep volcanic soils
Flowering period: March–April
Distribution: SE Turkey (Karacadağ), endemic; 1,500–1,650m
Cytology: 2n=22

This recently described species was correctly placed in *Colchicum*, even though it is most closely related to species formerly in *Merendera*. It finds its closest ally in *C. trigynum* and the two can easily be mistaken for one another. However, the relatively broad, oblong-elliptic mature leaves of *C. ignescens*, generally three or four per corm, as opposed to two or three in *C. trigynum*, and the more numerous bright rose-purple flowers of *C. ignescens*, are helpful diagnostic characters. Both *C. kurdicum* and *C. raddeanum* also bear a superficial resemblance to *C. ignescens* but have paler flowers with thicker, more substantial filaments and styles.

Colchicum ignescens in cultivation in Turkey.

Colchicum imperatoris-friderici

Colchicum imperatoris-friderici Siehe ex K. Perss., *Edinburgh J. Bot.* 56: 129 (1999)
Type: Turkey, Içel, Goksu river, Kirche to Thekla, 100m, Oct., Siehe *Fl. Orient.* 99 (holotype JE; isotype LE)
Syns *C. kotschyi* sensu C.D. Brickell, non Boiss.

Corm: Ovoid, often broadly so, 6–8cm long, tunics membranous to papery, often rather fragile, shiny brown, extended above into a stout neck, 6–12cm long
Cataphylls: Yellowish white, sometimes with purple marks towards the top, quite stout, to 16cm long
Leaves: 4–5, rarely more, appearing long after flowering, rather crowded close to the ground, narrow-oblong to oblong-lanceolate or oblong-ovate, 16–45 × 4.5–11cm, glaucous green, the young tip and base often yellowish, glabrous, very thick, flattish and somewhat pleated and twisted, apex obtuse, the margin narrowly cartilaginous
Flowers: 5–10, occasionally more, funnel-shaped, lustrous pale lilac to lilac-purple, almost translucent; perianth tube yellowish white, generally flushed with lilac towards the top, 2–8cm long
Tepals: Narrow-oblong, 3.5–6.5 × 0.7–1.7cm, shallowly and bluntly keeled on the back, apex obtuse to rounded, sometimes slightly cucullate, filament channels densely pubescent
Stamens: One third to half the length of the tepals, subequal, the outer slightly shorter than the inner; filaments white, usually flushed lilac in the upper part, and with a swollen yellowish base; anthers pale yellow to straw-coloured, mostly 8–10mm long, equalling or shorter than the filaments
Styles: White, often lilac- or purple-flushed towards the top, equalling or exceeding the stamens, curved or hooked; stigmas punctiform or shortly decurrent
Fruit capsules: Ellipsoid, 2.3–4cm long, located just above ground level
Habitat: Rocky places, gullies, pine woodland (*Pinus brutia*), oak scrub, maquis, generally over limestone
Flowering period: Late September–November
Distribution: S Turkey (Mersin); near sea level to 1,650m
Cytology: 2n=54

Colchicum imperatoris-friderici growing in a limestone habitat in southern Turkey.

This species has distinctively lustrous, lilac flowers with straw-coloured anthers. Like *C. dolichantherum* and *C. balansae* it has impressively large foliage, the leaves more glaucous and thicker than the former, and as leathery as the latter. The species has perhaps the largest corms of any wild species, the largest measuring up to 8cm in diameter. However, this corm size does not rival measurements of some cultivated specimens, such as *C.* × *byzantinum*, in which they can be almost 12cm in diameter.

It was first named by Walter Siehe (1859–1928), one of the most significant collectors of the Anatolian flora. Although his collection of this species was annotated with the current name, it was not formally published until 1999 (Persson 1999). The specific epithet commemorates Holy Roman Emperor Frederick I, Frederick Barbarossa (1122–1190) who drowned during the Third Crusade while crossing the Göksu river, which is adjacent to the location where the type specimen was collected.

RIGHT The flowers of *Colchicum imperatoris-friderici* emerging from between cyclamen leaves in southern Turkey.

BOTTOM RIGHT A pale-flowered variant of *Colchicum imperatoris-friderici* in cultivation.

Colchicum inundatum

Colchicum inundatum K. Perss., *Edinburgh J. Bot.* 56: 99 (1999)
Type: Turkey, Konya-Antalya, Gencek to Aydinkent, 1,200–1,250m, Apr. 1991, *K. Persson 505* (holotype GB, isotype E)

Corm: Ovoid to almost globose, 2.5–6cm long, with membranous, pale brown, rather fragile, veiny tunics and a fairly stout neck, to 8.5cm long
Cataphylls: Whitish, 4–12cm long
Leaves: 4–12, appearing long after flowering, crowded at ground level making a spreading rosette, mostly oblong-lanceolate, the upper leaves noticeably smaller than the lower, glaucous-green, glabrous, sometimes scabrid or somewhat ciliate at the margin
Flowers: 1–3, pale lilac to purplish lilac
Tepals: Thin and semi-translucent, rather fragile looking, elliptic-oblong or oblong-oblanceolate, 2.7–4.5cm long, the inner clearly shorter than the outer, filament channels puberulous to densely pubescent and ciliate along the margins
Stamens: A third to half the length of the tepals; filaments white, greenish yellow and slightly swollen at the base; anthers deep to golden-yellow, 4.5–7mm long, shorter than or more or less equalling the filaments
Styles: Slender, white, slightly hooked at the tip and somewhat exceeding the stamens; stigmas shortly decurrent
Fruit capsules: Not recorded
Habitat: Alluvial flats, areas subject to winter flooding, meadows
Flowering period: September–October
Distribution: S Turkey (SW of Konya), endemic; 1,200–1,300m
Cytology: 2n=54

Described in 1999, this southern Turkish endemic has its closest ally in *C. persicum*, a more widespread species which is found further east, extending into Iraq, Syria, Iran and Lebanon. It is very similar in leaf to *C. persicum* but the flowers are clearly different, being a paler, lilac-pink in *C. inundatum*, the tepals more oblong with the stamens attached close to the base in the throat of the flower, rather than in the top of perianth tube as in *C. persicum*. Both species have the same chromosome number.

Interestingly, *C. inundatum* was discovered in the same habitat as *Lagotis stolonifera*, a plant of the *Plantaginaceae*, which is otherwise only known in Turkey much further east, stretching into north-west Iran, Armenia and Georgia.

Colchicum inundatum typically has pale, lilac-pink flowers and oblong tepals.
INSET **Leaves of** *Colchicum inundatum* **in cultivation.**

Colchicum kesselringii

Colchicum kesselringii Regel, *Trudy Imp. S.-Peterburgsk. Bot. Sada* 8: 646 (1884), as '*kesselringi*'
Type: Tajikistan, 'Bukara orientalis infra Kulab,' 1,500ft, Apr. 1883, A. *Regel s.n.* (neotype LE)
Syn. *C. crociflorum* (Regel) Regel

Corm: Oblong-ovoid, 2–4 × 1–2cm, occasionally larger, tunics membranous to subcoriaceous, brown, reddish or blackish brown, extended into a neck, 1–4(8)cm
Cataphylls: Whitish, sometimes flushed violet-purple at the top, 6–15cm long
Leaves: (2–)4–7, occasionally more, generally visible at flowering and then 3–7.5cm long, mature leaves ascending to spreading, linear to linear-lanceolate or falcate, 5–15 × 0.3–1.2cm, bright or grey-green, glabrous, sometimes glossy, margins often flushed purple, flattish to shallowly channelled and often a little twisted, apex obtuse to subobtuse, margins thinly cartilaginous, occasionally slightly scabrid
Flowers: 1–3(–4), narrow funnel-shaped to more or less star-shaped (especially in strong sunshine), white, striped brown-crimson or -purple along the longitudinal veins, occasionally flushed pale purple-lilac or slate-blue, intensely fragrant; perianth tube greenish white to white, occasionally greyish, often with the darker colour running down from the tepal bases, often grooved and somewhat dilated at the top, not or only exceeding the cataphylls by up to 2.5cm
Tepals: Linear to linear-lanceolate or oblanceolate, 1.7–3 × 0.2–0.6cm, apex obtuse to subobtuse, rarely acuminate, shallowly ridged at the base, filament channels edged with short lamellae, especially on the inner three tepals
Stamens: Filaments white, greenish or pale yellow, the base swollen, deep green or buff; anthers basifixed, pale yellow, 7–11mm long, about twice the filament length
Styles: Fused in the perianth tube, pale green to yellowish green or yellowish white, more or less equalling the stamens, straight or slightly bent; stigmas punctiform
Fruit capsules: Ellipsoid-oblong, 1.5–2.5cm long, located at ground level, sessile or very shortly pedicellate to 1cm
Habitat: Stony slopes, passes, alpine meadows, moist depression, snow hollows, submontane valleys and depressions
Flowering period: February–July
Distribution: Restricted to the Syr Darya river valley and the mountains of Pamir-Alay and Tien Shan in central Asia, E and NE Afghanistan and NW India; (450–)900–3,800(–4,000)m
Cytology: 2n=54

Colchicum kesselringii, with its dark-striped and white flowers, is very different to the pink- and purple-flowered species of the genus.

Among a genus of primarily pink- and purple-flowered species, C. kesselringii is boldly different with its usually white flowers striped with brownish or crimson-purple. The flowers can also be pale yellow. The stripe on the underside of the tepals can be either elegantly thin or prominent, and from the palest violet to a lurid, deep purple-pink.

Despite the colour difference between this species and the only yellow-flowered species, C. luteum, these two Asian species are closely related and mark the easternmost extension of the genus. Colchicum kesselringii and C. luteum are close genetically and share a chromosome number, 2n=54. Their flowers appear to have the normal three, thread-like styles but they are united midway down perianth tube into a single style. This feature is only known in one other species, C. bulbocodium. In the latter species the style is solitary and three-branched near the top at the level of the anthers. Colchicum kesselringii also crosses in the wild with C. luteum, givinge a hybrid called C. × alberti.

Colchicum robustum also inhabits this easternmost distribution. Genetic studies show that it is closely related to C. kesselringii and C. luteum and yet it was formerly included in Merendera, whose species have no perianth tube and therefore separate tepals. Colchicum robustum has separate styles.

Regel recognised two varieties of C. crociflorum, a species now considered to be synonymous with C. kesselringii. One was var. typicum, which equates with the current species. The other was var. stenosepalum, said to have solitary or paired flowers with narrow tepals, and recorded from the Bokhara-Darwaz region. However, such distinctions can be found in many populations of C. kesselringii.

Cultivars of C. kesselringii include 'Boldness', 'Modesty', 'My Choice', 'Prosperity', 'Purple Star', 'Snow of Highland' and 'Yeti'.

RIGHT **In some variants of *Colchicum kesselringii* the flowers almost lack the dark stripes.**

FAR RIGHT **The leaves of *Colchicum kesselringii* are usually visible at flowering.**

BOTTOM RIGHT ***Colchicum kesselringii* is one of the easternmost species of the genus and is generally found in mountain habitats, as here on the Taldyk Pass in Kyrgyzstan.**

Colchicum kotschyi

Colchicum kotschyi Boiss., *Diagn. Pl. Orient. ser. 1* 13: 38 **(1854)**
Type: Iran, Tehran, Elburz, Passgala (Dareke, Aug. 1843, *Kotschy 655* (holotype G-BOISS; isotypes FI, LE, P, UPS)
Syns *C. candidum* var. *hirtiflorum* Boiss., *C. laetum* sensu Baker pro parte, non Steven, *C. obtusifolium* Siehe ex Hayek

Corm: Ovoid, 3–5cm long, with dark brown, membranous to papery, rather fragile tunics, extended above into a stout neck, up to 7(14)cm long, the inner tunics paler
Cataphylls: Whitish or yellowish, generally extending for up to 5cm beyond the neck of the corm
Leaves: 3–4(–5), appearing after flowering, ascending to spreading, elliptic-lanceolate to oblong-elliptic, 10–16 × 3–5cm, glabrous, slightly undulate at margin and often twisted, apex obtuse to subacute, the base truncate or somewhat cordate, borne somewhat above the ground on a short pseudostem, to 9cm long
Flowers: 3–7 or more, funnel-shaped, white, yellowish white to purplish pink in the upper part; perianth tube whitish or yellowish white, sometimes flushed purple towards the top, as long as the tepals or somewhat longer
Tepals: Oblong-elliptic to oblanceolate, mostly 3–5.5 × 0.6–1.2cm, apex obtuse to subacute, densely pubescent along the ridges of the filament channels
Stamens: One-third to half the length of the tepals; filaments white, base swollen, orange-yellow; anthers yellow, 6–12mm long, equalling or longer than the filaments, membranous along the margins
Styles: White, straight but curved at the apex, overtopping the stamens; stigmas decurrent for 1.5–4mm
Fruit capsules: Oblong-ovoid, 3–4.5cm long, glabrous, beaked
Habitat: Margins of and clearings in *Pinus* woodland, limestone cliffs
Flowering period: Late August–November
Distribution: NW Iran, NE Iraq, SC and SE Turkey; (750–)1,000–3,000m
Cytology: 2n=20

The pale flowers of *Colchicum kotschyi* adorning a clearing in pine woodland.

This species is often confused with *C. balansae*. An easy way of telling them apart in flower is by studying the filament channels towards the base of the tepals. In *C. kotschyi* they are pubescent along the bases of the ridges, whereas in *C. balansae* they are glabrous. The extended fibrous neck of the corm of the latter species is also very distinctive and conceals the cataphylls, which are exposed in *C. kotschyi* to some extent. These two species overlap in distribution in inner Anatolia. Another closely related species is *C. heldreichii*.

Colchicum kurdicum

Colchicum kurdicum (Bornm.) Stef., *Sborn. B'lghar. Akad. Nauk* 22: 42 (1926)
Type: Iraq, Arbil, Algurd Dağ (Mt Helgurd), 3,000m, Jun. 1893, *Bornmüller 1840* (lectotype JE; isotypes BM, E, FI, G, K, LD, LE, P, W, WU)
Syn. *Merendera kurdica* Bornm.

Corm: Oblong to ovoid-oblong, 2–4.5cm long, with a membranous, sometimes almost leathery, dark to reddish brown tunic, extended above into a short neck, not more than 3cm long
Cataphylls: White, very thin, 6–12cm long, soon dispersing
Leaves: (2–)3, present at flowering, then as long as or slightly overtopping the flowers, at maturity oblong-lanceolate to elliptic-oblong, 10–21 × 2–6cm, glossy grey-green, glabrous, apex subacute to subobtuse, margin narrowly cartilaginous
Flowers: Solitary or sometimes paired, broad funnel-shaped, pale to deep pinkish purple with a whitish base, sometimes extending for half the length of the tepals; tepals split to the base, claws white or yellowish white, 3–4.5cm long
Tepals: Oblong to obovate or oblanceolate, 2.6–4.3 × 0.5–1.7cm, apex obtuse to subobtuse, sometimes somewhat apiculate or cucullate, bases with interlocking filiform appendages, 1–3mm long
Stamens: With stout, whitish or yellowish, rarely purplish filaments and yellow, greyish, brownish or occasionally purplish black anthers, 3–5mm long, much shorter than the filaments
Styles: Thread-like, whitish, yellowish or greenish, equalling or slightly longer than the stamens
Fruit capsules: Oblong-ovoid or oblong-ellipsoid, 2.5–4.5cm long, apex pointed, located above ground on fruiting pedicels to 7cm long
Habitat: A snow-melt species of upper forest zones, open alpine slopes and snow hollows, sometimes in large colonies
Flowering period: March–early August
Distribution: NW Iran, N Iraq and SE Turkey; 1,650–3,100m
Cytology: 2n=20

This is the largest species among those formerly included in *Merendera*. The broad developing leaves are convoluted and ensheath the base of the flowers, preventing the tepals from parting, their cohesion is also helped by the presence of interlocking appendages at the base of the tepal limbs. Spectacular colonies can sometimes be seen in the wild, especially following a winter with plenty of snow.

The broad leaves of *Colchicum kurdicum* help to keep the flowers in shape.

Colchicum laetum

Colchicum laetum Steven, *Nouv. Mém. Soc. Imp. Naturalistes Moscou* 1: 262, t. 18 (1829)
Type: Russia Federation, Chechnya, Terek, Sep. 1922, *Steven s.n.* (holotype H 1085078; possible isotypes G-BOISS, GOET, P)

Corm: Ovoid, 2.5–3cm long, with a coriaceous, blackish brown tunic, this extended above into a long, slender neck
Cataphylls: White, 5–11cm long
Leaves: (3–)4, appearing after flowering, ligulate, 14–20 × 1.5–3.5cm, the lowermost rather broader than the upper, pale green, glabrous, apex obtuse, acute in uppermost leaves
Flowers: 1–3, funnel-shaped, lilac to purple, often rather pale; perianth tube white, 5–8cm long
Tepals: Lanceolate to elliptic, 2.8–4 × 0.8–1.6cm, apex obtuse, filament channels glabrous
Stamens: About half the length of the tepals; filaments white, somewhat swollen and yellowish at the base; anthers yellow, 6–8mm long
Styles: Very slender, greatly exceeding the stamens, somewhat thickened and curved at the top; stigmas shortly decurrent
Fruit capsules: Ovoid, 1.5–2cm long, apex acuminate, borne on a short pedicel at ground level or slightly above
Habitat: Meadows and steppes
Flowering period: August–September(–October)
Distribution: Caucasus (Dagestan and Chechnya in particular), SW and S Russia; *c.*1,500–2,500m
Cytology: 2n=42

A primarily Russian species, its distribution also stretches south-east into the northern Caucasus region from where it was first described. It probably finds its closest ally in *C. umbrosum* which is also found in the Caucasus region. *Colchicum laetum* has also been likened to a slimmer version of *C. speciosum* but, apart from obvious size differences, *C. laetum* differs in its ligulate leaves, lanceolate to elliptic tepals, shorter anthers and slender, long styles. Having said this, *C. laetum* is an understudied species deserving further investigation.

This species is occasionally seen in cultivation. It has, however, been much confused with a plant widely cultivated under the same species name which is now correctly known as *Colchicum* 'Pink Star'.

Meadows and steppes are the favoured habitat of *Colchicum laetum*, here growing in the Manych river basin in Stavropol Krai in the Caucasus.

Colchicum lagotum

Colchicum lagotum K. Perss., *Bot. Jahrb. Syst.* 127 (2): 197 (2007)
Type: Turkey, Erzurum, Çat to Bingöl, 1,900m, Jun. 2004, *Zetterlund et al. BATM 223* (holotype GB)

Corm: Ovoid to oblong-ovoid, with a short 0.5-1.5cm foot, the tunics membranous, transitory, dark brown, the neck absent or very short
Cataphylls: 5-8cm long
Leaves: 2, present at flowering, then more or less equalling perianth tube, at maturity erect to somewhat spreading, narrow-lanceolate to linear-oblong, to 18 × 2cm, but often narrower, glossy deep green, flushed brownish carmine when young, glabrous, channelled but flat, not convolute at the base
Flowers: Usually solitary, narrow goblet-shaped, purplish lilac; perianth tube whitish, 2.5-5cm
Tepals: Oblong to oblanceolate, 2-2.5 × 0.4-0.9cm, with narrow channels along the filament bases
Stamens: Filaments white or palest yellow, swollen and dark yellow at the base; anthers greyish yellow to black
Styles: White to palest yellow, slender, with punctiform stigmas
Fruit capsules: Oblong to ellipsoid, to 2.5cm long, located at or just below the soil surface
Habitat: Steep, unstable serpentine slopes
Flowering period: December-March
Distribution: NE Turkey (Artvin and Erzurum provinces, Çoruh river valley), endemic; c. 1,900m
Cytology: $2n = 14$

Colchicum lagotum is a pretty little species but it scarcely has the impact of the closely related *C. szovitsii* in its finest forms. Based on a collection from 2004, *C. lagotum* is distinguished by its thin, papery corm tunic that soon disintegrates, by the strictly paired leaves that are not furled around one another at the base, by the lack of perianth auricles, and by the more slender filaments and styles. In addition, the fruit capsules are subterranean in *C. lagotum*, but positioned at ground level or slightly above in *C. szovitsii*. The original discoverers likened the leaves to rabbit ears, hence the epithet *lagotum*, although *lagōs* is Greek for hare.

This species is rare in cultivation, although present in the extensive collection at Gothenburg Botanic Garden, Sweden.

Colchicum lagotum always has two, narrow leaves per corm, that more or less equal the flowers at flowering time.

Colchicum leptanthum

Colchicum leptanthum K. Perss., *Bot. J. Linn. Soc.* 135: 85 (2001)
Type: Turkey, Çoruh, SW of Yusufeli, 660m, Feb. 1996, cult, *Kerndorff & Pasche 96–02* (holotype GB)

Corm: Soboliferous, with short, thick soboles, about 10mm diameter, with thinner, stolon-like lobes, to 4cm long
Cataphylls: White, suffused green at the top, 6–9cm long, protruding above ground by up to 40mm
Leaves: 3, present at flowering and then generally overtopping the flowers, at maturity linear, to 20 × 0.2–0.35cm, deep green, sometimes flushed purple in the lower part, glabrous, deeply channelled but not keeled, margin thinly cartilaginous
Flowers: 1 or 2, narrowly funnel-shaped, white, occasionally flushed purplish pink towards the base; perianth tube white 1.5–4cm long
Tepals: Linear-elliptic, 13–19 × 1–3mm, apex subacute
Stamens: Filaments pale yellow; anthers yellow or greyish, 2.5–4mm long
Styles: Thread-like, yellowish, exceeding the stamens; stigmas punctiform
Fruit capsules: Not recorded
Habitat: Stony steppe
Flowering period: February–April
Distribution: NE Turkey (Erzurum province, Çoruh river valley), known only from the type locality; c. 650m
Cytology: 2n=18

This Turkish endemic is similar in appearance to both *C. minutum* and *C. munzurense*, both also soboliferous species and local endemics, although well isolated from one another. From *C. minutum*, *C. leptanthum* differs in its thicker, shorter soboles, and in its narrower, deeply channelled, non-keeled leaves with the cataphylls distinctly exserted above ground after flowering has ceased. In addition, the flowers tend to be paler in *C. leptanthum* and the filaments and styles are yellowish rather than white or purplish. *Colchicum munzurense* differs in its broader and flatter, more yellowish green leaves that are barely shorter than the flowers at anthesis, in the flatter flowers, often with somewhat recurved tepals, and its reddish brown or almost black anthers.

Colchicum leptanthum is one of only five species so far identified with a low chromosome count of 2n=18, the others being *C. hierosolymitanum*, *C. serpentinum*, *C. szovitsii* and some variants of *C. trigynum*.

The cataphylls of *Colchicum leptanthum* protrude from the ground at flowering.

Colchicum lingulatum

Colchicum lingulatum Boiss. & Spruner, *Diagn. Pl. Orient. ser.* 1 5: 66 **(1844)**
Type: Greece, Sterea Ellada, Attiki, Mount Parnes. May 1842, *Boissier* (holotype G-BOISS, isotype G)
Syns *C. sibthorpii* sensu B.L. Burtt., non Baker

Corm: Almost globose to ovoid, mostly 2.5–5cm long, with dark blackish brown, coriaceous tunics, the inner tunics paler, neck to 10cm long
Cataphylls: White to yellowish white, often greenish or purplish towards the tip, slender, to 10cm long, often half that length
Leaves: (3)4–5, sometimes with several additional much smaller and narrower leaves, appearing after flowering, spreading, often pressed close to the ground, oblong-lanceolate to strap-shaped, 2–15 × 1–3.3cm, occasionally longer, deep green to glaucous green and shiny, glabrous, channelled in the lower half, rather undulate and thinly cartilaginous at the margin
Flowers: 1–5(–9), funnel-shaped, often narrowly so, pale to deep purple-pink or purple-lilac, sometimes faintly tessellated; perianth tube white, sometimes flushed with pinkish lilac at the top, shorter or somewhat longer than the tepals, 1–5cm long
Tepals: Narrow oblong-elliptic to oblanceolate, 2.5–4 × 0.3–0.8(–1.2)cm, filament channels glabrous or sparsely papillose
Stamens: One-third to half the length of the tepals, the outer series inserted in the top of perianth tube; filaments white or yellowish, somewhat swollen and yellow at the base; anthers yellow, rarely greyish yellow, 4–8mm long
Styles: White, slender, overtopping the stamens, curved at the top; stigmas decurrent for 1–2mm
Fruit capsules: Oblong-ovoid, 1.5–2.5cm long, glabrous, apex blunt, located at ground level
Habitat: Stony and short grassy habitats, scrub, often with *Arbutus*, *Erica*, *Genista* and *Cistus*, maquis, pine woodland, generally over limestone, sometimes non-calcareous substrates
Flowering period: August–early October
Distribution: C Greece (Stereá Elláda and Euboea island); near sea level to 1,200m
Cytology: 2n=54

Colchicum lingulatum subsp. *rigescens* in its typical habitat of stony ground in south-western Turkey.

Closely related to *C. chalcedonicum*, there are obvious similarities in the corm, tunics and leaves. The prime differences are in the longer corm necks and cataphylls in *C. chalcedonicum*, and in the greater number of flowers that are rather narrower in profile in *C. lingulatum*, those flowers being generally paler and usually faintly tessellated, while the anthers are always yellow, which is rarely the case in C. chalcedonicum. Colchicum lingulatum subsp. lingulatum has quite a restricted distribution in eastern Greece and on the island of Euboea, but is apparently absent from other Greek islands. Records of its existence in the Peloponnese are almost certainly erroneous and are probably referable to C. pulchellum.

The description above is for the typical plant, C. lingulatum subsp. lingulatum, which is endemic to Greece. Turkish material of this species has been assigned to C. lingulatum subsp. rigescens (Persson 1998a).

Colchicum lingulatum subsp. rigescens K. Perss., *Candollea* 53: 409 (1998)
Type: Turkey, Muğla, Marmaris to Datca, 140m, Apr 1991, *K. Persson 515* (holotype GB)

> **Flowering period:** September–October
> **Habitat:** Stony places in pine forests and among *Erica* on ultramafic substrates, primarily serpentinite
> **Distribution:** SW Turkey (Muğla); 50–200m
> **Cytology:** 2n=54

This subspecies, which is endemic to south-west Turkey, differs in its subcoriaceous, dark reddish brown tunics and three or four leaves per corm which are thicker and channelled at least in the lower half. The tepals have glabrous filament channels and the anthers are yellow, often rather pale.

RIGHT *Colchicum lingulatum* growing on Mount Parnitha in central Greece.

FAR RIGHT The leaves of *Colchicum lingulatum* are often held close to the ground.

BOTTOM RIGHT *Colchicum lingulatum* subsp. *rigescens* differs from the typical species in features of the corm and leaves.

BOTTOM FAR RIGHT A pale-flowered variant of *Colchicum lingulatum* growing on Mount Parnitha in central Greece.

Colchicum longifolium

Colchicum longifolium Castagne., *Cat. Pl. Marseille*: 135 **(1845)**
Type: France, Bouche du Rhone, Montaud, Castagne (1838) (lectotype G; possible isotypes P, UPS)
Syns *C. castrense* De Laramb., *C. longifolium* var. *micranthum* Emb. & Maire, *C. neapolitanum* var. *castrense* (De Laramb.) Rouy, *C. neapolitanum* f. *micranthum* (Emb. & Maire) Maire & Weill., *C. provinciale* Loret

Corm: Oblong-ovoid, 1.8–3.5cm long, with a coriaceous, dark reddish brown tunic, extended above into a neck
Cataphylls: White, 4.2–5.8cm long
Leaves: 3(–4), appearing after flowering, in the later winter, ascending to spreading, linear-lanceolate to lanceolate, 12–28 × 1.2–2.5cm
Flowers: 1–2(–4), funnel-shaped, uniformly lilac to rose-purple; perianth tube whitish, flushed purple in the upper part, 5.2–7.2cm long, 2–3 times the length of the tepals
Tepals: Linear-lanceolate to lanceolate-oblong or oblong-oblanceolate, 2.6–3.8(–4.5) × 0.4–1cm
Stamens: One-third to half the length of the tepals; filaments whitish, slightly swollen at the base; anthers yellow, 4.5–5.5mm long
Styles: Filiform, slightly overtopping the stamens; stigmas short-decurrent
Fruit capsules: Not recorded
Habitat: Meadows and other grassy places, open scrub, woodland margins
Flowering period: August–September
Distribution: Algeria, Balearic Islands, Corsica, Sardinia, S France, NW Italy, Morocco; 1,250–1,400m
Cytology: $2n=144$

A relatively little-known species centred upon the western Mediterranean region, this is most closely related to members of the *C. autumnale* complex, particularly *C. lusitanum* and *C. neapolitanum*. It is included by some authors as a synonym under the last species.

Colchicum longifolium and *C. neapolitanum* are closely allied species, the former from the western Mediterranean, the latter from the central Mediterranean, the two only overlapping on the island of Corsica. In general, *C. neapolitanum* is a larger plant with rather broader leaves and relatively larger flowers. However, it has to be admitted that these taxa require further close study. Both have high chromosome numbers, with a count of $2n=90$ and 140 for *C. neapolitanum*.

Colchicum longifolium is a little-known species from the western Mediterranean.

Colchicum lusitanum

Colchicum lusitanum Brot., *Phytogr. Lusitan. Select.* (1816–1827) 2: 211, tt. 173, 174 (1827)
Type: Portugal, Alcantara by Lisbon and Extremadura, Brotero – destroyed (D'Amato 1955) (lectotype t. 173 in Brotero,, *Phytogr. Lusitan. Select.* 2 (1827))
Syns *C. actupii* Fridl., *C. algeriense* Batt., *C. autumnale* var. *algeriense* (Batt.) Batt. & Trab., *C. autumnale* var. *gibraltaricum* Kelaart, *C. fharii* Fridl., *C. fritillatum* Link ex Jahand. & Maire, *C. levieri* Janka, *C. texedense* Pau.

Corm: Ovoid to oblong-ovoid, 3-4.5cm long, adorned with a bright chestnut-brown coriaceous tunic, extended upwards into a stout neck, to 9cm long
Cataphylls: 6.5-11cm long
Leaves: 4-5, appearing after flowering, oblong-lanceolate to linear-lanceolate, 40-60 × 1-1.5cm, subacute to subobtuse, deep green, glabrous
Flowers: 1-4, chalice-shaped, pinkish purple to purple, generally faintly tessellated and whitish in the throat; perianth tube white, 5-8.5cm long, often suffused with pink or purple towards the top
Tepals: Elliptic-oblong to oblanceolate, 4-6 × 0.6-1.4cm, obtuse, glabrous
Stamens: Filaments white, swollen and yellowish green at the base; anthers purple-black to purplish pink, 6-8mm long, half the length of the filaments
Styles: Whitish, slender, overtopping the stamens, slightly curved at the top, the stigmas decurrent for 2-3mm
Fruit capsules: Oblong, 3-4.5cm long, apex apiculate, located well above ground level
Habitat: Meadows, woodland margins, waysides, generally over limestone
Flowering period: August-September (early October)
Distribution: Algeria, Balearic Islands, N and C Italy, Morocco, Portugal, Sardinia, Spain and Tunisia; 50-1,550m
Cytology: 2n=108

Often mistaken for *C. autumnale*, *C. lusitanum* has a decidedly south-western, Europe-North African distribution, while *C. autumnale* is a widespread, European (except the north), Russian and Ukrainian species. *Colchicum autumnale* is widespread in most parts of the Alps, while *C. lusitanum* is absent in all except the area of Brescia in Italy.

Colchicum lusitanum is primarily distinguished from *C. autumnale* by two characters – its slightly tessellated flowers and its black or purplish pink, rarely yellow, anthers, although the colour is quickly obscured by the yellow pollen.

Colchicum lusitanum growing in a meadow near Ronda in Spain and showing some tessellation.

Plants generally have four or five leaves, whereas there are normally four in *C. autumnale*. In addition, the flowers of *C. lusitanum*, although about the same size, generally have narrower tepals. There is a great disparity in chromosome numbers, with *C. autumnale* having 2n=38.

Because of its close resemblance to *C. autumnale*, *C. lusitanum* is seldom seen in gardens although it it is present in collections at both Royal Botanic Garden Edinburgh and Gothenburg Botanical Garden, Sweden.

RIGHT **A floriferous clump of *Colchicum lusitanum* at the trial at RHS Garden Hyde Hall.**

FAR RIGHT **The leaves of *Colchicum lusitanum* in the wild in Portugal.**

BOTTOM RIGHT **The usual dark stamen colour of *Colchicum lusitanum* is often obscured by the yellow pollen soon after the flowers open.**

Colchicum luteum

Colchicum luteum Baker, *Gard. Chron. n.s.* 2: 34 **(1874)**
Type: Kashmir, 5,000–7,000ft, 1848, *Thomson s.n.* (lectotype K; isotypes G, GOET, LD, M, P, S, W)

Corm: Oblong-ovoid, often flattened on one side, generally 2–5 × 1–2.5cm, with a several-layered, membranous, pale to mid-brown tunic, extended upwards into a neck, 1.5–7cm long
Cataphylls: White, generally flushed violet-purple towards the top, 6–17cm long
Leaves: (2)3–5(6), present at flowering, then to 5cm long but occasionally still hidden within the cataphylls, at maturity spreading to arcuate, oblong-lanceolate to elliptic-oblong or linear, 10–25 × 0.6–2cm, occasionally broader, rather bright green, shallowly channelled, and slightly undulate, apex obtuse to subobtuse, margin narrowly cartilaginous, scarcely visible, glabrous to minutely papillose
Flowers: 1–2(–4), narrow funnel-shaped but star-like and wide-spreading in strong sunshine, pale yellow to golden-yellow, plain, flushed bronze or streaked brownish purple towards the base, very fragrant; perianth tube slender, 0.5–4.5(–7)cm long, white to pale yellow, sometimes bronze-flushed or purple-brown streaked, occasionally deep maroon or maroon-purple or brown-purple, usually grooved in the upper part
Tepals: Linear to linear-lanceolate or narrow-oblong, 1.3–3.4 × 0.2–0.8cm, occasionally larger, apex obtuse to subobtuse, ridged and grooved and with short lamellae at the base
Stamens: Subequal; filaments linear, pale yellow, attached to the upper part of perianth tube, about half the length of the tepals; anthers basifixed, yellow or golden, 4.5–16mm long, narrow
Styles: Fused below, equalling or somewhat overtopping the stamens, yellow or golden, straight or bent; stigmas punctiform
Fruit capsules: Ovoid to oblong-ellipsoid, 2–3.7cm long, located at ground level or slightly above, sessile or with a short pedicel to 1.5cm long
Habitat: Alpine meadows and pastures, high passes, rocky and stony slopes, stream banks, open woodland, moist slopes, snow hollows, often near melting snow, sometimes in extensive colonies
Flowering period: March–June
Distribution: E and NE Afghanistan, Kazakhstan, Kyrgyzstan, N Pakistan east to Himachal Pradesh in India, Tajikistan, S Turkmenistan and S Uzbekistan; 600–3,900m
Cytology: 2n=50, 54

The dramatic yellow flowers of *Colchicum luteum*, here blooming in the Hazrati-Shoh mountain range in Tajikistan, are unique in the genus.

A startlingly beautiful spring-flowering species, the yellow flowers of *C. luteum* separate it from all other known species of *Colchicum*. Like its close cousin *C. kesselringii*, with white and purple-striped flowers, it is a denizen of high habitats in Central Asia and these species form the easternmost distribution of the genus. Both have their strongholds in the mountains of Central Asia, particularly the Pamir and Tien Shan, but extending down in one or two localities in the western Himalaya and the Hindu Kush of Afghanistan.

Colchicum luteum is quite a variable species with 2- or 3-leaved and solitary-flowered forms being found in the wild, often in colonies of the more typical forms. The flowers at first are goblet-shaped but open in strong sunshine to blunt-tipped stars. At a glance the flowers could be mistaken for a *Sternbergia*, however this genus belongs to the daffodil family, *Amaryllidaceae*, and all the yellow species of *Sternbergia*, except *S. vernalis*, are autumn-flowering.

Cultivars of *C. luteum* include 'Carrot Line'. 'Golden Baby', 'Minion', 'Vakhsh' and 'Yellow Empress'.

RIGHT *Colchicum luteum*, sharing an alpine pasture with emerging leaves of *Eremurus*, near the Alabel Pass in the Ala-Too range in Kyrgyzstan.

FAR RIGHT A cultivated specimen of *Colchicum luteum* originating from Urungachsai in Uzbekistan.

BOTTOM RIGHT Sheets of *Colchicum luteum* on the slopes above the Chychkan Gorge in Kyrgyzstan.

Colchicum macedonicum

Colchicum macedonicum Košanin, *Glas Srpske Kraljiveske Akademije* 85: 232, t. 6 (1911)
Type: Republic of North Macedonia, Mount Karad, Solunska to Jakupica, *Kosanin* (holotype BEOU)
Syn. *C. pieperianum* Markgr.

Corm: Ovoid, 1.5–2.2cm long, with a sub-coriaceous, dark brown tunic which is extended upwards into a slender neck, usually 3–4cm long
Cataphylls: White, slender, 3–5.2cm long
Leaves: 2–3, appearing after flowering, suberect, narrow-oblong to strap-shaped, 5–16 × 0.5–2cm, mid-green, not channelled, apex obtuse to subobtuse
Flowers: Funnel-shaped, deep rose-purple; perianth tube very slender, white, flushed purple in the upper part, 1.3–5.2cm long
Tepals: Narrow-elliptic, 1.2–2.7 × 3–7.5cm, apex subacute
Stamens: Half the length of the tepals, with pale yellow anthers
Styles: Very slender, slightly overtopping the stamens
Fruit capsules: Ovoid-oblong, 1.8–2.2cm long, located at ground level
Habitat: Alpine pastures on calcareous soils
Flowering period: June–August
Distribution: Albania, Republic of North Macedonia, Serbia; 1,500–2,300m
Cytology: 2n=54

Relatively little known, this species flowers before the leaves develop, but it often does so with the leaves and fruits of the previous season still intact or partly so.

Colchicum macedonicum has affinities with *C. pulchellum* but apart from distribution it differs in having rather larger and more deeply coloured, rose-purple flowers. In addition, the leaves are very different. *Colchicum pulchellum* has three or four upright and linear or linear-oblong leaves, with two suberect, while *C. macedonicum* has oblong-linear to almost ligulate leaves. Both species share similar chromosome numbers.

A specimen of *Colchicum macedonicum* collected in Albania under the synonymous name, *C. pieperianum*.

Colchicum macrophyllum

Colchicum macrophyllum B.L. Burtt, *Kew Bull.* 5(3): 433 (1951)
Type: Greece, Crete, *H.C. Baker s.n.*, cult. E.A. Bowles (holotype K)
Syns *C. latifolium* auct., non Sm., *C. latifolium* var. *longistylum* Pamp.

Corm: Rounded to ovoid, 5–7cm long, with a dark brown outer tunic extended into a long neck, up to 14cm long, often shorter, the inner tunics thinner and paler
Cataphylls: Stout, white or yellowish white, often flushed green or reddish purple at the top, to 25cm long, well exceeding the corm neck
Leaves: 3–4, appearing after flowering, in late winter and early spring, large, ovate to elliptic-oval, 24–40 × 11–16cm, strongly pleated, apex subacute to acuminate, glabrous
Flowers: 1–4, occasionally more, funnel- to star-shaped, pale lilac, tessellated lilac- to rosy-purple, or purplish violet, especially in the upper part, generally white in the throat, occasionally pure white overall; perianth tube white, sometimes streaked purple, up to twice the length of the tepals, 2.5–10.5cm long
Tepals: Subequal, elliptic to oblong-elliptic, 4.5–7 × 1.5–2.5(–3)cm, apex subobtuse to acute, glabrous, filament channels glabrous or puberulous
Stamens: About half to two-thirds the length of the tepals; filaments white, slightly incurved, yellow and scarcely enlarged at the base; anthers purple-brown, 8–11mm long, the pollen green or greyish purple
Styles: White, often flushed purple towards the top, overtopping the stamens, sometimes as long as the tepals, curved at the tip; stigmas decurrent for 1.5–2.5mm
Fruit capsules: Ovoid to ellipsoid-oblong, to 5cm long, glabrous
Habitat: Dry, open, grassy and rocky places, coniferous woodland (*Abies* and *Pinus*), olive groves, phrygana, occasionally in abandoned fields
Flowering period: September–October
Distribution: Greece (Crete, Euboea and East Aegean Islands (Kos to Rhodes)), SW Turkey (Muğla); 200–750m
Cytology: 2n= 54

A highly distinctive species, this is noted for its tessellated flowers and for its large and handsome, pleated leaves that are reminiscent of those of a *Veratrum*. In the wild the leaves have been mistaken for those of the widespread and often abundant sea squill, *Charybdis maritima*, but those of the latter are more leathery, duller and unribbed. Although it can be compared to other tessellated species, *C. bivonae* and *C. variegatum* for instance, the foliage is so distinct that it is unlikely ever to be confused with them.

Colchicum macrophyllum in cultivation, showing its pale lilac, tessellated flowers and purple-brown anthers.

While this species is available in the horticultural trade and present in private and public collections, it does not always perform well in gardens, often failing to flower to its full potential. This is because it needs a drier, warmer place in the garden than most colchicums. A site that suits *Sternbergia lutea* also suits this species. It was noticeable how well the corms flowered after the hot summer of 2018. Despite this, *C. macrophyllum* does on occasion set seed in the open garden.

Cultivars of *C. macrophyllum* include 'Anopolis', 'Hora Sfakion' and white-flowered 'Cretan White'.

RIGHT **If grown in the open garden,** *Colchicum macrophyllum* **prefers a dry, warm site.**

FAR RIGHT **The large, pleated leaves of** *Colchicum macrophyllum* **are distinctive.**

BOTTOM RIGHT **Native to dry habitats in Greece and Turkey,** *Colchicum macrophyllum* **often thrives better in a pan under glass than in the open garden.**

Colchicum manissadjianii

Colchicum manissadjianii (Azn.) K. Perss., *Bot. Jahrb. Syst.* 127(2): 202 (2007)
Type: Turkey, Amasya, Tavchan Dağ, near Merzifoun, 1,000–1,400m, May 1907, *Mannisadjian 6* (holotype G, isotype LE)
Syn. *Merendera manissadjianii* Azn.

Corm: Ovate-oblong, 2-3cm long, with a coriaceous, dark brown tunic, this extended above into a short neck, up to 3cm long
Cataphylls: 4.5-8cm long
Leaves: 3-4, partly developed at flowering and then shorter than the flowers, at maturity ascending to spreading, linear-lanceolate, 12-17 × 0.6-1cm, glabrous channelled, apex obtuse or subobtuse, margin thinly scabrid
Flowers: 1-2, funnel-shaped to starry, lilac or violet; tepals split to the base, the claws filiform, slightly widened towards the top, equalling or slightly longer than the tepal limbs
Tepals: Narrow elliptic to elliptic-oblong, 2.2-3 × 0.5-0.9cm, with a pair of short auricles at the narrowed base
Stamens: A third to half as long as the tepal limb; filaments white, slender, very slightly expanded at the base; anthers greenish yellow, 3-3.5mm long
Styles: Whitish, slender, equalling the stamens; stigmas punctiform
Fruit capsules: Not recorded
Habitat: Stony places, meadows
Flowering period: April-May
Distribution: N Turkey (Kastamonu to Samsun), endemic; 1,000-1,400m
Cytology: 2n=22

Described as a species of *Merendera* more than 100 years ago, this little-known species is endemic to a small area in northern Turkey centred on Amasya. It has been associated in literature with *Colchicum raddeanum* (syn. *Merendera raddeana*), a species of eastern Turkey, Caucasus and northern Iran. The two species are clearly closely related, differing in the generally fewer fowers and greenish yellow anthers of *C. manissaddjianii*, versus 1–5 flowers and dark grey, blackish or olive-brown anthers in *C. raddeanum*.

Colchicum manissadjianii is cultivated by a few enthusiasts.

Colchicum maraschicum

Colchicum maraschicum Özhatay & E. Kaya, *Turkiye Geofitleri* 3: 539 **(2014)**
Type: S Turkey, Kahramanmaraş, Başkonuş Dağı, 1,516m, 8 May 2008, *E. Kaya* 4609 (holotype ISTE 104800)

Corm: Ovoid-globose, 4-5cm long, extended above into a stout tunicated neck 6-10cm long; tunic dark blackish brown, coriaceous, the inner tunics paler and more memranous
Cataphylls: Whitish, included within the corm neck
Leaves: 4-6, appearing long after flowering, elliptic-lanceolate, the inner rather narrower than the outer, all greyish green, glabrous, erect to half-spreading
Flowers: 1-8, white, funnel-shaped; perianth tube white, equalling the cataphyll or exserted by 1-2cm
Tepals: Linear-lanceolate, 3.5-4cm long, the inner three equalling the outer but somewhat narrower, glabrous
Stamens: Half to two-thirds the length of the tepals, with white filaments, slightly swollen at the base; anthers yellow
Styles: White, filiform, slightly broadened and curved at the tip; stigma punctiform
Fruit capsules: Ellipsoid, 3-4cm long, located at ground level
Habitat: Stony places
Flowering period: July-August
Distribution: S Turkey, Kahramanmaraş, Taslik alanlar, endemic; 1,300-1,500m
Cytology: 2n=36

A recently described and little-known species, this is known to date only from its type locality of Kahramanmaras province where nine species of *Colchicum* have been recorded. It finds its allies in species such as *C. balansae* and *C. kotschyi*, differing in details of the corm and leaves and in its distinct chromosome number.

The species is named after its geographical origin, Kahramanmaraş, shortened in Turkey to Maraş.

A small, white-flowered species, *Colchicum maraschicum* is from a region of Turkey rich in colchicums.

Colchicum micaceum

Colchicum micaceum K.Perss., *Edinburgh J. Bot.* 56: 95 **(1999)**
Type: Turkey, Izmir, Ödemis, Boz Dağ, 1,700–1,800m, Aug. 1950, *Davis 18189* (holotype E; isotype GB)

Corm: Ovoid to almost globose, 1.7–4cm long, with a membranous tunic extended into a neck, up to 6cm long, occasionally longer
Cataphylls: White, to 8–9cm long
Leaves: Mostly 3, appearing after flowering, narrow-oblong to lanceolate-oblong, 6–10 × 1–2.7cm, occasionally larger
Flowers: 1–3, funnel-shaped, pinkish purple to rose-purple; perianth tube whitish, yellowish green towards the top usually, about twice as long as the tepals
Tepals: Narrow-oblong to oblanceolate, mostly 1.5–3.5 × 0.3–0.9cm, the filament channels glabrous or minutely puberulous
Stamens: Half to two-thirds the length of the tepals; filaments whitish yellow, bright yellow at the base and somewhat swollen; anthers yellow, 3.5–7mm long
Styles: Slender, white, equalling or slightly shorter than the stamens, somewhat thickened and hooked at the tip; stigma punctiform to shortly decurrent for up to 0.8mm
Fruit capsules: Not recorded
Habitat: Meadows and rocky terrain, often on schist and in areas where snow lingers late in the season
Flowering period: August–mid September
Distribution: W Turkey (Baba Dağ and Boz Dağ), endemic; 1,500–1,800m
Cytology: 2n=54

This recently described species has a limited distribution in western and west-central Turkey, focussed on the mountains of Baba Dağ and Boz Dağ.

It finds its closest ally in *C. micranthum*, which is confined to East Thrace and north-western Turkey, principally in the Bosphorus region. That species differs in its narrower leaves, not more than 1cm wide, and in the paler coloured, generally rather smaller flowers. While *C. micranthum* is a plant of relatively low altitudes up to 500m, *C. micaceum* is essentially a snowmelt plant of higher altitudes, certainly above 1,000m. *Colchicum chalcedonicum* subsp. *punctatum*, like *C. micaceum*, favours mica schist substrates, although is found further east in Turkey and at lower altitudes, up to 1,000m.

A cultivated specimen of *Colchicum micaceum* derived from a collection made on Boz Dağ in Turkey.
INSET Leaves of *Colchicum micaceum* in cultivation.

Colchicum micranthum

Colchicum micranthum Boiss., *Fl. Orient. [Boissier]* 5(1): 162 (1882)
Type: Turkey, Istanbul, Chodscha–dasch, Bujukdere, Aug. 1872, *Janka s.n.* (holotype G-BOISS; isotypes FI, GOET, JE, P, WU)

Corm: Small, subglobose to ovoid, to 1.5 × 1cm, with membranous, reddish brown tunic and a short, thin neck, often rather obscure
Cataphylls: 4-7cm long
Leaves: (2-)3, appearing after flowering, linear, 10-20 × 0.3-1cm, glabrous, apex obtuse
Flowers: Generally 1-2, funnel-shaped to starry, white to pale pink; perianth tube greenish white, 2-3 times as long as the tepals
Tepals: Elliptic-oblong to linear-elliptical, 1-2.5(-3) × 0.3-0.6(-1)cm
Stamens: Filaments white; anthers pale yellow, 3-4mm long
Styles: White, straight, distinctly curved at the tip, equalling or slightly longer than the stamens; stigmas decurrent for about 1mm
Fruit capsules: Oblong-ovoid, about 2.5cm long, glabrous
Habitat: Dry rocky places, open woods, maquis (*Erica* and *Arbutus*), oak scrub, meadows
Flowering period: September–October
Distribution: NW Turkey (north in the Bosphorus region); 480-500m
Cytology: 2n=54

This species is dainty and rather delicate looking in flower, with extremely slender perianth tubes. Its distribution is restricted to the Bosphorous region of Turkey.

It has clear affinities with the northern Anatolian *C. umbrosum*, which extends to Romania and Crimea. Both species have similarly coloured flowers and yellow anthers but the latter produces more leaves, generally three to five instead of two or three, which are strap-shaped rather than linear. Looking at the corms, those of *C. micranthum* have membranous, reddish brown tunics and a short or scarcely developed neck, while those of *C. umbrosum* are dark blackish brown and leathery, with a well-developed neck. There are also clear affinities with the little known, relatively recently described *C. micaceum*.

The very slender perianth tubes of *Colchicum micranthum* are one of its main features.

Colchicum minutum

Colchicum minutum K. Perss., *Edinburgh J. Bot.* 56(1): 90 (1999)
Type: Gündoğmuş to Manavgat, 1,000m, Apr. 1987, *K. Persson 431* (holotype GB)
Syns *C. hiemale* Siehe, nom. nud., non Freyn, *C. issicum* Siehe, nom. nud., *C. psaridis* sensu C.D. Brickell, non Heldr. ex Halácsy

Corm: Soboliferous, not more than 10mm wide, generally with 1–2 shoot-bearing lobes, occasionally more, these 2–5cm long and not more than 7mm wide, inclined to horizontal, with a yellowish brown or somewhat reddish brown, membranous tunic, scarcely with a neck
Cataphylls: White, delicate, slender and flexuous, to 10cm long
Leaves: 3(–4), well-developed at flowering time, then equalling or exceeding the flowers, at maturity linear to linear-lanceolate, 8–20 × 0.3–0.8cm, occasionally somewhat wider, mid-green, sometimes somewhat glaucous, glabrous, generally crimson-purple at the apex and along the margin, channelled, keeled below, tapering to the apex, the margin finely cartilaginous
Flowers: 1–2(–4), funnel-shaped, appearing starry from above, white, generally with a flush of pinkish purple, especially towards the tips; perianth tube very slender, white, sometimes flushed pinkish purple, 1.5–5cm long
Tepals: Linear to narrow-oblong, 1.5–3 × 0.15–0.5cm, glabrous, faintly channelled, apex acute to subacute
Stamens: To half the length of the tepals, the outer clearly shorter than the inner; filaments white, sometimes flushed with pinkish purple, very slender, slightly swollen and orange-yellow at the base; anthers buff-yellow to greyish yellow or dark grey, very slender, 2.5–4mm long
Styles: White, sometimes purple-flushed, very slender, straight or curved, long-overtopping the stamens; stigmas punctiform
Fruit capsules: Ellipsoid, green, flushed with carmine-purple, rarely seen
Habitat: Sparse oak scrub, abandoned fields, dolines, on deep terra rossa over limestone
Flowering period: January–March
Distribution: S Turkey (Antalya to Adana); *c.*1,000–1,400m
Cytology: 2n=44

The slender filaments and grey anthers of *Colchicum minutum* can be clearly seen in these plants in southern Turkey.

A quietly attractive little species, this is patch-forming by means of its branching soboliferous rootstock, often producing far more leaves than flowers. From a gardener's point of view it is easy to propagate. Both the renewal buds and the reserve buds branch out horizontally, eventually spreading plants in all directions.

As its name implies, *C. minutum* has small flowers, indeed they are probably the smallest recorded in the genus, and in this respect at least it appears to come close to *C. munzurense*.

A Turkish endemic, it was first collected by Walter Siehe in 1911 (Siehe 87) on 'the S coast of Asia Minor, above Anamas by the river, the source of the Tarmar'. Siehe recognised it as a distinct species (*C. hiemale* nom. nud.) but unfortunately he was unaware that the name had already been used for a Cypriot species, this latter now considered to be a synonym of *C. pusillum*. Brickell (1984) allied *C. minutum* with *C. psaridis*, an endemic of the Greek Peloponnese. The latter generally bears leaves in twos and has deeper coloured flowers.

RIGHT *Colchicum minutum* typically has three leaves per corm, which can overtop the flowers by flowering time.

FAR RIGHT At maturity the leaves of *Colchicum minutum* can reach 20cm long.

BOTTOM RIGHT The tendency to produce more leaves than flowers is evident in this colony of *Colchicum minutum*.

Colchicum montanum

Colchicum montanum L., *Sp. Pl.*: 342 **(1753)**
Type: Clusius, *Rar. Stirp. Hisp. Fig.* p. 157 (1576) (lectotype)
Syns *Bulbocodium autumnale* Lapeyr., *B. broteroi* Welw., *B. colchicoides* Nyman, *B. lusitanicum* Heynh., *B. montanum* (L.) Heynh, *Colchicum bulbocodioides* Brot., *C. hexapetalum* Pourr. ex Lapeyr., *C. pyrenaicum* Pourr., *Merendera bulbocodioides* Willd., *M. bulbocodium* Ramond, *M. montana* (L.) Lange, *M. pyrenaica* (Pourr.) P. Fourn.

Corm: Regular, ovoid to ovoid-oblong, 2–3cm long, with a mid-brown, membranous tunic, extended above into a short neck
Cataphylls: Whitish or greenish, 3–5cm long
Leaves: 3–4, appearing after flowering or sometimes then with the tips showing, at maturity narrow-linear to linear-lanceolate, 11–15 × 0.5–0.9cm, occasionally longer, dark or bluish green, glabrous, channelled, apex acute to subacute
Flowers: 1–2, rosy purple or violet-purple, funnel-shaped but opening to wide, flat stars, 3–5cm diameter; tepals split to the base
Tepals: Linear-elliptic to linear-oblanceolate, mostly 25–40 × 2–5mm, apex obtuse, narrowed below into the claw
Stamens: Filaments whitish; anthers yellow, 5.5–8mm long, equalling or longer than the corresponding filaments
Styles: Filiform, white, extending beyond the stamens
Fruit capsules: Oblong, 1.5–2.5cm long, glabrous, located at or just above ground level
Habitat: Grassy hillsides, rocky meadows, open scrub, often in rather dry places
Flowering period: July–October
Distribution: S France (C Pyrenees), C, N and NW Spain; 900–2,500m
Cytology: 2n=54

An attractive little plant, this can be seen in large numbers at some localities, studding the ground with small purple stars. It can start into flower in some areas in mid-summer.

Colchicum montanum is probably the best known of the species formerly in *Merendera* and makes an excellent pan plant in an alpine house or cold frame. It will succeed equally well in an open position on a raised bed in the garden.

A cultivar with slightly deeper coloured flowers, *C. montanum* 'Norman Barratt', is sometimes offered for sale, although it is scarcely distinguishable from many of those seen in the wild.

Colchicum montanum growing near Gavarnie in the Pyrenees.

Colchicum multiflorum

Colchicum multiflorum Brot., *Fl. Lusit.* 1: 597 (1804)
Type: Portugal, Beira, Sep.–Oct., *Brotero* l.c. (not located)
Syn. *C. guadarramense* Pau

Corm: Ovoid to almost globose, 2.5–4cm long, adorned with a coriaceous or subcoriaceous, chestnut-brown tunic, extended above into a short neck
Cataphylls: White, 9–14cm long
Leaves: 2–4(5), appearing after flowering, spreading, oblong-lanceolate to lanceolate, 15–35 × 2.5–4.5cm, glabrous, apex obtuse
Flowers: 1–3(4), funnel-shaped, pale lilac to purple or rose-purple, rarely white; perianth tube whitish, often flushed lilac or purple towards the top, up to twice the length of the tepals, 8–16cm long
Tepals: Oblong-linear, 3–6 × (0.5–)0.7–1.4cm, glabrous, apex subacute to obtuse
Stamens: One-third the length of the tepals; filaments white, 8–16mm long; anthers yellow, 8–10mm long
Styles: Slender, whitish, overtopping the stamens, straight or curved; stigmas decurrent for 1.5–3mm
Fruit capsules: Oblong to narrow-ovoid, 2.5–5cm long, located at ground level or just above
Habitat: Rocky places and meadows
Flowering period: September–November
Distribution: Portugal and C Spain; 600–1,200(–1,800)m, occasionally at lower elevations
Cytology: 2n = 140. 2n=144 was recorded by Persson (2007)

An Iberian endemic, this is closely related to *C. lusitanum* and in the complex of species that encompasses *C. autumnale*. It differs from *C. lusitanum* primarily in its uniformly coloured, purple or lilac-purple flowers without a white throat, and in the yellow, rather than purple-black to purplish pink anthers. Many references to *C. lusitanum* in Spain and Portugal of plants with yellow anthers are probably referable to *C. multiflorum*. Both species have dark chestnut, leathery corm tunics but differences can be seen in the leaves: *C. multiflorum* bears two to four leaves per corm, these at maturity 2.5–4.5cm wide, while those of *C. lusitanum* are four or five and generally narrower, 1.5–4cm wide. The flowers are about the same size although *C. multiflorum* has narrower tepals, 0.4–1cm wide, versus 0.9–1.5cm.

Records of *C. multiflorum* on the island of Sardinia are referable to *C. neapolitanum*.

Colchicum multiflorum has long, whitish perianth tubes and evenly purple flowers.

Colchicum munzurense

Colchicum munzurense K. Perss., *Edinburgh J. Bot.* 56(1): 86 (1999)
Type: Turkey, Tunceli to Ovacik, 950m, Apr. 1990, *Kammerlander et al. 90–193* (holotype GB)

Corm: More or less narrow- to broad-ovoid, 1.2–2cm long, producing pencil-thick horizontal lobes from both renewal and reserve buds, the tunic membranous, brown or yellowish brown, generally without a neck
Cataphylls: White, 3–6cm long, not appearing above ground
Leaves: 3, partly developed at flowering, then shorter than the flowers, at maturity rather flat, spreading to arcuate, sometimes spiralled at the apex, linear to linear-lanceolate, 13–20 × 0.7–1cm, glossy, yellowish green, often flushed with purplish brown at the tip and along the margins, glabrous, flattish with a central furrow above and a shallow keel beneath, margin obscurely cartilaginous
Flowers: Generally just 1–2, star-shaped, white to purplish lilac, fragrant; perianth tube 2.5–5cm long
Tepals: Narrow-oblong to somewhat oblanceolate, 1.8–3 × 0.25–7.5cm, distinctly parallel-veined, spreading to recurving
Stamens: Filaments pale yellow or yellowish white; anthers greyish purple to reddish brown or almost black, 2.5–3mm long
Styles: Usually pale yellow or greenish yellow
Fruit capsules: Ellipsoid to almost globose, to 1.5cm long, located at ground level
Habitat: Rocky places and screes, cliff ledges, open oak forests, snow patches, on limestone
Flowering period: February–April
Distribution: Turkey (Tunceli province), endemic; 950–1,000m
Cytology: 2n=24

Like the related *C. minutum*, this Turkish endemic flowers the moment the snow begins to recede in late winter and early spring. In most respects *C. munzurense* is a rather more substantial plant, although still with small flowers, but with leaves less developed at flowering time, more spreading, star-shaped flowers with distinctly veined tepals, and yellowish or greenish rather than basically white filaments and styles.

Colchicum munzurense, named after the Munzur River valley where it was discovered, is endemic to central Anatolia, an important centre of biodiversity. Other unique plants of the area include *Origanum munzurense*, *Ranunculus munzurensis* and *Stachys munzurdagensis*.

Colchicum munzurense has spreading, star-shaped flowers and three leaves per corm that are often purple-flushed at the margins.

Colchicum nanum

Colchicum nanum K. Perss., *Bot. Jahrb. Syst.* 127(2): 206 (2007)
Type: France, Corsica, Col de la Vaccia, 1,180m, May 1978, *K. Persson 349* (holotype GB)

Corm: Rounded to broad-ovoid, with a short foot and a membranous tunic extended upwards into a short neck, to 4cm long
Cataphylls: About the same length as, and included within, the corm neck
Leaves: 3, appearing after the flowers, linear to linear-lanceolate, 6–13 × 0.5–1.4cm, glabrous, channelled
Flowers: Generally solitary, funnel-shaped, pinkish purple; perianth tube yellowish, 2–4cm long
Tepals: Oblong to oblanceolate, 1.2–2.1 × 0.2–0.6cm, distinctly veined
Stamens: Filaments whitish with swollen yellow bases; anthers yellow, 2–3mm long
Styles: White, overtopping the stamens, slightly curved at the tip and with a decurrent stigma, for 0.6–1.5mm
Fruit capsules: Narrow ellipsoid, about 1–1.5cm long, pointed, located just above ground level
Habitat: Subalpine meadows
Flowering period: Late July–September
Distribution: Corsica (Bocca di Verde, Col de la Vaccia, Monte Cinto and Monte Coscione, Zicavo), Sardinia, SW Alps, the Apennines and Sicily; 1,180–2,000m
Cytology: 2n=54

A little-known and recently described species, this is closely related to *C. alpinum*, both being among the smallest species of *Colchicum* known. *Colchicum nanum* differs primarily in being a generally smaller plant with three, spreading, rather than two, ascending leaves. *Colchicum gonarei*, found just on Sardinia, differs in rather minor characters of which the more numerous and larger leaves, smaller flowers and longer, decurrent style are perhaps most significant. In many details, however, it is difficult not to regard these as island forms of *C. alpinum*.

Colchicum nanum growing in the wild in Sardinia.

Colchicum neapolitanum

Colchicum neapolitanum (Ten.) Ten., *Fl. Neap. Prod. App.* 5: 11 **(1826)**
Type: Italy, Campania, Monte S. Angelo di Castellamare, Oct., Tenore, not located
Syns *C. kochii* Parl., *C. multiflorum* sensu Arrigoni, non Brot.

Corm: Ovoid to ovoid-oblong, 1.8–2.5cm long, with a chestnut brown subcoriaceous tunic, this extended above into a long neck, to 7cm
Cataphylls: 5–11cm long
Leaves: (2–)3–4, appearing after flowering, linear-lanceolate, 14–26 × 1.0–3.0(–4)cm, mid-green, erect to half-spreading
Flowers: 2 or 3, chalice- to funnel-shaped, pink to lilac-purple; perianth tube white, sometimes flushed pink towards the top, 5–8cm long
Tepals: Narrow-elliptic to oblong-elliptic, 2.7–4.5 × 0.6–1.2cm, the inner somewhat shorter than the outer
Stamens: About half the length of the tepals; filaments white, glabrous; anthers yellow, 3–4mm long
Styles: Whitish, overtopping the stamens usually, somewhat curved; stigmas decurrent for 2–3mm
Fruit capsules: Ovoid-oblong, 2–3cm long, glabrous, apex pointed, located above ground level
Habitat: Meadows and other grassy places, open scrub, woodland margins
Flowering period: August–September(–early October)
Distribution: S Alps (France – Alpes Maritimes, Drôme, Alpes-de-Haute-Provence and Var departments; Italy – Biella and Vicenza only), Corsica, Slovenia, Sardinia and Sicily; generally below 700m
Cytology: 2n= 90, 140

Belonging to the *C. autumnale* complex, *C. neapolitanum* has a particularly Mediterranean distribution, concentrated from Italy and Slovenia westwards to encompass south-eastern France. Despite these differences in geography, *C. neapolitanum* can be readily confused with *C. autumnale*, especially small-flowered forms of this species. There are, however, differences worth stressing. *Colchicum neapolitanum* has smaller flowers, the tepals mostly in the range of 3–4.5 × 0.7–1.2cm, versus 40–6 × 1.0–1.5cm in *C. autumnale*. Both species have erect to ascending leaves, those of *C. autumnale* being broad-lanceolate and 2–5cm wide, occasionally up to 7cm, while those of *C. neapolitanum* are relatively longer and narrower, linear-lanceolate and mostly 1–3cm wide. There is great

Colchicum neapolitanum is a small-flowered species with a central Mediterranean distribution.

disparity in their chromosome numbers, with *C. autumnale* having 2n=36 or 38.

Plants apparently recorded from Spain and Portugal as *C. neapolitanum* are referable to either *C. lusitanum* or *C. multiflorum*.

Because of its small stature, *C. neapolitanum* has won few plaudits in the garden, with *C. autumnale* and *C. speciosum* and their cultivars having far greater impact. However, especially in its deeper coloured variants, *C. neapolitanum* is ideal for cultivation in pans or troughs, or planted in small groups on a raised bed.

RIGHT & FAR RIGHT **Two or three flowers can be produced by one corm of *Colchicum neapolitanum*.**

BOTTOM RIGHT **A small group of *Colchicum neapolitanum* makes a good subject for a raised bed.**

Colchicum osmaniyense

Colchicum osmaniyense Özhatay & E. Kaya, *Istanbul Üniv. Eczac. Fak. Mecm.* 46(2): 126 (2016)
Type: S Turkey, Osmaniye, Hasanbeyli, Kemiklikayatepe district, *E. Kaya C8001* (holotype ISTE 104799)

Corm: Asymmetrically ovoid, to 2.5cm long, with a dark greyish brown tunic, not extended above into a neck
Cataphylls: Yellowish white, purple-flushed at the top and just appearing above ground, 4-6cm long, becoming brown and membranous
Leaves: 3-4(-6), part-developed at flowering but then shorter than the flowers, dark matt green, erect at first, later spreading, linear to linear-lanceolate, 10-15mm wide, channelled, glabrous
Flowers: 2-5, occasionally solitary, narrowly chalice-shaped, opening more widely at maturity, pink, deeper rose at the base; perianth tube, slender, white or pinkish, 2-3 times longer than the tepals
Tepals: Narrow elliptic-oblanceolate, 12-15mm long, the inner somewhat shorter than the outer
Stamens: Filaments pale yellowish green or white, slightly expanded and orange-yellow at the base; anthers purple or purple-brown, 2-2.5mm long, with dark yellow pollen
Styles: White, straight, equalling or overtopping the stamens; stigma punctiform
Fruit capsules: Ellipsoid, located just below the soil surface
Habitat: Rocky areas
Flowering period: November–December
Distribution: S Turkey (Osmaniye province), endemic; 600-900m
Cytology: 2n=54

A recently described Turkish endemic, this species is known to date from several localities in the vicinity of Osmaniye province in Turkey. It is winter-flowering and is related to *C. stevenii* which is known from further west. Both species share the same chromosome number of 2n=54 but differ not only in flowering time but in the shape and colour of the flowers and details of corm shape.

Cultivated plants of *Colchicum osmaniyense* grown from the type collection, made at the locality for which the species was named.

Colchicum parlatoris

Colchicum parlatoris Orph., *Atti Congr. Bot. Firenze 1874*: 32 (1876)
Type: Greece, Peloponnese, Sep. 1873, *Orphanides 4010* (lectotype G-BOISS)
Syn. *C. neapolitanum* subsp. *parlatoris* (Orph.) K. Richt.

Corm: Ovoid to ovoid-oblong, 2–4 × 2.5–3cm, with a dark brown or blackish brown, leathery tunic and a long slender neck, to 10cm long
Cataphylls: White, slender, 5–9cm long
Leaves: Generally (4–)5–11, occasionally more, appearing shortly after flowering, spreading, linear, 7–15 × 0.1–0.4 (–0.6)cm, glabrous, shallowly channelled, apex acute, rarely sparsely ciliate along the margin
Flowers: 1–2(–4), funnel-shaped, pale lilac to purple-pink, whitish in the throat; perianth tube white, about twice as long as the tepals
Tepals: Oblong to narrow elliptical, 3–4.5 × 0.4–0.8(–1.2)cm, indistinctly veined, apex obtuse to subobtuse
Stamens: Filaments white, glabrous, sometimes greenish yellow towards the base; anthers yellow
Styles: White, straight or slightly bent at the apex, equalling or somewhat longer than the stamens; stigmas punctiform
Fruit capsules: Ovoid-oblong, 1.5–2cm long, glabrous, located at ground level
Habitat: Dry stony fields, banks, short grassy places in open scrub, olive groves, terraces, rocky pathways, often in deep terra rossa
Flowering period: late August–November
Distribution: Greece (S and W Peloponnese and islands of Cephalonia, Sapientza and Zakynthos), endemic; sea level to 900m, often close to sea level
Cytology: 2n=54

An attractive and floriferous Greek species, with rather larger flowers than its closest allies, this deserves to be more widely grown. The large number of narrow leaves is a very useful diagnostic character, as are the yellow anthers and leathery corm tunics. Although the leaves appear after flowering, they may be partly developed in the occasional late-flowering specimen.

Rather localised in the wild and rarely occurring in large numbers, *C. parlatoris* is sometimes found in close proximity to other species, notably *C. psaridis* and *C. sfikasianum*, but also at times in association with other geophytes such as *Crocus boryi* and *C. niveus*. The closely related *C. arenarium*, from central Europe, has fewer, generally three to five, broader, linear-lanceolate to strap-shaped leaves, to 1.7cm wide. From the similar looking *C. pusillum*, *C. parlatoris* differs most

A dark purple variant of *Colchicum parlatoris* growing near Monemvasia in Greece.

obviously in its yellow anthers, which are brown, black or grey in *C. pusillum*.

It is sometimes seen in bulb collections where it makes a desirable plant for pan, trough or bulb frame culture. It has performed well at various botanic gardens including Gothenburg Botanical Garden, Sweden, and Royal Botanic Gardens, Kew. It is unaccountably absent from many private collections of autumn-flowering bulbs.

RIGHT ***Colchicum parlatoris* succeeds well in cultivation.**

BOTTOM RIGHT **The linear leaves of *Colchicum parlatoris* can appear towards the end of flowering in late-flowering specimens.**

Colchicum parnassicum

Colchicum parnassicum Sartori, Orph. & Heldr., *Diagn. Pl. Orient. ser. 2* 4: 122 (1859)
Type: Greece, Sterea Ellada, Levadia, Mount Parnassus, 5,000–7,000ft, Jul. 1854, Orphanides Fl Graec. Exsicc. no. 465 (lectotype G-BOISS; isotypes ATIU, BM, FI, GB, JE, K, LD, P, S, UPS, WU)
Syn. *C. lingulatum* var. *parnassicum* (Sartori, Orph. & Heldr.) Stef.

Corm: Ovoid to subglobose, 2.5-5 × 2-4cm, occasionally larger, with or without a basal foot up to 2cm long, with a membranous, yellowish or reddish brown tunic, extended into a 4-11cm long neck
Cataphylls: White, relatively stout, 7-12cm long
Leaves: 3-4(5), appearing after the flowers, usually in late winter and early spring, ascending to more or less erect or arcuate, oblong to elliptic-oblong, mostly 9-23 × 1.8-5.5cm, bright glossy green, glabrous, not twisted, apex obtuse
Flowers: 1-4, rose- to deep lilac-purple, faintly but obviously tessellated, with a white median stripe in the lower third of the tepals; perianth tube cream or white, two to three times the tepal length
Tepals: Narrow-oblong to elliptic-oblong, 2.5-4.7 × 0.5-1.2cm, occasionally larger, apex obtuse to subobtuse, filament channels glabrous or minutely puberulous
Stamens: Filaments white, sometimes flushed purple in the upper part, the inner distinctly longer than the outer; anthers yellow, 5-10mm long
Styles: White, sometimes flushed purple, longer than the stamens; stigmas usually decurrent for 2-5mm
Fruit capsules: Oblong to ellipsoid, 2-3cm long, located just above ground level
Habitat: Rocky slopes and stony meadows, often with spiny *Astragalus angustifolius* and *Daphne oleoides* in particular, screes, on limestone
Flowering period: Late July-early September
Distribution: SC Greece (known only from Giona, Helicon and Parnassus to date; 1,300-2,300m
Cytology: 2n=54

This species is generally likened to *C. lingulatum* although the flowers of *C. parnassicum* tend to be smaller and clearly, although not strongly, tessellated. Colchicum lingulatum can be readily distinguished by its subcoriaceous corm tunics and the larger number of linear leaves that are appressed to the ground.

Colchicum parnassicum is closely related to *C. graecum*, but these species do not grow on the same mountains, despite the fact that their distributions overlap.

Colchicum parnassicum **is a narrow endemic with faintly tessellated flowers.**

Colchicum paschei

Colchicum paschei K. Perss., *Edinburgh J. Bot.* 56: 110 (1999)
Type: Turkey: Adiyaman, Nemrut Daği, 2,000m, Jun. 1986, *Pasche & Richter 86–01* (holotype GB)

Corm: Globose to ovoid, 2.5–3.5cm long, adorned with rather membranous, mid-brown tunics, the inner more shiny and yellow-brown, extended above into a relatively short neck, 2.5–5cm long
Cataphylls: Yellowish white, extending 2–3cm above the corm neck
Leaves: 3(–4), appearing after flowering, clustered at the soil surface, spreading to arcuate, narrow-oblong to oblong-oblanceolate, 12–16 × 1.6–4.5cm, mid-green, rather glossy, shallowly channelled and with a shallow keel beneath, glabrous overall, the margin finely cartilaginous
Flowers: 1–2, narrow funnel-shaped, white, occasionally with a faint mauve flush; perianth tube white, quite stout, 4–6cm long
Tepals: Linear to oblong-oblanceolate, 2.5–3.5 × 0.4–0.7cm, obtuse to subobtuse and often somewhat hooded at the apex, filament channels glabrous or minutely papillose
Stamens: To about half the length of the tepals; filaments white with a whitish or pale yellow base; anthers yellow, shorter than the filaments
Styles: White, equalling or overtopping the stamens, curved at the tip; stigmas decurrent for 0.5–1mm
Fruit capsules: Ellipsoid, 2.8–3.6cm long, undotted, glabrous, with a broad beak
Habitat: Stony and rocky mountain steppe
Flowering period: Late July–August
Distribution: SE Turkey (Nemrut Daği in Adiyaman province); *c.*2,000m
Cytology: 2n=48

A little-known species, this is known only from Nemrut Daği in Turkey, a mountain famous for its 2,000-year-old limestone statues. Described in 1999 and named in honour of bulb collector Erich Pasche, *C. paschei* was originally collected many years previously in 1883 as a fruiting specimen but it remained unrecognised.

This species is similar in many ways to *C. kotschyi*, but *C. paschei* is generally smaller with the leaves crowded directly at ground level. Furthermore, its leaves have an attenuated rather than a cordate or truncated base, and they are not markedly twisted and undulate. In *C. kotschyi* the filament channels are pubescent and the anthers are longer than the corresponding filaments.

Colchicum paschei is a white-flowered species known only from one mountain in Turkey.

Colchicum peloponnesiacum

Colchicum peloponnesiacum Rech. f. & P.H. Davis, *Oesterr. Bot. Z.* 95: 427 (1949)

Type: Greece, N Peloponnese, Achaia, Diakofto to Megaspelion, 30–100m, Oct. 1939, *Davis 1007* (holotype K; isotype G)

Corm: Ovoid, sometimes rather oblong, 1.5–2.5cm long, with a brown, generally membranous tunic, which is extended above into a 2–7.5cm long neck
Cataphylls: White, extending beyond the corm neck
Leaves: 4–6(–7), present at flowering, then short and sometimes rather inconspicuous, up to 5cm long, at maturity erect to suberect to arcuate, linear, mostly 6–15 × 0.1–0.25cm, rather dull green, shallowly channelled, occasionally with a few hairs along the veins above, margin scabrid or ciliate
Flowers: 1–5, goblet-shaped, pale to bright rose-pink, often with darker veins, paler in the centre; perianth tube white, 1.5–2.5cm long
Tepals: Narrow-oblanceolate to narrow-oblong, 1.7–2.8 × 0.4–0.8cm, apex obtuse to subacute, the inner tepals slightly smaller than the outer
Stamens: Filaments white, sometimes with a flush of pink, glandular-pilose in lower half, white, yellowish or greyish at the slightly dilated base; anthers greyish yellow, brownish or nearly black, 2–5mm long
Styles: White, straight, somewhat shorter than the stamens; stigmas punctiform
Fruit capsules: Narrow-ovoid to more or less oblong-ellipsoid, 1.3–2cm long, green, sometimes flushed purple, browning at maturity, located at ground level
Habitat: Sheltered, dry, open scrub, grassy places over non-calcareous gravels, pine (*Pinus halepensis*) woodland, often on north-facing slopes
Flowering period: Late September–November
Distribution: Greece (Achaia and Argolis in the Peloponnese), endemic; near sea level to 800m
Cytology: 2n=44

Leaf and corm details help separate *Colchicum peloponnesiacum* from similar species.

This species is restricted in distribution to the northern and north-eastern part of the Peloponnese although it can be locally common. It belongs to a group of small-flowered species found in southern Greece, Turkey and in the Middle East which includes *C. cretense*, *C. parlatoris*, *C. pusillum* and *C. stevenii*. Botanists are not generally agreed on the status of these various taxa, citing leaf development and anther colour as being either reliable or unreliable characters. Among these, however, both *C. cretense* and *C. parlatoris* are hysteranthous species, with leaves developing after flowering, while the others are synanthous species, with at least some leaf development at flowering time. *Colchicum parlatoris* has consistently yellow anthers. The chromosome numbers are not consistent within the group: *C. cretense* (2n=36), *C. parlatoris* (2n=54), *C. peloponnesiacum* (2n=44), *C. pusillum* (2n=54) and *C. stevenii* (2n=54).

Colchicum peloponnesiacum probably finds it closest ally in *C. stevenii,* found further to the east, which has a longer perianth tube and narrower tepals that spread widely apart to give a star rather than open goblet shape, and glabrous filaments. *Colchicum stevenii* has yellow anthers in its Middle East populations but they are reddish brown in western Turkey. The other species have anther colours primarily greyish yellow to virtually black or reddish brown.

Colchicum peloponnesiacum is sometimes confused in the field with *C. psaridis*, which occurs further south in the Peloponnese. The two species do not overlap in distribution. While the corms are clearly very different, with those of *C. peloponnesiacum* being regular and those of *C. psaridis* soboliferous, the flowers can look very similar at a glance. The prime differences are the leaves being four to six in *C. peloponnesiacum* and rarely more than 2.5mm wide, as opposed to two, or rarely three and mostly 2–10mm wide in *C. psaridis*, and filaments hairy at the base and white to yellowish or greyish yellow, as opposed to glabrous and white in *C. psaridis*. Out of flower, the leaves are sometimes difficult to spot, especially when plants are growing in grassy places.

The description of *C. peloponnesiacum* was based upon a Peter Davis collection of the late 1930s made along the railway line which runs from the north coast of the Peloponnese to the monastery at Megaspelion.

The species is rare in cultivation but has been maintained for a number of years at Gothenburg Botanic Garden, Sweden.

RIGHT **Dark tepal veins are clear in this plant of** *Colchicum peloponnesiacum* **growing near Sikea in the northern Peloponnese.**

FAR RIGHT *Colchicum peloponnesiacum* **usually has leaves present at flowering time.**

BOTTOM RIGHT *Colchicum peloponnesiacum.* **growing near Diakopto in the northern Peloponnese.**

Colchicum persicum

Colchicum persicum Baker, J. Linn. Soc., Bot. 17: 430 (1879)
Type: Iran, Luristan, 1856, *Loftus s.n.* (holotype BM)
Syns *C. halophilum* Freyn & Bornm., *C. haussknechtii* Boiss.

Corm: Asymmetrically ovoid to somewhat ellipsoid, mostly 5–7.5cm long, with reddish brown or orange-brown, membranous tunics, paler and more yellowish inside, extended at the top into a stout neck, 8–30cm long
Cataphylls: Substantial, white, equalling the corm neck or slightly longer, 10–34cm long
Leaves: 5–11, occasionally more, appearing long after flowering, crowded on a short pseudostem up to 10cm above ground, lowermost spreading, others ascending, linear-lanceolate to lanceolate or somewhat oblong, mostly 9–28 × 1–10cm, grey-green to yellowish green, sometimes crimson-purple flushed when young, leathery, flat or shallowly channelled, usually somewhat twisted, margin undulate, sometimes finely so, cartilaginous and moderately to densely scabrid
Flowers: 2–7, occasionally more, widely funnel-shaped, pinkish purple to lilac- or violet-purple, occasionally faintly tessellated, ribbed basally; perianth tube white or yellowish, quite stout, 1–6cm long, ribbed above
Tepals: Elliptic to oblanceolate, 2.5–6(–6.5) × 0.5–1.5(–1.9)cm, the outer and inner scarcely differing in length, often grooved above along the median line, apex acute to subobtuse, often acuminate, filament channels glabrous
Stamens: Half to two-thirds the length of the tepals, the outer three shorter than the inner three and often inserted in the top of perianth tube; filaments white or yellowish white, often with a purple flush, pale yellow and slightly swollen at the base; anthers yellow, sometimes greyish or brownish violet, shorter than the filaments, usually slightly curved, mostly 5–10mm long
Styles: White, yellowish or purplish, just exceeding the stamens, curved or slightly hooked at the apex; stigmas punctiform or shortly decurrent for 0.5–1mm
Fruit capsules: Ellipsoid-oblong, 2.5–5(–7)cm long, apex pointed to beaked, brown-dotted, especially when dried, located at or just above ground level
Habitat: Dry stony and rocky slopes, sandy and clay soils, saline flats, often in semi-desert
Flowering period: Late August–November
Distribution: W Iran, Lebanon, Syria, SC Turkey (Gaziantep and Şanlıurfa); sea level in the west, under 800m in Turkey, but otherwise mainly 1,000–3,300m
Cytology: 2n=54

Curved anthers and styles and ribbing on the flowers are features of Colchicum persicum.

A highly variable species, the leaf dimensions in particular differ greatly between populations and sometimes within populations. Plants can be large and robust, especially in the Zagros regions of western Iran, but elsewhere plants can be smaller. Larger specimens can be mistaken for *C. speciosum* but *C. persicum* can be distinguished by its larger number of more leathery leaves and brown-dotted fruit capsules. In its smaller manifestations, however, especially in semi-desert areas, it approaches *C. tunicatum*, known from Israel, Jordan, Palestine and south Syria. However, that species has very pale white or pinkish flowers while its leaves, six to nine in number, are no more than 1.5cm wide and form a flat rosette on the ground.

RIGHT *Colchicum persicum* is a variable species in flower as well as leaf.

FAR RIGHT The number and texture of the leaves help identify *Colchicum persicum*.

BOTTOM RIGHT Some smaller and paler variants of *Colchicum persicum* can only be correctly identified by looking at their leaves.

Colchicum polyphyllum

Colchicum polyphyllum Boiss. & Heldr., *Diagn. Pl. Orient. ser.* 2 4: 121 **(1859)**
Type: Turkey, Içel, Mersina, Sep. 1855, *Reinert 3024* (lectotype G-BOISS)
Syns *C. decaisnei* var. *cilicica* Siehe, *C. lockmannii* Siehe ex Stef.

Corm: Globose to ovoid, somewhat asymmetric, 2.5–4.5cm long, with a membranous, reddish brown, somewhat lustrous tunic, extending above into a short neck, to 2cm long, or neck absent
Cataphylls: White, slender, 9–18cm long
Leaves: 10–30, appearing after flowering, spreading to form a procumbent rosette, linear, 10–32 × 0.2–0.8(1.1)cm, bluish green, glabrous, channelled, apex acute
Flowers: 7–25, funnel-shaped, pale pink to purplish pink, rarely faintly tessellated, the tepals often with a thin white median line and recurving in strong sunshine; perianth tube white, equalling or somewhat longer than the tepals
Tepals: Linear-oblong to narrow-elliptic or narrow-oblanceolate, 2.3–5 × 0.3–0.9cm, occasionally larger, apex obtuse to subobtuse, sometimes slightly apiculate, filament channels somewhat papillose to finely pubescent
Stamens: Outer stamens rather shorter than the inner ones; filaments slender, white, sometimes flushed purple towards the top, longer than the anthers; anthers pale yellow, 3–6mm long
Styles: White, straight, usually shorter than the stamens, curved but not thickened at the tip; stigmas punctiform or decurrent for no more than 1mm
Fruit capsules: Not recorded
Habitat: Stony slopes, field margins and open scrub, roadsides and streamsides, often dominated by *Quercus coccifera*, generally on deep terra rossa soils
Flowering period: October–November
Distribution: N Syria, S and SE Turkey (Adana, Mersin); sea level to 1,500m
Cytology: 2n=22

The large number of leaves of this species make this a very useful character to separate it from most of the other small-flowered, eastern species. It is difficult to distinguish from *C. hierosolymitanum* in flower, although the tepals in *C. polyphyllum* can often have a thin, central white stripe. However, the two are very different in leaf. *Colchicum polyphyllum* has 10 to 30 linear leaves that spread out horizontally over the ground, whereas *C. hierosolymitanum* has no more than nine ascending, broader leaves. In reality the two species do not overlap in the wild except in a small area of north-west Syria.

Although narrow tepals and pale yellow anthers are characters of *Colchicum polyphyllum*, its large number of leaves after flowering help confirm it.

Colchicum psaridis

Colchicum psaridis Heldr. ex Halácsy, *Consp. Fl. Graec.* 3(1): 274 (1904)
Type: Greece, Peloponnese, Taigetos, Sep. 1874, *Psarides* 982 (lectotype W; isotypes B, BM, E, FI, GB, JE, LD,m,m, P, S, UPS)
Syn. *C. zahnii* Heldr.

Corm: Soboliferous, small, 1.2–5cm long, not more than 1cm wide, often as little as 2–5mm, with a dark red-brown, papery, neckless tunic, and often with one or several, horizontal, shoot-bearing lobes
Cataphylls: White, slender, 4–9cm long
Leaves: 2(–3), present at flowering, then to 3–9cm, at maturity strongly arcuate to more or less procumbent, linear-lanceolate, to 15 × 0.6–1.4cm, deep green, channelled, apex acute to subobtuse, margin glabrous or ciliate
Flowers: 1–4, funnel-shaped at first but soon star-shaped, pale pink to pinkish purple or rose-purple, with darker veins, whitish in the centre; perianth tube equalling or longer than the tepals, greenish white to pale yellow, sometimes pink-flushed, 2–6cm long
Tepals: Linear-oblong to linear-elliptical, sometimes very narrow, 1.4–2.5 × 0.2–0.5cm, rarely larger, apex obtuse to acute, occasionally cucullate
Stamens: Filaments white; anthers purplish black or purple-brown, rarely greyish yellow, 1.5–2.5mm long
Styles: Straight, equalling or overtopping the stamens; stigmas punctiform
Fruit capsules: Oblong-ovoid to ellipsoid, 1–1.8cm long, glabrous or sometimes partly hirsute, located at, or partly below ground level.
Habitat: Stony slopes and fields, phrygana and olive groves, generally in terra rossa over limestone
Flowering period: September–December
Distribution: Greece (S Peloponnese, particularly the Mani Peninsula and NW of Monemvasia, Kythira island); sea level to 900m
Cytology: 2n=54

A dainty and rather charming Greek endemic, this has small flowers that open into little stars. It is known only from the extreme southern Peloponnese where it is locally common, particularly in the northern Mani Peninsula and foothills of the southern Taygetos mountains but also in the region to the north-west of Monemvasia. The appearance of the slender young leaves with the flowers in autumn is a very useful diagnostic character.

The small-flowered, autumn-flowering species of *Colchicum* are readily

The narrow leaves of *Colchicum psaridis* are usually present at flowering.

confused in the wild. *Colchicum psaridis* can be easily mistaken for *C. cupanii*, especially when plants have three leaves or when *C. cupanii* plants just have two leaves. The leaves of *C. psaridis* are generally thinner and paler and the tepals elliptic or oblong rather than oblanceolate but the only definitive way to confirm identity is to observe the corms. *Colchicum psaridis* has small horizontal soboles, whereas *C. cupanii* has a more conventional, upright corm.

Colchicum psaridis was included in *Flora of Turkey* (Brickell 1984) but Turkish material was later assigned to a separate species, *C. minutum*, by Persson (1999b). Both share a soboliferous habit but Turkish plants generally have three or four leaves, corms with yellowish brown rather than dark brown tunics, and rather spidery looking flowers with very narrow, obscurely veined tepals just 1.3–4mm wide that are fairly pallid and white to pale pinkish purple. In addition, *C. minutum* does not come into flower until January.

The soboliferous habit makes *C. psaridis* a very attractive plant in cultivation, the corms quickly building up in numbers, making propagation easy. It is best grown within the confines of an alpine house or bulb frame as, at more northerly latitudes, it is not particularly hardy.

The original specimens of this species collected in 1874 were sent to Theodor von Heldreich by Elias Psarides, after whom the species was named, and von Heldreich gave it the provisional name *Colchicum psaridis*. A more detailed and validating description was made by Eugen von Halácsy in 1904. Heldreich also described *C. zahnii* in 1900, based on specimens with rather wider leaves than average, but this fits in comfortably with the circumscription of *C. psaridis* as understood today, and so is included as a synonym here.

RIGHT **A pale-flowered *Colchicum psaridis* growing near Paliopyrgos on the Mani Peninsula in the Peloponnese.**

FAR RIGHT ***Colchicum psaridis* being visited by a hoverfly.**

BOTTOM RIGHT ***Colchicum psaridis* growing near Areopoli on the Mani Peninsula in the Peloponnese.**

Colchicum pulchellum

Colchicum pulchellum K. Perss., *Willdenowia* 18: 30 (1988)
Type: Greece, Peloponnese, Corinth, Mount Killini, south-west of Ano Trikala, 1,520m, Sep. 1974, *K. Persson 237* (holotype GB; isotypes C, K)

Corm: Ovoid to almost globose, 1.5–2.5(–3)cm long, with a membranous to sub-leathery, brown tunic and a slender neck up to 5cm long, occasionally longer, corms often producing one or several droppers, sometimes up to twice as long as the mother corm
Cataphylls: White, slender, 6–9cm long
Leaves: 3–4(5), appearing after flowering, in the late winter, ascending, linear to linear-oblong, mostly 7–12 × 0.5–1.2cm, mid to deep green, channelled, glabrous or minutely scabrid on the margins, sometimes also underneath at the base
Flowers: 1–3, funnel- to chalice-shaped, pinkish to rose-purple, often with darker veins, the tepals with a median white stripe in the lower two-thirds; perianth tube white, greenish or greenish yellow, often flushed pink towards the top, one to two times as long as the tepals
Tepals: Elliptic to elliptic-oblanceolate or elliptic-oblong, 1.7–3.1 × 0.4–1cm, occasionally somewhat larger, apex obtuse or subobtuse, the filament channels glabrous
Stamens: Filaments white, yellowish and somewhat swollen at the base; anthers yellow, generally 3.5–6mm long
Styles: White, equalling or somewhat longer than the stamens; stigmas punctiform to slightly decurrent
Fruit capsules: Ellipsoid-oblong, 1.4–2cm long, glabrous, often present but withered when the next season's flowers appear, located just above ground level
Habitat: Stony and rocky slopes, grassy, gravelly and sandy places, low scrub
Flowering period: August–September
Distribution: Greece (restricted to the Taygetos Mountains and Killini in the Peloponnese); 1,400–1,850m
Cytology: 2n=54

Colchicum pulchellum in one of its typical habitats of stony slopes.

This species appears to be fairly closely related to *C. macedonicum* but the latter is larger in all its parts and the flowers are a deep rose-purple. In addition, the leaves are very different, being three or four in number and upright and linear or linear-oblong in *C. pulchellum*, but just two and suberect and oblong-linear to almost ligulate in *C. macedonicum*. Interestingly, dry capsules and leaves are often still present when plants of *C. pulchellum* come back into flower in the autumn, a feature exhibited by *C. macedonicum* in which the previous years fruit capsules can still be green, but the latter species flowers in mid-summer.

It is closely related to *C. rausii* but the corms of C. *pulchellum* are more regular and ovoid with thicker, darker and more persistent tunics.

Colchicum pulchellum was collected twice on Killini in the 1930s but not recognised as distinct at the time. It was not until the 1970s when it was collected several times by Jimmy and Karin Persson, both on Killini and on the Xerovouna ridge of the Taygetos Mountains, that it was recognised as a new species and named in 1988. Although other mountains in the Peloponnese and also in the Štereá Elláda have been thoroughly searched, the species has not been found elsewhere.

This species is rare in cultivation, although it has been grown at Gothenburg Botanical Garden, Sweden.

RIGHT **Grassy places are another favoured habitat of *Colchicum pulchellum*, here on Killini in the Peloponnese.**

FAR RIGHT **Up to five linear leaves is a feature that separates *Colchicum pulchellum* from similar species.**

BOTTOM RIGHT **A slightly darker flowered variant of *Colchicum pulchellum* in the Peloponnese.**

Colchicum pusillum

Colchicum pusillum Sieber, *Flora* 5(1): 248 (1822)
Type: Greece, Crete, Chania, Akrotiri Peninsula, Maleka, Oct., *Sieber* (holotype not located; Sieber sheet marked 'Creta' at PRC could be an isotype)
Syns *C. andrium* Rech. f. & P.H. Davis, *C. hiemale* Freyn, *C. montanum* var. *pusillum* (Sieber) Fiori

Corm: Ovoid to subglobose, to 2 × 2cm, with a dark brown, papery to almost leathery tunics and a long, slender neck, to 6cm
Cataphylls: White, slender, exceeding the neck of the corms but not appearing above ground
Leaves: (2-)3-6(-8), present at flowering but then rarely exceeding 3cm, at maturity spreading to recurved, often flat on the ground, linear to filiform, up to 20 × 0.3-0.5cm, deep green and rather shiny, glabrous, apex acute to subobtuse, occasionally ciliate at the margin
Flowers: 1-4(-5), small, funnel-shaped to starry, pale pink to lilac-pink, pale mauve, often with darker veins, or white; perianth tube white or greenish white, equalling or up to twice as long as the tepals
Tepals: Narrow-elliptic to oblong-elliptic, 1-3 × 0.2-0.6 (-0.9)cm, often with 5-7 veins only, apex obtuse to subobtuse
Stamens: Filaments white, often orange-yellow and somewhat swollen at the base; anthers purplish black to greyish or brownish, 1.5-3mm long
Styles: White, straight, shorter than the stamens; stigma punctiform
Fruit capsules: Small, ovoid, 0.3-0.5cm long, glabrous
Habitat: Rocky places, open places in fields, grassless tracks and road margins, low phrygana, clay dolines, cultivated land, locally common
Flowering period: September-early December
Distribution: Cyprus, Greece (Crete, Cyclades, Karpathos, Kythira, East Aegean islands including Rhodes and Symi); sea level to 1,400m
Cytology: 2n=54

A dainty, small-flowered species of the eastern Mediterranean islands, it forms scattered, occasionally extensive, colonies. The very narrow leaves characteristically splay out flat on the ground at maturity. In its broader tepalled and deep pink versions this is a very pretty little species.

Colchicum pusillum is the most widespread colchicum in Crete, occasionally forming colonies mixed with *C. cretense*. It belongs to an aggregate of rather similar species that includes *C. cretense*, *C. stevenii* and perhaps also *C. fasciculare*.

Colchicum pusillum on Crete where it is a widespread species.

The degree to which the leaves are developed at flowering time in this aggregate is very variable. Plants from higher altitudes tend to have very little leaf development at flowering while those from lower altitudes or in early wet autumns may have considerable leaf development at anthesis. Anther colour can vary a lot between and within populations. *Colchicum cretense*, restricted to the high mountains of Crete, develops leaves after flowering and has rather larger, yellow or grey anthers. *Colchicum stevenii*, from the eastern Mediterranean, has pale yellow anthers, while the leaves are undeveloped or very short at flowering time. *Colchicum fasciculare*, native to Jordan and Syria, generally has three to five leaves, partly developed at flowering time, and dark grey-green anthers. Interestingly the latter species also extends the top of its cataphylls above ground, a useful diagnostic character.

    Colchicum pusillum looks similar to *C. parlatoris*, also from Greece, but they differ in anther colour.

RIGHT **The narrow leaves of *Colchicum pusillum* are usually visible at flowering time.**

FAR RIGHT **Cultivated specimens of *Colchicum pusillum* from Crete.**

BOTTOM RIGHT ***Colchicum pusillum* on the Greek island of Symi.**

Colchicum raddeanum

Colchicum raddeanum (Regel) K. Perss, *Fl. Iran.* [*Rechinger*] 170: 22 (1992)
Type: Iran, Azerbaijan Sharqi, Kuh-i Sabalan, 20 Jun. 1880, *Radde 146* (lectotype LE)
Syns *C. caucasicum* var. *raddeanum* (Regel) Stef., *Merendera raddeana* Regel

Corm: More or less ovoid to ovoid-ellipsoid, 1.5–3cm long, with membranous, shiny, reddish or yellowish brown tunics, extended above into a very short neck, or neck absent
Cataphylls: White, 4–8cm long, occasionally longer
Leaves: 2(–3), upper, much smaller leaf, present at flowering, then equalling or overtopping the flowers, at maturity linear to narrow-oblong or oblong-oblanceolate, 8–21 × 0.6–2.5cm, mid to dark green, often purplish crimson flushed along the margins and on the under-surface, especially when young, glabrous, shallowly channelled
Flowers: 1, funnel-shaped to globular-campanulate, lilac-pink to purple, with darker veins, whitish in the centre, often as a basal band; tepals split to the base, narrowed into a whitish or yellowish claw, 1–5cm long
Tepals: Oblong-oblanceolate to narrow-obovate, 1.5–3 × 0.3–0.9cm, with filiform, basal auricles at least on the inner tepals, these up to 4mm long
Stamens: Filaments white or yellowish, sometimes purple-flushed; anthers dark yellow or greyish, 2–4mm long
Styles: Equalling or overtopping the stamens, greenish or yellowish white
Fruit capsules: Ellipsoid-oval, 1.5–3cm long, located at ground level
Habitat: Meadows and alpine slopes and other places often close to the snow line
Flowering period: Late March–June
Distribution: Armenia, Azerbaijan, Georgia, N and NW Iran, E Turkey; (1,750–)2,200–3,200m
Cytology: 2n=20

This species is closely allied to *C. trigynum* and sometimes included in it. However, *C. raddeanum* has thinner and paler, more membranous corm tunics, generally two, more upright leaves, and more globular flowers with more distinct veins and with longer basal auricles on the tepal limbs.

Globular flowers and two purple-flushed leaves per corm are features of *Colchicum raddeanum*.

Colchicum "ramonensis"

Colchicum "ramonensis" – name not formally published.

Corm: Not recorded
Cataphylls: Not recorded
Leaves: 3, partly developed at flowering, then equalling or longer than the flowers, at maturity curving and spreading close to the ground, linear to linear-lanceolate, 0.3–0.6cm wide, dark green, glabrous, channelled, apex acute
Flowers: 2–7, white to pale pink with deeper veins, funnel-shaped to rather starry, small, 1.5–2cm across; perianth tube whitish, about equalling the tepals or slightly longer
Tepals: Narrow-elliptic to narrow-oblong
Stamens: Filaments slender, white, much longer than the brownish grey anthers
Styles: Filiform, solitary, slightly longer than the stamens; stigma punctiform
Fruit capsules: Not recorded
Habitat: Dry sandy places
Flowering period: December–January
Distribution: Israel (upper Negev desert), rare, endemic; *c.* 900m

This small, recently discovered and rare desert species is unique in that each flower has just a single style instead of three. It appears to be closely related to *C. schimperi*. It was discovered near Mount Ramon in the Negev Desert, Israel, in 2009 by Amiel Vasl and Avi Shmida who were conducting research on the adaptive role of nectarial appendages. The name *C. ramonensis* was proposed by Vasl & Shmida (2015), along with a photograph, but has not been formally published. Photographs of this species can also be found in Peri (2015) under the name *C. ramonensis*.

Colchicum "ramonensis" is an undescribed species from Israel distinguished by its single style.

Colchicum rausii

Colchicum rausii K. Perss., *Pl. Syst. Evol.* 217: 58 (1999)
Type: Greece, Thessaly, Kardistsa, Agrithea-Mouzakion, 1,500–1,600m, Sep. 1980, *Binder, Hagemann, Hempel & Raus 489* (holotype B, isotype GB)

Corm: Soboliferous, irregular in shape, often rather oblique, up to 5 × 1cm, with a very pale brown, transient, membranous tunic and a very short or no neck
Cataphylls: White, very slender, 3.5–7cm long, just reaching the soil surface
Leaves: 2–3(–4), appearing after flowering, in the late winter and early spring, linear-oblong to oblong-oblanceolate, often hooded at the tip, to 22 × 2.2cm, mid to deep green, with a distinct keel beneath
Flowers: Funnel-shaped, pinkish purple to pale mauve
Tepals: Oblong to oblanceolate, 3.3 × 0.9cm, but to as little as 0.3cm wide, darker veined, the inner tepals distinctly shorter than the outer, filament channels glabrous, sometimes bordered by minute lamellae, especially on the inner tepals
Stamens: Filaments white with golden yellow, swollen bases
Styles: Whitish, equal to or somewhat exceeding the stamens; stigmas punctiform to slightly decurrent
Fruit capsules: Not recorded
Habitat: Short alpine turf, on limestone
Flowering period: September–October
Distribution: Greece (S Pindus mountains), in a very restricted area as far as it is known; c.1,500–1,650m
Cytology: $2n=54$

This species is confined in the wild to just a small area in the Pindus mountains of Greece. An obvious feature is the disparity in size between the tepals, the inner ones generally noticeably shorter than the outer, often by several millimeters. As with some of the other soboliferous species, vegetative growth is slow and rather restricted and it is rare in cultivation.

Colchicum rausii has been likened to both *C. micranthum* and *C. pulchellum*. All three are autumn-flowering, with the flowers appearing in advance of the leaves. Examination of the corms shows that *C. micranthum* and *C. pulchellum* bear more regular, ovoid corms with thicker, darker and more persistent tunics.

Colchicum rausii **growing on Mount Karáva, its only locality in the wild.**
INSET **Leaves of** *Colchicum rausii* **in cultivation.**

Colchicum ritchii

Colchicum ritchii R. Br., *Narr. Travels Africa [Denham & Clapperton]*, App.: 241 (1826)
Type: Libya, in desert near Tripoli, *Walter Oudney s.n.* (holotype BM)
Syns *C. aegyptiacum* Boiss., *Hermodactylus ritchii* R.Br. ex Steud.

Corm: Ovoid, 2.5–3.5cm long, with membranous, brown tunics, extended above into a slender neck, 2.5–5cm long
Cataphylls: White, 7–10cm long, not appearing above ground
Leaves: 3(4), partly developed at flowering, then shorter or exceeding the flowers, rarely absent at flowering, at maturity spreading and curved, linear to narrow-lanceolate, 15–23 × 1–3(–4)cm, glaucous green, margins often flushed with purple, glabrous, channelled, sometimes somewhat twisted, apex acute to subacute
Flowers: 2–8, squat, broad funnel-shaped, white to pale, or sometimes deep pink; perianth tube white, sometimes flushed pink, shorter than the tepals
Tepals: Oval to obovate to narrow-elliptic, 1.8–3.2 × 0.7–1cm, rounded to obtuse to subacute at the apex, filament channels adorned on both sides with parallel lamellae in the form of small, lanceolate teeth
Stamens: Half to two-thirds the length of the tepals; filaments white or pinkish; anthers grey-green to dark purple, occasionally yellow
Styles: White or pinkish, straight, shorter than, or slightly overtopping the stamens; stigmas punctiform
Fruit capsules: Ellipsoid, 2–3.5cm long, apex acute, located at ground level
Habitat: Sandy and stony desert places on various soils
Flowering period: Late December–February
Distribution: Egypt (including the Sinai Peninsula), S Jordan, Libya, Palestine, Tunisia, particularly common in the Judean Desert; 350–900m
Cytology: 2n=14

This is a small Saharan and Arabian species that often forms little, pale posies in its stony desert habitats. The leaves are generally partly developed at flowering time when they generally exceed the flowers, although this is not always so.

Colchicum ritchii is not common in cultivation, but Edward B. Anderson (1973) records growing it in a bulb frame in a well-drained, limy soil. He described it as 'A rather tender, bunch-flowered species from Palestine, Egypt and Tripoli. Flowers pink or white' and sometimes 'recorded as a semi-desert plant'.

Colchicum ritchii in Jordan in its typical desert habitat.

Colchicum robustum

Colchicum robustum (Bunge) Stef., *Sborn. B'lghar. Akad. Nauk* 22: 24 **(1926)**
Type: Uzbekistan, Nasarbai–Chuduk, Apr. 1842 (in fr.!), A. Lehmann s.n. (lectotype P; isotype LE)
Syns *C. hissaricum* (Regel) Stef., *Merendera robusta* Bunge

Corm: Ovoid-oblong to oblong, 3–6.5cm long, the tunics pale to dark or blackish brown, coriaceous to membranous, multi-layered, with pronounced longitudinal veins, extended above into a neck, up to 15cm long, occasionally longer
Cataphylls: White, sometimes flushed purple or violet-purple at the top, mostly 6–23cm long, with an oblique mouth, extending just above the soil surface
Leaves: 2–7(–10), present at flowering, at maturity erect to ascending, the upper part often recurving, linear to linear- or oblong-lanceolate, 7–25 × 0.3–3.5cm, pale to mid-or grey-green, sometimes purple suffused, matt or slightly shiny, shallowly channelled or flat, occasionally somewhat undulate, apex acute to attenuate, margin inconspicuously cartilaginous, glabrous to minutely papillose or scabrid
Flowers: 1–4(–6), white, sometimes flushed with rose-, violet- or grey-purple; tepals split to the base, claws white, sometimes flushed purple towards the top, 2–3cm long, sometimes scarcely exceeding the cataphylls
Tepals: Linear to narrow oblong or lanceolate, occasionally oblong-obovate, 1.5–3.5(–4.5) × 0.15–0.8(–1)cm, glabrous, attenuate at the base, apex acute to slightly hooded, shallowly ridged towards the base, sometimes absent from outer tepals
Stamens: Rather unequal, up to half the length of the tepals; filaments white or greenish, with a prominent, swollen green base, sometimes flushed purple towards the top; anthers basifixed, pale yellow to greyish or greenish black, generally longer than the filaments, often slightly twisted
Styles: Green, often rather pale, sometimes purple-flushed at the top, shorter than or equalling the stamens, straight or curved at the apex; stigmas punctiform
Fruit capsules: Ellipsoid to ovoid or oblong, 1.3–4cm long, apex apiculate to beaked, located at ground level or elevated above ground for 5cm or more
Habitat: Sandy and rocky habitats on plains and hills, loess slopes, saline flats, fixed sand dune scrub, bushy places and mountains, often by melting snow
Flowering period: (January–)February–June
Distribution: Afghanistan, N and NE Iran, Kashmir region, S Kazakhstan, Pakistan, Tajikistan, Turkmenistan and S Uzbekistan; (200–)1,200–3,500(–4,500)m
Cytology: 2n=54. 2n=36 was recorded by Zakhariyeva and Makushenko (1969)

Colchicum robustum is an eastern species related to *C. kesselringii* and *C. luteum.*

This species presents a complex picture, both geographically and morphologically, that would repay a more detailed investigation. It has a broad distribution that encompasses much of Central Asia. In addition to this, the species has an extremely wide altitudinal range from 200m above sea level to 4,500m, a range not achieved by any other species in the genus. It often forms clumps of leaves and flowers, due to the development of several corms within the existing 'mother' tunic. At maturity, the cluster of flowers often forms an interlocking mass as the tepals fall into partial disarray, although the leaf bases and short tepal claws help to hold the flowers together.

Colchicum robustum as presently interpreted includes a wide range of variation, especially in details such as leaf number, leaf width and stance, in the number, size and colour of the flowers, and in the colour of the anthers. Even within small geographical areas there is considerable variation in flower number and size, the flowers varying in colour from white to pale rosy purple and sometimes darkening with age (Grey-Wilson, pers. obs.). High-altitude representatives are rather small, with membranous tunics and fewer, broader leaves with glabrous or minutely scaberulous margins, and small, often solitary flowers with short anthers. Plants from lower altitudes and drier habitats have tougher and darker tunics, and many narrower, scabrous leaves, and up to four, larger flowers. The pattern of variability, however, cannot be interpreted in any meaningful way. As a result of this, the taxon has accumulated an interesting number of synonyms in *Colchicum* but also in *Merendera* and *Bulbocodium*. Persson (1992) emphasises that much of the variation reflects the variety of habitats and altitudes and concedes that it is not really possible to distinguish infraspecific units.

Colchicum robustum shows a close affinity to both *C. kesselringii* and *C. luteum*, these three forming the easternmost extension of the genus. On morphological criteria the latter two differ in their entire perianth tubes and in the well-developed necks to the corms, as well as differences in flower colour. All three species bear basifixed anthers which are considerably longer than in most other *Colchicum* species.

Colchicum hissaricum was described from Tajikistan in 1882 and based on a collection by Eduard August von Regel. It is probably a high altitude variant, from 3,000m and above, of *C. robustum* and is regarded as a synonym here.

RIGHT & FAR RIGHT *Colchicum robustum* grown from low-altitude collections made in north-eastern Iran.

BOTTOM RIGHT The flowers of *Colchicum robustum* are held together by the leaf bases and cataphylls.

Colchicum sanguicolle

Colchicum sanguicolle K. Perss., *Edinburgh J. Bot.* 56(1): 92 (1999)
Type: Turkey, Antalya, Ak Dağ, near Yeşilgöl, 1,650–1,800m, Sep. 1976, *T. Baytop & Leep ISTE 36226* (holotype ISTE; isotype GB)

Corm: Ovoid to somewhat oblong, 3–4.5cm long, often with a well-developed foot, up to 3cm long, adorned with a pale or reddish brown, membranous tunic, extended above into a slender neck, 3.5–7.5cm long
Cataphylls: Extending beyond the corm neck, whitish below, heavily stained crimson-purple in the upper part, protruding above the soil surface
Leaves: 3(–4), appearing long after flowering, ascending to wide-spreading, strap-shaped, 20–28 × 3.5–4cm, glabrous, flattish with an obtuse or somewhat truncated apex, the basal sheaths split for most of their length
Flowers: 2–4, funnel-shaped, pinkish purple to rich rose-purple, occasionally violet-purple, whitish towards the base; perianth tube white, 2.5–5cm long, about equalling the tepals
Tepals: Oblanceolate to narrow-obovate, 3.3–4.6 × 0.8–2cm, apex obtuse or subobtuse and often somewhat cucullate, filament channels glabrous but covered in short, filiform lamellae
Stamens: About half the length of the tepals; filaments, slender, white, yellowish at the slightly swollen base; anthers yellow, 5–8mm long
Styles: White, straight or slightly kinked, overtopping the stamens and sometimes as long as the tepals; stigmas punctiform or decurrent for just 0.4–0.6mm
Fruit capsules: Oblong-ellipsoid to almost ovoid, 1.7–3cm long, apex blunt or short-pointed, located at ground level
Habitat: Margins of cedar forests, open slopes and meadows
Flowering period: September–October
Distribution: SW Turkey (Antalya and Muğla); 1,200–1,800m
Cytology: 2n=22

A distinctive, autumn-flowering species, this is sometimes compared to *C. cilicicum*. However, the dark purple cataphylls and toothed lamellae along the filament channels at the base of the tepals of *C. sanguicolle* distinguish it from all other autumn-flowering species.

Colchicum sanguicolle has a limited distribution in the wild, being found principally on Akdağ, Baba Dağ and Tahtalı Dağı.

The deeply coloured cataphylls of *Colchicum sanguicolle* are clearly visible in this wild specimen.

Colchicum schimperi

Colchicum schimperi Janka ex Stef., *Sborn. B'lghar. Akad. Nauk.* 22: 31 **(1926)**
Type: Saudi Arabia, Jebel Kora, Dec. 1835, *Schimper 870* (lectotype W; isotypes E, G, GOET, LE, M, P, WU)
Syns *C. deserti-syriaci* Feinbrun, *C. jesdianum* Czern., *C. palmetorum* Czern., *C. szovitsii* var. *cornigerum* Schweinf., *C. velutinum* Bornm. & Kneuck.

Corm: Oblong-ellipsoid to broad ovoid, mostly 2–3.5cm long, often rather narrow, with a dark chestnut-brown, papery several-layered tunic, extended upwards into a long, slender, split neck, to 5cm long, occasionally longer
Cataphylls: White, extending well beyond the neck of the corm, 5–10cm long, not appearing above ground
Leaves: 3(–5), present at flowering time, then as long as or somewhat overtopping the flowers, at maturity ascending to spreading or arcuate, close to the ground, linear to linear-lanceolate, 13–30 × 0.5–3cm, mid-glaucous-green, sometimes reddish or purplish flushed towards the apex or along the margins, glabrous, shallowly channelled, often twisted, apex acute, the margin finely cartilaginous, rarely shortly ciliate
Flowers: 2–7, rarely more, broadly goblet-shaped to almost starry, white to pale or mid-pink, occasionally deeper, especially outside in the lower half; perianth tube slender, white, 1–5cm long, equal to or slightly longer than the tepals
Tepals: Narrow elliptic to elliptic-oblanceolate or oblong, 1.8–3.5(–4.5) × 0.3–0.9(–1.2)cm, apex subobtuse to subacute, filament channels without lamellae
Stamens: About half to two-thirds the length of the tepals, filaments slender, white or straw-coloured, with a swollen dark brown or yellow-brown base; anthers dark grey, purplish brown or sometimes yellow, 2.5–5mm long
Styles: White or straw-coloured, straight, generally exceeding the stamens; stigmas punctiform
Fruit capsules: Oblong to ovoid, 1.5–3cm long, apex pointed or somewhat apiculate, located at ground level or slightly above on a short pedicel
Habitat: Sandy and rocky areas in semi-desert, slopes on rocky or clay substrates, field boundaries and waste places
Flowering period: December–early April
Distribution: NE Egypt (including Sinai Peninsula), C and S Iran, Iraq, Lebanon, Palestine, Saudi Arabia and Syria; 1,450–2,600m, almost to sea level in lowland Iraq
Cytology: 2n=14

Colchicum schimperi growing in the Negev Desert, Israel, its typical arid habitat.

One of several desert species, this has a wide distribution across the Middle East and parts of North Africa. It is closely related to *C. ritchii*, with which it overlaps in distribution in the Middle East, although the distribution of the latter extends further into North Africa. *Colchicum ritchii* can be distinguished by its squat, broad, funnel-shaped flowers, with a perianth tube that is shorter than the tepals, and by the presence of tiny lanceolate teeth at the base of the filaments.

RIGHT **The leaves of *Colchicum schimperi* are present at flowering time, usually overtopping the flowers.**

BOTTOM RIGHT **A white variant of *Colchicum schimperi* growing in the Negev Desert, Israel.**

Colchicum serpentinum

Colchicum serpentinum Woronow ex Miscz., *Fl. Caucas. Crit.* 3(4): 114 (1912)
Type: Turkey, Artvin, cult in Tbilisi Bot. Gard., *Woronow s.n.* (holotype LE)
Syns *C. crociflorum* Schott & Kotschy, *C. falcifolium* sensu C.D. Brickell in P.H. Davis, non Stapf, *C. stevenii* subsp. *taurii* (Stef.) Thiébaut, comb. inval., *C. tauri* Siehe ex Stef.

Corm: Ovoid to oblong, 1.5–4 × 1.5–2cm, with papery to subcoriaceous, reddish brown or maroon tunics, without or with a very short neck
Cataphylls: White, slender, 7–10cm long
Leaves: (2–)3–4, occasionally up to 6, generally partly developed at flowering time, then up to 6cm long, at maturity spreading, linear, 9–15 × 0.1–0.7(–1)cm, channelled, glabrous or with the margins and dorsal surface scabrid to densely hispid, apex acute to subobtuse
Flowers: 2–7, occasionally more, funnel-shaped to starry, white to pink or purplish pink; perianth tube whitish, very slender, much longer (3–4 ×) than the tepals
Tepals: Narrow-elliptic to narrow-oblanceolate, 2–2.5 × 0.2–0.4(–0.6)cm, apex usually acute
Stamens: Filaments whitish, slender, glabrous, slightly swollen and orange-yellow at the base; anthers black, greenish black, purplish brown or brownish black, 2.5–4.5mm long, about a quarter the length of the filaments
Styles: Straight, white, equalling the stamens; stigmas punctiform
Fruit capsules: Subglobose to narrow-ovoid, 1–2.5cm long, apex shortly apiculate, glabrous or hispid
Habitat: Montane steppe and bare limestone areas, generally by melting snow
Flowering period: January–April, occasionally later
Distribution: C and E Turkey (Amasya, Diyarbakir, Erzincan, Hatay, Mersin, Kahramanmaraş, Mardin, Tokat, Trabzon), endemic; 250–1,800m
Cytology: 2n=18

In *Flora of Turkey* (Brickell 1984), *C. serpentinum* is listed as a synonym of *C. falcifolium*. Persson (2007) pointed out that the type specimen (held at WU) of *C. falcifolium* from Gilan, Iran, is not a *Colchicum* at all but *Iris pseudocaucasica*. *Colchicum serpentinum* is actually a good species. Plants recorded from Iran as *C. falcifolium* are probably referable to *C. varians*.

Closely related to *C. serpentinum* is the recently described *C. erdalii* which is native to eastern Turkey. They differ in their chromosome numbers (2n=14 in *C. erdalii*), but more significantly the flowers of *C. erdalii* are squatter, with a

Colchicum serpentinum growing near Aladağ in south central Turkey.

stouter but short perianth tube while the tepals are acuminate. Differences can also be noted in the broader mature leaves of C. erdalii. Many of these smaller Turkish species require further and more detailed investigation, both to define their variabilty as well as their distribution in the wild.

Colchicum serpentinum is in cultivation where it is mostly grown by enthusiasts, usually in an alpine house. Selections with deeper pink flowers, or those with white flowers and deep pink markings on the outside are usually favoured.

RIGHT *Colchicum serpentinum* growing wild in Turkey.

FAR RIGHT Some selections of *Colchicum serpentinum* have contrasting purplish pink markings.

BOTTOM RIGHT An attractively coloured selection of *Colchicum serpentinum* from near Develi in central Turkey.

Colchicum sfikasianum

Colchicum sfikasianum Kit Tan & Iatroú, *Rock Gard.* 24(3): 255 (1995)
Type: Greece, Peloponnese, Malea Peninsula, W of Monemvasia, Oct. 1986, *Iatrou 3233* (holotype UPA; isotypes C, E)
Syn. *C. polymorphum* Orph., nom nud.

Corm: Ovoid to almost globose, 2–5 × 1.7–4.5cm, with a subcoriaceous to rather membranous, dark brown or reddish brown tunic, the inner layers more orange-brown and quite shiny, extended upwards into a neck, up to 6.5cm long, occasionally longer
Cataphylls: Yellowish white, occasionally flushed purple towards the top, slender, to 10cm long, exceeding the corm neck
Leaves: 3–4(–6), appearing after flowering, usually in late winter, grey-green, narrow oblong to lorate, 5–15 × 0.6–2cm, occasionally larger, generally somewhat undulate and twisted, glabrous or with a minutely scabrid margin
Flowers: 1–3, narrow funnel-shaped, white, lightly and sparsely tessellated in the upper part with purplish lilac to purple, sometimes purple-striped along the veins; perianth tube white, 2.5–7.5cm long
Tepals: Narrow-elliptic to oblong-lanceolate, 2–5 × (0.3–)0.5–1.2cm, apex acute to subobtuse, filament channels glabrous
Stamens: Up to a half the length of the tepals; filaments pale yellow, slightly expanded at the pale yellow base; anthers pale to mid-yellow, 4–7mm long
Styles: Yellowish white, equalling or exceeding the stamens, slightly curved at the tip; stigmas decurrent for 1.5–2.5mm
Fruit capsules: Narrow oblong to ellipsoid, 1.5–3cm long, located at ground level
Habitat: Dry open rocky slopes, scrubby areas of maquis, pine woodland and coastal phrygana
Flowering period: September–early November
Distribution: S Greece (Attica, Mount Hymettus, Ionian Islands (mainly Cephalonia but also Ithaca and Zakynthos), SE Peloponnese); sea level to 500m
Cytology: 2n=54

This is a pale-flowered, lightly tessellated species found in a handful of scattered localities in southern Greece and the Ionian Islands. It inhabits generally dry, open places.

Colchicum sfikasianum is named in honour of Greek botanist George Sfikas who has written books on the Greek flora, including Crete. The taxon had in fact been known since as long ago as 1874 when Theodhoros Orphanides published

Levels of tessellation on *Colchicum sfikasianum* can be quite varied.

the name *C. polymorphum*, but as no formal description was presented the name is considered invalid. Lafranchis & Sfikas (2009) thought that the species was endemic to the Peloponnese, but it has outlying colonies in the Ionian Islands and on Mount Hymettus near Athens. It is very different in character to the other tessellated species found in Greece, *C. bivonae*, *C. macrophyllum* and *C. variegatum*, both in its smaller, pale flowers and in its narrow leaves.

RIGHT *Colchicum sfikasianum* growing near Molaoi in the Peloponnese.

FAR RIGHT A white-flowered variant of *Colchicum sfikasianum*.

BOTTOM RIGHT *Colchicum sfikasianum* grown from seed collected north of Neapoli Voion in the south-eastern Peloponnese by Jim and Jenny Archibald.

Colchicum sieheanum

Colchicum sieheanum Hausskn. ex Stef., *Sborn. B'lghar. Akad. Nauk.* 22: 47 (1926)
Type: Turkey, Içel, Fundukbunar, 1,400m, Siehe Fl. Orientalis no. 92 (holotype B; isotypes JE, LE, W)

Corm: Ovoid to subrounded, 1–2.5cm long, with 1 or 2, short, peg-like projections, the tunics mid-brown, thin and evanescent, extended above into a slender neck, 1.5–5cm long
Cataphylls: White, very slender, 7–11cm long
Leaves: 3–4, just appearing at flowering, linear, 7–12 × 0.2–0.6cm
Flowers: 1–2, narrow funnel-shaped, violet-purple, concolourous or slightly paler in the throat; perianth tube very slender, purplish, 2.5–5cm long
Tepals: Narrow oblong to oblong-oblanceolate, 3.5–5 × 0.5–1cm, apex subobtuse, the filament channels glabrous
Stamens: About one-third the length of the tepals; filaments very slender, filiform, whitish flushed with purple, somewhat swollen and yellow at the base; anthers yellow
Styles: Slender, more or less equalling the stamens, white; stigmas punctiform
Fruit capsules: Not recorded
Habitat: Open clearings in pine forests on deep terra rossa over limestone
Flowering period: September–October
Distribution: S Turkey (Mersin); 1,000–1,400m
Cytology: Not determined

A small species, this has delicate flowers borne on seemingly thread-like perianth tubes. It looks very similar to the recently described *C. chlorobasis*. Both species have corms with one or two short lobes and narrow leaves, not more than 12mm wide, but *C. sieheanum* has deep violet-purple, usually concolorous flowers, occasionally paler in the centre, while *C. chlorobasis* has bright rosy lilac flowers with a prominent white centre. The filaments of *C. sieheanum* are yellow at the base but green in *C. chlorobasis*. Both are Turkish endemics, with *C. sieheanum* from the south-east and *C. chlorobasis* from the south, and not known to overlap.

Colchicum sieheanum has been in cultivation on and off for more than 100 years. Material was received at Royal Botanic Gardens, Kew, in 1903 but it did not persist. Later introductions included live material also being sent to Kew, in 1987, by Prof. Turhan Baytop. Despite this, *C. sieheanum* remains a little-known species.

Colchicum sieheanum in cultivation in Ukraine, derived from material collected near Mersin in Turkey.

Colchicum soboliferum

Colchicum soboliferum (C.A. Mey.) Stef., *Sborn. B'lghar. Akad. Nauk* **22:44 (1926)**
Type: Iran, Azerbaijan, Khoi, Jun. 1828, *Szovits 424* (holotype LE; isotypes G, H, K)
Syns *Bulbocodium bastulatum* Friv., *Merendera bastulata* (Friv.) Baker, *M. sobolifera* Fisch. & C.A. Mey.

Corm: Soboliferous, 2–8cm long, with slender, horizontal soboles and a membranous, mid-brown tunic
Cataphylls: Whitish, 2–5.5cm long
Leaves: 3, present at flowering, then up to 5cm long, ascending at first, later spreading, at maturity narrow lanceolate to linear, 10–18(–21) × 0.3–0.7(–1.3)cm, green, channelled, glabrous, apex acute or subacute
Flowers: 1–2, funnel-shaped, white, sometimes pale pink or pale purple;
Tepals: Narrowed into slender claws about the same length as the limb, sometimes rather longer, linear to narrow-elliptic, 20–30 × 2.5–5mm, occasionally larger, usually with a pair of hair-like auricles located at the base of the limbs
Stamens: Filaments white, flushed greenish yellow at the base; anthers blackish violet, 1.5–3.5mm long, about half the length of the filaments
Styles: Very slender, white, projecting beyond the stamens
Fruit capsules: Oblong-ovoid, 1.5–2cm long, glabrous, located at ground level
Habitat: Rocky and stony habitats, fields, snow hollows, flowering as the snow recedes
Flowering period: March–May, occasionally into June
Distribution: Afghanistan, Armenia, Azerbaijan, Bulgaria, NE Greece, Iran, Jordan, North Macedonia, Palestine, Romania, Syria, Tajikistan, Turkey and Turkmenistan; 1,000–2,400m
Cytology: 2n=54

A locally common, pretty little species, this is sometimes confused with *C. atticum* although the two species only overlap in distribution in Bulgaria, Greece and parts of Turkey. They can, however, be readily separated by the soboliferous, patch-forming character of *C. soboliferum* and the presence of small, hair-like appendages at the base of the tepal limbs in this species, not found in *C. atticum*. These hair-like appendages prevent the delicate flowers from falling apart.

The patch-forming habit of *Colchicum soboliferum* results in dense colonies when grown in a pan.

Colchicum speciosum

Colchicum speciosum Steven, *Nouv. Mém. Soc. Imp. Naturalistes Moscou* 1: 265, t. 15 **(1829)**
Type: Georgia, in leaf, D. Wilhelms, May 1826 (lectotype Steven Herbarium H 1085097)
Syns *C. illyricum* auct., non Stokes, *C. lenkoranicum* (Miscz.) Grossh.

Corm: Large, oblong-ovoid, sometimes broadly so, 4–8 × 2.5–4cm, often with a short, obtuse-triangular foot, to 1.5cm long, sometimes absent, tunic matt or somewhat lustrous, mid to orange brown, sometimes slightly yellowish, membranous to sub-leathery, the inner tunic more reddish and papery, extended upwards into a substantial neck, to 12cm long, occasionally longer
Cataphylls: White or yellowish, somewhat exceeding the corm neck and just reaching the soil surface
Leaves: (3–)4–5, appearing long after flowering, ascending to suberect on a stout pseudostem, oblong-lanceolate to narrow-elliptic, 16–25(–30) × (3–)5–9.5cm, mid to deep green, glabrous, bluntly keeled below and with several, poorly marked longitudinal folds above, apex obtuse, margin obscurely cartilaginous
Flowers: 1–2(–3), narrow to broadly chalice- or funnel-shaped, pale to deep rose-purple, rarely more intense reddish or violet-purple, plain or sometimes with a white or yellowish white throat extending a third way up the tepals from the base; perianth tube stout, 5–12cm long, sometimes longer (especially in cultivated specimens), greenish white or whitish, sometimes flushed purple towards the top
Tepals: Elliptic to oblanceolate, mostly 4.5–8 × 1–2.7cm, apex obtuse to subacute, filament channels finely pubescent to puberulous along the ridges
Stamens: Half to two-thirds the length of the tepals; filaments greenish or yellowish, scarcely swollen at the base; anthers yellow, occasionally purple or purple-brown, 6–11mm long
Styles: White to yellowish, rarely purple flushed, equalling or exceeding the stamens, straight, curved or hooked at the top; stigmas decurrent for 0.5–4mm
Fruit capsules: Ellipsoid to obovoid, 4–5.5cm long, apex often long-beaked, pale brown at maturity and sometimes darker dotted, glabrous, borne well above ground on leaf clumps 30–60cm tall, very shortly pedicellate
Habitat: Meadows and banks, grassy slopes, stream sides, gullies, open maquis, open woodland
Flowering period: August–October
Distribution: Caucasus, N Iran, N Turkey; 600–3,000m
Cytology: 2n=38, 40, 42, 44

Colchicum speciosum forming a bold clump near the Zigana Pass in north-eastern Turkey.

The elegant, goblet-shaped flowers of this species, some of the largest in the genus, are a distinctive feature of grassy meadows in the Pontic Mountains of north-eastern Turkey, the high mountains of north-western Iran (primarily in the Elburz and Talysh ranges) and the Caucasus in the early autumn. It is highly variable in the wild, with some forms having rather narrow tepals, and others having green perianth tubes or prominent white throats to the flowers. Such variations fit comfortably within the natural variation found within and between the populations. This species has the habit of multiplying quickly due to the plants developing new corms from the flowering shoots (axillary to the lowermost leaf) and also the reserve bud (axillary to the second leaf), located within the tunics of the mother corm in fruiting specimens.

Colchicum speciosum was introduced to gardens in the west around 1850 and is without question the best of the autumn-flowering species. It has given rise to a number of distinctive named selections and is the parent of many fine cultivars. The exact origin of some its hybrids is obscure because either no record of the cross was made or, more usually, hybrids have occurred spontaneously in gardens, but C. *speciosum* is often apparent in their parentage. Cultivars of C. *speciosum* itself include 'Album', 'Atrorubens', 'Dombai', 'Maximum', 'Naeisanum', 'Ordu', 'Paul Furse', 'Paul Furse Early', 'Rubrum' and 'Van Tubergen's'. In addition, mostly in botanic gardens, but also in a few private collections, there are some plants that have retained their original collector's numbers from wild gatherings.

Colchicum liparochlamys, a name that was given by Woronow without a validating description, based on plants collected at 1,800m in Abkhazia, west Caucasus, is said to differ enough from C. *speciosum* to warrant recognition as a separate species. Persson (2007) equates it with C. *woronowii*, pointing out that C. *liparochlamys* is a nomen nudum. This is discussed further under C. *woronowii*. The prime characters as described are: corms obcordate, borne on 'an obtusely rostrate offshoot', the tunic membranous and cinnamon brown, shiny, extended above into a short neck, to 3.5cm long, flowers 2–4, relatively small, the narrow tepals 3–5cm long, the anthers linear, *c.* 6mm long.

Colchicum bornmuelleri and C. giganteum

Despite natural variation in the species, there are two names, C. *bornmuelleri* Freyn and C. *giganteum* hort. ex Stef., that require further consideration. The first was based on a 1889 collection by Joseph Fredrich Nicholaus Bornmüller at 1,800m on Ak Dağ mountain in Amasya province, Turkey. Bornmüller (1862–1948) was a German botanist who graduated in botany from Potsdam. He then worked for several years at Jevremovac Botanical Garden in Belgrade which allowed him to travel to Greece botanising and collecting samples. Later he travelled extensively in Turkey, the Middle East and North Africa. *Colchicum bornmuelleri* is said to differ from C. *speciosum* primarily in its paler, more funnel-shaped flowers, the purplish brown as opposed to yellow colour of its

RIGHT **The leaves and fruit capsules of** *Colchicum speciosum* **are held above ground on short pseudostems.**

FAR RIGHT **Variants of of** *Colchicum speciosum* **with deep pink flowers can be found in the wild.**

BOTTOM RIGHT *Colchicum speciosum* **can flower as early as August in cultivation.**

anthers, and in the length of the decurrent stigmas which are 0.5–1.5mm as opposed to 2–4mm. It has a chromosome count of 2n=42. Brickell (1984) in *Flora of Turkey* states that: '*C. bornmuelleri*, although very closely related, is distinguished by the purple or purple-brown (not orange-yellow) anthers, the distinctly swollen style apex with a short stigmatic surface of 0.5–1(–1.5)mm (style apex not or only slightly swollen, stigmatic surface of 2–4mm in *C. speciosum*) and in the proportionately narrower leaves'. It is clear from observations made by Persson (1992), Brickell (1984) and Grey-Wilson (pers. obs.) that although there is a marked tendency for plants with the purple-brown or purple anthers linked to *C. bornmuelleri* to be found at the western end of the range of *C. speciosum*, this character is found elsewhere, in mixed populations of *C. speciosum*, including at the eastern end of the range of this complicated taxon. Persson (2001a) treats *C. bornmuelleri* simply as a synonym of *C. speciosum*. However, it is a distinct entity, especially in cultivation, so the present authors prefer to treat it as a cultivar group, *C. speciosum* Bornmuelleri Group. *Colchicum speciosum* Bornmuelleri Group is rare in cultivation, but is certainly present in the collections at Royal Botanic Garden Edinburgh. It should also be noted that some plants sold in the horticultural trade as *C. bornmuelleri* are almost certainly hybrids involving *C. speciosum*, and do not correspond with *C. bornmuelleri* discussed above, and hence can not be assigned to *C. speciosum* Bornmuelleri Group. These have been given the cultivar name *Colchicum* 'Joseph' in this book to avoid further confusion. Cultivars related to 'Joseph' include 'Artur Klark', 'Fabergé's Silver', 'Harlekijn', 'Jarka', 'Redgrave' and 'World Champion's Cup'.

Colchicum giganteum was described from cultivated plants, presumably of Turkish origin, although material that could be accepted as representing the type has not been traced. It is characterised by having larger, broader, more funnel-shaped, pale flowers with a white centre, white perianth tube and pale yellow anthers. They are generally earlier to come into flower than *C. speciosum*. Plants showing characteristics of *C. giganteum* appear in populations of *C. speciosum* in eastern and north-eastern Turkey and more commonly in Iranian populations. Plants corresponding to *C. giganteum* are also seen in cultivation, some having been grown for many years. Because it is a distinct entity in cultivation, the present authors prefer to treat *C. giganteum* a cultivar group, *C. speciosum* Giganteum Group. The cultivars 'Chequers' and 'Revelation' belong to Giganteum Group.

Further research might reveal that *C. bornmuelleri* and *C. giganteum* are distinct taxa in the wild, overlapping with *C. speciosum*, and giving hybrid or introgressed intermediates, or that *C. speciosum* is just one very variable taxon showing variation west to east. At present, it is better to recognise one species, *C. speciosum*, following the research made by Persson (2001a, 2007). However, in gardens, plants can be found corresponding to all three taxa, hence our solution of two cultivar groups under *C. speciosum*. See the Cultivars chapter for further discussion and images of the two cultivar groups.

RIGHT *Colchicum speciosum* multiplies quickly, a useful feature in cultivation.

BOTTOM RIGHT The white throat of *Colchicum speciosum* can extend a considerable way up the tepals.

BOTTOM FAR RIGHT Some variants of *Colchicum speciosum* have narrower tepals.

Colchicum stevenii

Colchicum stevenii Kunth, *Enum. Pl. [Kunth]* 4: 144 (1843), as 'steveni'
Type: Syria, Latakia, Labillardière (lectotype H; isolectotype? G-BOISS)

Corm: Subglobose to oblong-ellipsoid, 1.5–3 × 1.5–2.3cm, with matt, dark bown, membranous to sub-leathery, somewhat shiny tunics, these extending upwards into a slender neck, up to 6cm long, occasionally longer
Cataphylls: White, 5–10cm long, mostly covered by the corm neck
Leaves: 4–8, occasionally as many as 12, partly developed at flowering time, then sometimes very short, at maturity ascending to spreading and often recurving, linear, 8–15 × 0.2–0.5(–0.8)cm, occasionally longer, channelled and with a distinct midvein, apex obtuse to subacute, glabrous or somewhat ciliate on the margins
Flowers: (1–)2–5 occasionally more, funnel- to star-shaped, pale to bright purplish pink, with a white throat and often with a median white stripe, occasionally almost pure white, fragrant; perianth tube white, 10–12cm long
Tepals: Narrow elliptic-oblong to oblanceolate, 1.5–3 × (0.2–)0.4–0.6(–0.8)cm, the inner slightly shorter than the outer, all obtuse to subacute at the apex
Stamens: Filaments white, glabrous, slightly swollen and yellowish orange at the base; anthers pale yellow, *c.* 3mm long
Styles: Straight or slightly curved at the top, white, usually slightly exceeding the stamens; stigmas punctiform.
Fruit capsules: Oblong-ovoid to ellipsoid, 1–1.8cm long, pale brown, apex shortly acuminate
Habitat: Rocky places and slopes, field banks, sometimes in profusion, avoiding very arid areas
Flowering period: (September–)October–December
Distribution: NE Cyprus (rare), SE Greece (only on Kastelorizo and Strongyli), Israel, Jordan, Lebanon, Palestine, W Syria and S Turkey; sea level to 900m
Cytology: 2n=54

An attractive, small, autumn-flowering species, this can form large, multi-flowered clumps in the wild, each clump representing a close cluster of corms. It can sometimes be found in large, dispersed colonies. The leaves vary considerably in how developed they are at flowering time, with those from higher altitudes only having their leaf tips revealed by then. During especially dry autumns some flowers may appear in advance of the leaves.

Colchicum stevenii is closely related to *C. pusillum* and *C. peloponnesiacum*, neither of which overlap in the wild with the former. *Colchicum pusillum* has

Colchicum stevenii is often found in rocky places in the wild.

rather smaller flowers that are pinkish lilac to white and purplish black or greyish brown anthers. In addition, the leaves spread out over the ground. *Colchicum peloponnesiacum*, which is confined to the northern Peloponnese, is characterised by having more lanceolate leaves and rather larger, bright purplish pink flowers, while the filaments are pilose in the lower part. The recently described *C. osmaniyense* is also similar but differs in flowering time and the shape and colour of the flowers.

In cultivation *C. stevenii* is an easy and very satisfactory species for pan culture or a bulb frame. Plants vary considerably in the colour, size and substance of the flowers. The capacity of corms to produce both renewal and secondary corms in cultivation means that a quick increase can be expected.

TOP RIGHT
The flower colour of *Colchicum stevenii* can vary quite widely in the wild.

TOP FAR RIGHT
A dense-flowered clump of *Colchicum stevenii* growing near the Israeli coast.

BOTTOM RIGHT At flowering, the extent of development of the notably linear leaves of *Colchicum stevenii* is variable.

BOTTOM FAR RIGHT
Colchicum stevenii is easy to grow in a pan.

Colchicum szovitsii

Colchicum szovitsii Fisch. & C.A. Mey., *Index Seminum [St.Petersburg (Petropolitanus)]* 1: 24 (1835)
Type: Armenia, Arekligeduk, Jun. 1829, in fruit, *Szovits 361* (holotype LE; isotypes G, K)
Syns *C. armenum* B. Fedtsch, *C. bifolium* Freyn & Sint., *C. diampolis* Delip. & Cheshm., *C. gohariae* Gabrieljan, *C. ninae* Sosn., *C. syriacum* Siehe ex Stef., *Merendera nivalis* Stapf

Corm: Ovoid, 1.5–4 × 2–3cm, with a papery to subcoriaceous, blackish or reddish brown tunic, without an obvious neck
Cataphylls: Whitish, 4.5–9cm long, occasionally longer
Leaves: 2–3, present at flowering, then 2–6cm long, generally not overtopping the flowers, at maturity ascending, linear-lanceolate to ligulate, up to 22 × 3.5(–4.5)cm, mid-green and somewhat glossy, glabrous, sometimes twisted, apex usually subacute and often slightly hooded, margin flat or sometimes undulate
Flowers: 1–5, occasionally more, chalice- to funnel-shaped, pink, pale to deep purplish pink or white, sometimes greenish at the base within or stained purple or purple-brown; perianth tube quite stout, white, straw-coloured or greenish white, sometimes pink-flushed at the top, 1.5–6cm long, furrowed
Tepals: Narrow-elliptic to oblanceolate, occasionally lanceolate, mainly 2–3.5 × 0.4–1(–1.3)cm, more or less equal, generally with basal tooth-like or thread-like auricles, apex acute to obtuse
Stamens: About one third the length of the tepals; filaments whitish or greenish white, with a swollen yellowish or greenish yellow base, the inner stamens somewhat longer than the outer; anthers greyish yellow to chocolate-brown, purplish or greenish black, 3–4mm long
Styles: Fairly stout, straight, whitish to greenish or yellowish green, slightly shorter than or equalling the stamens; stigmas punctiform
Fruit capsules: Subglobose to ellipsoid, 2.5–5cm long, glabrous, apex shortly apiculate, located at ground level or slightly above on short pedicels, to 5cm long
Habitat: Moist or wet meadows, alpine pastures, stream sides, margins of coniferous forests, by melting snow, sometimes in substantial colonies
Flowering period: February–May, occasionally later
Distribution: Bulgaria, Caucasus, N and NW Iran (Azerbaijan, Gorgan, Fars), N Iraq (Kurdistan), C and E Turkey (Inner and S Anatolia) and S Turkmenistan; 200–3,250m
Cytology: 2n=18, 36

Colchicum szovitsii with particularly pronounced chalice-shaped flowers growing on the Soğanlı Pass in central Turkey.

This is a widespread species of moist or wet grassy habitats, often flowering shortly after snowmelt and sometimes occurring in considerable numbers. Larger, more vigorous plants are to be found in the wetter habitats. It is a very variable species and has, as a result, accumulated a large number of synonyms. The only variant that can be upheld on current evidence is *C. szovitsii* subsp. *brachyphyllum* described below. It has been argued by some that the Bulgarian *C. diampolis*, which is included in synonymy here, differs in some respects. However, taking into consideration the wide range of extremes found in *C. szovitsii* over its distribution in the wild, it is difficult to uphold this particular taxon. Another is *C. ninae*, described from Armenia, which inhabits a similar habitat in the wild, namely wet, boggy ground. It is said to differ in that the flowers open out more or less flat and are star-like at maturity, but again we regard it as a synonym.

Because of its furrowed perianth tubes and basal auricles, *C. szovitsii* has sometimes been mistaken in the past for a *Merendera* (now included in *Colchicum*) which has the perianth tube split to the base (i.e. with free tepals). Closely related to *C. szovitsii* is the recently described *C. lagotum*.

In its various guises *C. szovitsii* is a fine plant for a bulb frame or pan culture under cold glass. Cultivars of *C. szovitsii* include 'Erich Pasche', 'Snowwhite', 'Tivi', 'Vardahovit' and 'Zigana'.

RIGHT An attractive selection of *Colchicum szovitsii* with deep pink flowers from Tunceli province in eastern Turkey.

FAR RIGHT A white-flowered variant of *Colchicum szovitsii*.

BOTTOM RIGHT As far as the eye can see, *Colchicum szovitsii* dominates alpine pasture in the mountains of Kahramanmaraş province in Turkey.

Colchicum szovitsii subsp. brachyphyllum (Boiss. & Hausskn.) K. Perss., *Bot. Jahrb. Syst.* 127(2): 221 (2007)

Type: Syria, Halab, Jan 1867, *Haussknecht 923* (lectotype G-BOISS)
Syns *C. brachyphyllum* Boiss. & Hausskn, *C. fasciculare* var. *brachyphyllum* (Boiss. & Hausskn.) Stef., *C. hydrophilum* Siehe, *C. libanoticum* Ehrenb. ex Boiss.

> **Flowering period:** Mid October–February
> **Habitat:** Heavy, damp, often basaltic soils in the mountains, stony ground, meadows, woodland margins
> **Distribution:** NW Jordan, Lebanon, N Palestine, Syria and SE Turkey; c. 900m
> **Cytology:** 2n=18

Although similar to *C. szovitsii* subsp. *szovitsii*, this subpsecies is a more vigorous, robust-looking plant, with three or four, narrow-ovate to lanceolate leaves that are scarcely developed at flowering time, and it has pale to deep pink, rather broader, oblanceolate, acute, pointed tepals. The anthers are dark grey or brownish purple. *Colchicum szovitsii* subsp. *brachyphyllum* replaces subsp. *szovitsii* in the south and south-east of the range.

Under the name *C. fasciculare* var. *brachyphyllum* this subspecies received an RHS Preliminary Commendation on 1 December 1959, when shown by Eliot Hodgkin. The bulbs had been sourced in Lebanon by Professor W.A. West, who

advised that the flower colour varied from white to pale pink, with April the peak time for blooming. It had been encountered on mountain slopes up to 1,850m, growing in places where in spring the ground was often saturated or actually flooded with water from the melting snows.

Edward B. Anderson (1973) mentions that in his bulb frames, raised 15cm above ground level on bricks or breeze blocks, and in a well-drained, limy soil, he grew *Colchicum szovitsii* subsp. *brachyphyllum*. He described it as 'A very handsome species whether white- or pink-flowered' from Lebanon.

RIGHT **The more acute tepals of *Colchicum szovitsii* subsp. *brachyphyllum*.**

FAR RIGHT **A white-flowered variant of *Colchicum szovitsii* in the Alborz Mountains of northern Iran.**

BOTTOM RIGHT **Wet habitats tend to be favoured by *Colchicum szovitsii*.**

Colchicum trigynum

Colchicum trigynum (Stevens ex Adams) Stearn, *J. Bot.* 72: 344 (**1934**)
Type: Georgia, no collector named but labelled '*Bulbocodium trigynum* m' (holotype LE; isotype K)
Syns *Bulbocodium trigynum* Stevens ex Adams, *Colchicum eichleri* (Regel) K. Perss., *C. greuteri* (Gabrieljan) K. Perss., *C. mirzoevae* (Gabrieljan) K. Perss., *Merendera candidissima* Miscz. ex Grossh., *M. caucasica* M. Bieb., *M. eichleri* Boiss., *M. ghalghana* Otsch., *M. greuteri* Gabrieljan, *M. mirzoevae* Gabrieljan, *M. trigyna* (Stevens ex Adams) Stapf.

Corm: Oblong-ovoid, 1.5–3.5cm long, adorned with a blackish brown, coriaceous, often multi-layered tunic, extended above into a relatively short neck, rarely much more than 3cm long
Cataphylls: Whitish, exceeding the corm neck by up to 6cm
Leaves: (2–)3(–4), present at flowering, then to 4cm long, not overtopping the flowers, at maturity spreading to recurving, linear to linear-lanceolate, 12–17 × 0.4–0.8cm, sometimes broader, pale to glaucous green, sometimes flushed crimson-purple at the margins, especially when young, glabrous, channelled, apex acute, margins thinly cartilaginous and scabrid
Flowers: 1–3(–5), funnel- to chalice-shaped, white to mauvish, purplish pink, occasionally bright purple; tepals split to the base, the claw greenish white, equalling or somewhat longer then the limb to 4cm long, occasionally longer
Tepals: Oblanceolate, generally narrowly so, 1.8–3.3 × 0.2–0.8(–1)cm, the base of the limb with interlocking auricles holding the tepals together, these denticulate or filiform, rarely more than 2mm long
Stamens: Filaments whitish or greenish yellow; anthers versatile, olive-brown, dark grey, purplish- or greenish-black, 2–3.5mm long, shorter than the filaments
Styles: Filiform, whitish, equalling the stamens or slightly longer
Fruit capsules: Ovoid to ellipsoid, 1.5–3.2cm long, located at ground level or slightly above
Habitat: Meadows, pastures, stony places, often by melting snow patches, gullies
Flowering period: (January–)February–June
Distribution: Armenia, Azerbaijan, N and NW Iran, SW, C, N and NE Turkey, 750–3,400m
Cytology: $2n=18, 22$

The tepals of *Colchicum trigynum* are split to the base.

This small, spring-flowering species exhibits a range of flower colour in the wild, and some of this is reflected in cultivated plants. It is closely related to *C. ignescens* and *C. raddeanum*.

Colchicum eichleri (syn. *Bulbocodium eichleri*) from central and eastern Caucasus is considered by some to be distinguishable from *C. trigynum*, citing its larger corms, longer, more falcate leaves that overtop the flowers and more numerous pale (white or palest mauve) linear, loosely connected tepals that readily fall apart. However, there is a considerable overlap in some of these characters and both are recorded from the same general vicinities in the Caucasus Mountains. In view of this, the present authors think that *C. eichleri* is best included in *C. trigynum* awaiting further detailed field studies.

RIGHT *Colchicum trigynum* growing at Çam Geçidi pass in Ardahan province, Turkey.

BOTTOM RIGHT & BOTTOM FAR RIGHT The typical pink-flowered *Colchicum trigynum* as well as white-flowered variants are cultivated.

Colchicum triphyllum

Colchicum triphyllum Kunze, *Flora* 29: 755 **(1846)**
Type: Spain, Andalucia, Sierra de Yunquera, 6,000–7,000ft, 1844, *Willkomm* 824 (holotype LZ – destroyed; isotype W)
Syns *C. ancyrense* B.L. Burtt, *C. biebersteinii* Rouy, nom illeg., *C. bulbocodioides* M. Bieb., non Brot., *C. catacuzenium* Heldr. ex Stef., *C. clementei* Graells, *C. holoophum* Coss. & Dur., *C. montanum* auct., non L.

Corm: Ovoid to oblong-ovoid, 1.5–2.5 × 1–1.5 (–2)cm, with a membranous, chestnut brown or yellowish brown tunic, neck short or absent
Cataphylls: Whitish, 5–9cm long mostly
Leaves: 3(–4), present at flowering, then shorter than the flowers, erect at first, then spreading, at maturity linear-lanceolate, 12–20 × 0.5–1.2(–1.9)cm, dull deep green, sometimes purplish at the tip, channelled, apex acute to subacute, margin minutely scabrid or smooth
Flowers: 1–5, occasionally more, globose-campanulate, pale lilac-pink to purplish pink, flushed purple towards the base, sometimes very pale, occasionally white, delicately veined; perianth tube white or cream, generally flushed purple towards the top, up to twice the length of the tepals
Tepals: More or less equal, the outer elliptic, the inner oblanceolate, all 1.5–3 × 0.5–0.9(–1.2)cm, glabrous, concave, especially in the upper half, often with filiform lamellae at the base, especially of the inner tepals, sometimes with basal auricles
Stamens: Filaments white or cream, glabrous, with a swollen yellow or orange-yellow base; anthers dark grey, purplish green or purplish black
Styles: Straight, equalling or slightly exceeding the stamens; stigmas punctiform
Fruit capsules: Ovoid to ovoid oblong, 1.5–3cm long, spongy, glabrous, apex apiculate, subterranean; seeds 4–5mm, narrowly keeled
Habitat: Mountain habitats, banks, open ground, roadsides, often on stony, gravelly or sandy substrates and near melting snow
Flowering period: Late February–April
Distribution: Algeria, Bulgaria, Crimea, Greece (mainly C and N Peloponnese and S Pindus mountains), North Macedonia, Morocco, NW Iran, Romania, W Russia, Sicily, C and S Spain, Tunisia, Turkey (W, C and S Anatolia), Ukraine; 700–3,000m, but close to sea level in Romania and Crimea
Cytology: 2n=60, 62

The goblet-shaped, flowers of *Colchicum triphyllum* adorn the bare ground at Karamanbey Pass in southern Turkey.

This is a delightful, neat little species with beautifully formed, goblet-shaped flowers and generally with three leaves, which give the species its epithet. It is irregularly widespread throughout the Mediterranean, extending north into the mountains close to the Black Sea. Along with *C. autumnale* it is probably one of the most wide-ranging of all the species in the genus. Although occasionally found at lower altitudes, *C. triphyllum* is essentially montane in its preferences.

Colchicum triphyllum has been in cultivation since 1938, although it has never been widely available. Edward B. Anderson (1973) records growing it in a bulb frame in a well-drained, limy soil. He described it as 'a winter-flowering species from the mountains of Spain and west North Africa. Flowers pale rose-pink, globular, marked with mauve outside'. In its finest forms it is a wholly delightful species for pan culture under cold glass or in a raised bed. The corms are slow to increase and tend to come into growth in late winter. They are probably at peak flowering in cultivation in late February and early March. There is a fine collection of different forms at Gothenburg Botanic Garden, Sweden, but it also appears quite often on the show benches of the Alpine Garden Society around the UK.

RIGHT & FAR RIGHT
Although it **has been in cultivation for 80 years**, *Colchicum triphyllum* **tends to only feature in specialist collections.**

BOTTOM RIGHT
Colchicum triphyllum **can produce a good number of flowers per corm.**

Colchicum troodi

Colchicum troodi Kotschy, in Unger, F. & Kotschy, C.G.T. *Ins. Cypern*: 190 (1865)
Type: Cyprus, 4,000ft, May 1862, *Kotschy 904* (lectotype W; isotypes JE, K, P, S)

Corm: Ovoid, 3–6cm long, with a blackish brown, papery-membranous, shiny tunic, extended above into a persistent neck, 5–9cm long, the inner tunics paler, reddish brown
Leaves: 3–6(–8), appearing after flowering, ascending to spreading, strap-shaped, mostly 12–20 × 1.5–4.5cm, dark green, glabrous, occasionally thinly pilose, flattish, not or scarcely channelled but with a distinct midrib, apex obtuse to subacute, ciliate along the margin
Flowers: 2–8, sometimes more, appearing in succession, narrow funnel-shaped at first, becoming starry, white to pale purplish pink; perianth tube white, slender, two to three times as long as the tepals, to 12cm long
Tepals: Narrow elliptic to narrow oblong-lanceolate, 2.5–4.5 × 0.4–1.2cm, glabrous, apex obtuse to subacute, filament channels glabrous or minutely pubescent
Stamens: One-third to half the length of the tepals, occasionally more; filaments very slender, white, glabrous or slightly hairy at the base; anthers slender, yellow, 6–8mm long
Styles: White, straight or somewhat curved, equalling or slightly overtopping the stamens; stigmas punctiform or very shortly decurrent for no more than 0.5mm
Fruit capsules: Ellipsoid, 1.5–3cm long, glabrous or minutely puberulous, apex beaked
Habitat: Maquis, phrygana, *Pinus* woodland, *Corylus* (hazel) groves, often in open sunny places
Flowering period: September–November, occasionally into December
Distribution: Cyprus (most prominent in the Troodos Mountains), endemic; near sea level to 1,620m
Cytology: 2n=54

The flowers of *C. troodi* vary enormously in overall size and in the relative width of the perianth segments. This variation is largely due to the vigour of individual plants and can be seen even within a single population, as can variation in the flower colour, from white to pink. The leaves are decidedly less variable, retaining much the same proportions in lush or impoverished plants. Likewise, there is little noteworthy variation in stamens and styles.

It is quite difficult to distinguish *C. troodi* from *C. decaisnei*, a mainland species found from northern Israel to southern Turkey. In *Flora of Cyprus*, Meikle (1985)

In it native Cyprus, pine woodland is a favoured habitat for *Colchicum troodi*.

comments that, 'Having examined a very wide range of specimens, I am convinced that the supposed differences between *Colchicum troodi* and *C. decaisnei* are illusory, and that, far from being an endemic as generally supposed, the former has a wide distribution in the eastern Mediterranean area'. However, Persson (1999b) argues that the differences between *C. decaisnei* and *C. troodi* are real and outlines the characters that separate them. These are outlined in this book under the former species.

RIGHT **The tepals of** *Colchicum troodi* **can be obtuse at the apex.**

FAR RIGHT **Flower colour variation is a characteristic of** *Colchicum troodi.*

BOTTOM RIGHT **A fine purplish pink variant of** *Colchicum troodi.*

Colchicum tunicatum

Colchicum tunicatum Feinbrun, *Palestine J. Bot., Jerusalem Ser.* 6: 87 (1953)
Type: Israel, Negev, N of Avdat, Sep. 1951, *D. Zohary s.n.* (lectotype HUJ)

Corm: Ovoid, (1.5)2-4cm long, adorned with a dark brown, coriaceous, many-layered tunic which is extended above into a thick neck, 3-7cm long
Cataphylls: White, 3-8cm long, almost completely enveloped by the corm neck
Leaves: (5-)6-9, appearing long after flowering and borne in a flat rosette, linear to lanceolate, the inner leaves progressively narrower, 7-15 × 0.3-1.2cm, channelled, slightly undulate, apex obtuse
Flowers: 2-6, often appearing with the remnants of the previous year's leaves still present, funnel-shaped, white to palest pink; perianth tube white, sometimes flushed pink at the top, shorter than the tepals
Tepals: Narrow-elliptic to elliptic-oblanceolate or almost oblong, 2.4-4 × 0.3-0.6cm, apex subobtuse to subacute, sometimes slightly apiculate
Stamens: About a third to half the length of the tepals; filaments white; anthers yellow, 3-4mm long
Styles: White, straight and relatively thick, generally exceeding the stamens, sometimes much longer, curved at the top; stigmas punctiform
Fruit capsules: Oblong, 2.5-3.5cm long, covered in tiny brown dots
Habitat: Rocky, gravelly and sandy places, steppe, bare desert hills, loess, *Artemisia* brush
Flowering period: September-mid October
Distribution: Israel, Jordan, Palestine, S Syria; 370-900m
Cytology: 2n=54

A small, desert species, this is easily overlooked when not in flower. It is noted particularly for its many-layered protective corm tunics that result from several years of accumulation. Leaf development is triggered by the onset of the winter rains.

It could be mistaken for smaller forms of *C. persicum* where they overlap in range, but that species has darker flowers and broader, more numerous, ascending leaves.

Colchicum tunicatum is often a denizen of bare, sandy places.

Colchicum turcicum

Colchicum turcicum Janka, Öesterr. Bot. Z. 23: 242 (1873)
Type: Turkey, Istanbul, Buyukdere, Sep. 1872, *Janka s.n.* (holotype CL; isotypes G, GOET, WU)

Corm: Ovoid to subglobose, with a coriaceous, dark blackish brown tunic, this extended above into a neck, up to 9cm long, occasionally more
Cataphylls: Whitish, 7–11cm long
Leaves: 5–9, appearing after flowering, ascending, narrow, linear-lanceolate to ligulate, 12–15 × 1.5–3cm, occasionally larger, glaucous-green, glabrous, pronouncedly twisted and undulate, apex subacute to obtuse, margin thinly cartilaginous and usually finely ciliate
Flowers: 3–8, funnel-shaped to more or less chalice-shaped, bright reddish purple, rarely paler, faintly tessellated; perianth tube white flushed purple towards the top, 3.5–6cm long
Tepals: Elliptic to narrow-obovate, 3–5(–6) × (0.35–)0.5–1.3cm, apex subacute to obtuse, the filament channels pubescent along the ridges
Stamens: One-third to half the length of the tepals, with glabrous, white or purple-flushed filaments and yellow anthers, 5–8mm long
Styles: Whitish, often flushed purple in the upper part, curved and slightly thickened at the apex, otherwise straight, overtopping the stamens; stigmas decurrent for 3–4mm
Fruit capsules: Oblong-ovoid to ovoid, 2–3cm long, glabrous, apex acute, located at ground level of slightly above
Habitat: Open oak woodland, coppices, wet meadows, fields, waste land
Flowering period: (mid August–)September–October
Distribution: SE Bulgaria, NE Greece and NW Turkey (in the region of the Bosphorus); low altitudes, sea level to 200m
Cytology: 2n=54

A fine, reddish purple selection of *Colchicum turcicum* propagated from the plants at Royal Botanic Garden Edinburgh.
INSET **Leaves of *Colchicum turcicum* in cultivation.**

This species is characterised by its relatively small flowers of an intense, deep reddish purple. These are normally lightly tessellated, although this may be difficult to discern. The glaucescent leaves with finely ciliate margins are also a useful diagnostic character. Its intense colour may have contributed to cultivars such as 'Benton End' and 'E.A. Bowles', but this has yet to be substantiated.

This species makes a useful garden plant, being particularly suited to pockets on a rock garden where it flourishes in a moist yet well-drained soil. Fine examples can be seen on the rock garden at Royal Botanic Garden Edinburgh.

Colchicum tuviae

Colchicum tuviae Feinbrun, *Palestine J. Bot., Jerusalem Ser.* 6: 79 (1953)
Type: Palestine, Wadi Tin near Mount Hordos, Nov. 1941, *Kushmir s.n.* (holotype HUJ)

Corm: Ovoid, 1.5-2.5cm long, with dark brown, coriaceous tunics, these extended upwards into a narrow neck, 2-5cm long and covering the lower third of the cataphylls
Cataphylls: White, often greenish towards the top, 5-10cm long
Leaves: 4-6(-9), partly developed at flowering, then shorter than the flowers, occasionally scarcely showing, at maturity linear-lanceolate, 10-18 × 0.3-0.5(-0.8) cm, glabrous or sometimes sparsely to quite densely pubescent on the dorsal or both surfaces, apex subacute to acute
Flowers: 2-5, funnel-shaped, but opening widely to star-shape, white or palest rose; perianth tube white, 2-4.5cm long
Tepals: Elliptic or oblong-elliptic, 1.5-2.2 × 0.3-0.5(-0.8)cm, apex subobtuse, sometimes shortly apiculate, ridged towards the base where the filaments merge with the tepals, and adorned along the basal ridges with prominent, dense, hair-like lamellae
Stamens: Filaments white, slender, about half the length of the tepals; anthers dark brown or brownish purple, *c.*2mm long
Styles: White, straight, equalling or exceeding the stamens; stigmas punctiform
Fruit capsules: Oblong, *c.*2.5cm long, located at ground level
Habitat: Calcareous steppes, often in *Artemisia-Phlomis* associations
Flowering period: November–January
Distribution: C Israel and Palestine (endemic to the Judaean and Negev Deserts); 370-520m
Cytology: $2n=14$

Colchicum tuviae is a delightful species, only 8–12cm tall overall. It is one of several in which the filament ridges are adorned with exceptionally well-pronounced, long and filamentous lamellae which are readily observed forming a hairy mass in the centre of the flower. It was named in honour of Tuvia Kushnir, a botanical student and talented young artist who was tragically killed in the Arab-Israeli War in January 1948.

Colchicum tuviae is closely related to *C. guessfeldtianum* and *C. ritchii*, the former found in the Sinai Peninsula and Jordan, the latter with a wider distribution that encompasses Egypt, Israel, southern Jordan, Palestine and Tunisia.

The hair-like lamellae at the base of the tepals of *Colchicum tuviae* are visible on this plant in the Judaean Desert.

Colchicum umbrosum

Colchicum umbrosum Steven, *Nouv. Mém. Soc. Imp. Naturalistes Moscou* 1: 264, t. 14 (1829)
Type: Ukraine, Crimea, Jaltam, Sep., *Steven* (lectotype H)
Syns *C. arenarium* var. *umbrosum* Ker-Gawl., *C. trapezunticum* Boiss. ex Baker

Corm: Ovoid to subglobose, 1.5–2.5 × 1–1.5cm, with a leathery or sub-leathery, blackish brown tunic and a long, slender neck
Cataphylls: White, slender, 5–11cm long
Leaves: 3–5, appearing after flowering, strap-shaped to narrow lanceolate, 8–15 × 1.3–2.2cm, glabrous, quite thick, apex obtuse
Flowers: 2–6, occasionally solitary, funnel- to star shaped, white to pale purplish pink; perianth tube white, shorter than, or equalling, the tepals
Tepals: Linear-elliptic to elliptic-oblanceolate, 1.6–3 × 0.2–0.6cm, the inner somewhat shorter than the outer giving the flower a slightly uneven appearance, apex subobtuse, somewhat incurved and hooded at the top
Stamens: Less than half the length of the tepals; filaments white; anthers pale yellow with a hyaline margin, 2.5–4mm long
Styles: White, shorter than or equalling the stamens, slightly curved at the tip; stigmas shortly decurrent for not more than 0.5mm
Fruit capsules: Elliptic-oblong, 2–4cm long, glabrous, apex acuminate
Habitat: Shady and part-shaded meadows, moist woodland (*Abies* and *Fagus*), shrubby and rocky places
Flowering period: August–September
Distribution: Caucasus and Transcaucasus, Crimea, S Russia, N Turkey (Anatolia); 150–1,400m
Cytology: 2n=24

First described from the region of Yalta in Crimea, this is a small species. It probably finds its closest ally in *C. arenarium* which is native to eastern and south-eastern Europe and has larger flowers with a punctiform stigma and narrower leaves. It is also related to *C. micranthum*, but there are differences in corm and leaf.

Colchicum umbrosum is of little horticultural value but present in collections at some botanic gardens. In recent years a new alkaloid compound has been extracted from the seeds of *C. umbrosum* and identified as 4-hydroxycolchicine.

The shorter inner tepals are noticeable on this plant of *Colchicum umbrosum*.

Colchicum varians

Colchicum varians (Freyn & Bornm.) Dyer, *Index Kew., Suppl.* 2: 45 (1904)
Type: Iran, Markazi, Kom to Sultanabad, Rahjerd, 1,600m, Mar. 1892, *Bornmüller 4729* (lectotype JE)
Syns *C. bifolium* var. *pleiophyllum* Bornm., *C. bakhtiaricum* Matin & Iranshahr.

Corm: Ovoid, often rather oblong, 2.5–5cm long, usually flattened on one side, covered in several layers of dark brown or reddish brown, thin or rather thick leathery tunics, extended upwards into a narrow neck, 1–6cm long
Cataphylls: Slender, white, to 12cm long, occasionally longer
Leaves: 4(–5), present at flowering and then up to 7cm long, not overtopping the flowers, at maturity spreading to recurved, linear to linear-lanceolate, 9.5–15 × 0.5–2cm, occasionally larger, grey-green, often flushed crimson-purple to the top, glabrous, channelled to flattish, margins finely undulate, sometimes slightly scabrid at the base
Flowers: 2–5, funnel- to somewhat chalice-shaped, white to pale pinkish purple, often flushed violet-purple in the lower part; perianth tube white, sometimes flushed with pinkish purple towards the top, 0.5–5cm long
Tepals: Oblanceolate to oblong-oblanceolate, 2–3.8 × 0.4–1cm, apex subobtuse to obtuse and often slightly cucullate, the inner at least, with auricled lamellae at the base along the filament channels
Stamens: To four-fifths the length of the tepals, more or less equal in length; filaments greenish or yellowish green, occasionally white, with a swollen green or yellow-green base; anthers buff or yellow, 2.5–5mm long
Styles: Straight, white to greenish or yellowish, equalling or slightly overtopping the stamens; stigmas punctiform
Fruit capsules: Ovoid-oblong, 1.4–2.5cm long, apex pointed or apiculate, mostly located at, or just below, ground level
Habitat: Stony slopes and shale and sandy areas, plains and hills, sometimes on cultivated, fallow or wasteland
Flowering period: February–May
Distribution: W and S Iran (Fars, Hamādan, Qom provinces); 2,300–2,800m
Cytology: 2n=46

The violet-purple flushing often seen in *Colchicum varians* is spectacular in this cutlivated specimen.

A western Iranian endemic species, this is primarily found in dry, sandy habitats and it generally has rather small, pale flowers.

At one time this species was confused with, and included in, *C. szovitsii* which is found further north in Iran, but the two are readily separated. *Colchicum szovitsii* has usually two leaves that enfold the flower perianth tubes for most of their length, these more erect at maturity and less spreading or recurved than in *C. varians*. In addition, *C. szovitsii* is a plant of damper habitats, streamsides and by melting snow at relatively higher altitudes, mostly at 1,600–3,600m in Iran at least. In both species the presence of small, tooth-like or filiform auricles can be observed at the base of the tepals, these sometimes confined to the inner tepals.

Christopher Grey-Wilson has observed *C. varians* in the wild (Grey-Wilson & Hewer 22) between Deh Bid and Shiraz in Fars province in southern Iran at 1,980m. There it was growing in considerable numbers in a dispersed colony in poor, very dry, sandy terrain. All the plants seen had white or very pale pink flowers.

RIGHT **Colchicum varians growing in western Iran.**

BOTTOM RIGHT **Colchicum varians in cultivation in Ukraine, grown from material collected in Iran.**

BOTTOM FAR RIGHT **A white variant of Colchicum varians in cultivation in Ukraine, also grown from Iranian material.**

Colchicum variegatum

Colchicum variegatum L., *Sp. Pl.* 1: 342 (1753)
Type: R. Morison, *Plantarum Historiae Universalis* 2: t.3, sect.4, fig.4 (1680), 'Colchicum Chionense floribus Fritillariae instar tessulatis, foliis undulatis' (lectotype), see Persson (2007)
Syns *C. agrippinum* auct., non hort. Angl. ex Baker, *C. chionense* Haw. ex Kunth, *C. parkinsonii* Hook. f., *C. tessellatum* Salisb., *C. variegatum* subsp. *parkinsonii* (Hook. f.) K. Richt, *C. variegatum* var. *desii* Pamp.

Corm: Ovoid to subglobose, 2–4 × 2–2.5cm, with a leathery, dark brown tunic, the long persistent neck, up to 15cm long
Cataphylls: Relatively stout, 8–22cm long
Leaves: 3–4, appearing in winter, patent, spreading close to the ground and forming a rosette, linear-lanceolate to ligulate, 9–15 × 1–2(–2.5)cm, glaucous, glabrous, undulate, often markedly so, apex obtuse to subacute, margin cartilaginous
Flowers: 1–3, star-shaped, white, heavily tessellated and flushed with violet-purple or deep red; perianth tube 4–7cm long
Tepals: Elliptic-lanceolate to elliptic, 4.5–7 × 0.8–2cm, the inner somewhat narrower than the outer, apex acute
Stamens: Half the length of the tepals; filaments whitish, spreading; anthers purplish black, 6–8mm long
Styles: Slender, white, flushed purple towards the top, equalling or slightly longer than the stamens, sometimes almost equalling the tepals, curving outwards, the tip slightly swollen, often curved; stigmas decurrent for 1.5–2mm
Fruit capsules: Oblong-ovoid, about 2cm long, glabrous, apex obtuse
Habitat: Woodland, scrub (*Erica* and *Juniperus*), maquis, *Pinus* or *Abies* woodland, rocky places on terra rossa
Flowering period: September–November
Distribution: S and SE Greece (S Peloponnese, Cyclades and Aegean islands) and SW Turkey; 150–1,450m
Cytology: 2n=54

A very striking species, this has large, starry, prominently tessellated flowers which set it apart from all the other autumn-flowering colchicums. The large flowers are spectacular, especially in the deep-coloured forms.

Colchicum variegatum is rarely seen in gardens because it requires a hot summer to ripen the corms. It is a fine subject, however, for pot culture or a bulb

The exotically tessellated blooms of *Colchicum variegatum* growing among rocks on the island of Chios in Greece.

frame where these conditions are more likely to be achieved. Despite this, the species has a long history of introduction and cultivation. It is a likely parent of C. × *agrippinum* which has also been cultivated for a long time. The hybrid is somewhat similar in appearance to C. *variegatum* but smaller in all its parts.

Reginald Farrer, who knew it under its synonym of C. *parkinsonii*, said in *The English Rock Garden* (1928) '*Colchicum parkinsonii* is often advertised, but the true plant hardly ever seen. It may, however, easily be recognised, for when, in spring, its leaves appear, they are but few in number, long and notably narrow, and they lie quite flat on the ground, undulating and wavy at the edge. Then, from the bare fields of Delos, Chios, and Naxos, there spring the flowers in early autumn, beautifully wide cups of deep lilac-rose vividly chess-boarded with squares of white. It may possibly only be a variety of C. *variegatum*'. He gives C. *variegatum* equal praise, saying 'It is the parent of C. *Parkinsonii*, having the same fantastic chequered blooms in autumn, but of a lighter pink, and rather larger, while the narrow leaves are held erect. It is a variable species of which forms sometimes appear as species – an instance being C. *chionense*'.

RIGHT **In cultivation *Colchicum variegatum* requires hot summers to ripen the corms.**

BOTTOM RIGHT **The long undulating leaves of *Colchicum variegatum* spread close to the ground.**

BOTTOM FAR RIGHT **This plant of *Colchicum variegatum* in Antalya province in Turkey has relatively narrow tepals.**

Colchicum wendelboi

Colchicum wendelboi K. Perss., *Fl. Iranica [Rechinger]* 170: 19 (1992)
Type: Iran, Fars, Shiraz to Kazerun, Feb. 1971, *Grey-Wilson & Hewer 35* (holotype K; isotypes E, GB, W)
Syns *C. caucasicum* sensu Stefanov, non. (M. Bieb.) Spreng., *Merendera caucasica* sensu. Boiss., non M. Bieb., *M. raddeana* sensu Boiss., non Regel, *M. wendelboi* (K. Perss.) Oganezova

Corm: Oblong to oblong-ovoid, 2–4cm long, covered in pale brown or reddish brown, membranous tunics, extended upwards into a slender neck, 3–7.5cm long
Leaves: 3(–4), appearing with the flowers but not overtopping them, at maturity ascending to somewhat arcuate, linear to linear-lanceolate, up to 9 × 0.3–1.1cm, rather pale to mid-green, channelled, margin often stained purplish crimson, narrowly cartilaginous, sometimes minutely scabrid, otherwise glabrous
Flowers: 2–4, occasionally solitary, goblet-shaped, rose-purple to pinkish lilac, rarely whitish overall; tepals split to the base, claws white, mostly 0.5–5cm long, interlocked at the top by small auricles, sometimes extending well up the blades, c. 2.5mm long
Tepals: Oblong to oblanceolate, 2–3.7 × 0.25–0.8cm, apex subacute or slightly acuminate
Stamens: Filaments white or yellowish white, sometimes flushed purple in the upper half, slightly swollen and pale yellow at the base; anthers greyish yellow to purplish grey, 3–4mm long
Styles: Filiform, straight, equalling or slightly overtopping the stamens; stigmas punctiform
Fruit capsules: Not recorded
Habitat: Rocky slopes in the mountains and valleys, often in damp spring soils, sometimes on alluvial soils or by melting snow
Flowering period: February–March
Distribution: W and S Iran (Fars and Kermanshah provinces, in or near the Zagros Mountains), endemic; 2,000–3,000m
Cytology: 2n=24

This species has been likened to *C. schimperi* in general habit and habitat preferences but is clearly separate on account of its split perianth tube. Among those species with split perianth tubes it comes closest to *C. trigynum*, but *C. wendelboi* has more membranous corm tunics and a more pronounced neck.
    Specimens of this species collected in 1971 by Christopher Grey-Wilson and

Colchicum wendelboi in the Zagros Mountains of Iran.

Professor Tom Hewer in southern Iran, between Shiraz and Kazerun, were used as the type specimen of *C. wendelboi* by Karin Persson when she described this species. Plants were growing in a wide, scattered community in very dry, rocky ground, flowering at the end of February.

RIGHT **The leaves of *Colchicum wendelboi* appear with the flowers.**

FAR RIGHT **Cultivated *Colchicum wendelboi* derived from material collected in the Zagros Mountains in Iran.**

BOTTOM RIGHT **A scattered colony of *Colchicum wendelboi* on a precarious substrate in the Zagros Mountains.**

Colchicum woronowii

Colchicum woronowii M.R. Bokeriya, *Bot. Zhurn. (Moscow & Leningrad)* 75(2): 201 (1990)
Type: Abkhazia, Tzebelda, Amtkel, 800m, Oct. 1986, *Bokeriya* 29 (holotype TBI; isotype LE)
Syns *C. liparochlamys* Woron. (nom. nud.), *C. liparochiadys* Woron. ex Czerniak. (nom. inval.)

Corm: Obcordate, 5 × 3.5–4cm, borne on an obtusely rostrate offshoot, with a thin, membranous, cinnamon-brown, lustrous tunic
Cataphylls: White, 8–13cm long
Leaves: Similar to those of *C. speciosum*
Flowers: 2–4, narrowly goblet-shaped, lilac-pink with a white throat and faint tessellations; perianth tube white, relatively slender, 3–3.5cm long
Tepals: Narrow, elliptic-obovate, obtuse, 3–5cm long
Stamens: Filaments white, slightly swollen and greenish yellow at the base; anthers linear, 6mm long
Styles: Linear, white or pink-flushed, equalling or slightly overtopping the stamens; stigmas decurrent
Fruit capsules: Not recorded
Habitat: Beech and chestnut woods and woodland margins
Flowering period: Late August–October
Distribution: W Caucasus (Abkhazia, Georgia); 300–1,700m
Cytology: $2n=42, 48$

An early rendition of this species was as a watercolour from 1908 held in the Leningrad herbarium labelled as *C. liparochlamys* Woron., but there is no validating description for this name. The name was then repeated, albeit as *C. liparochiadys*, by Czerniakovskaya in Komarov (1935) where there is a Russian description but no validating Latin description. It was finally named *C. woronowii* in 1990 (Persson 2007).

This species has been frequently placed as a synonym under *C. speciosum*. Apart from the slight disparity in chromosome number, which is $2n=38–44$ for *C. speciosum*, morphological differences have been noted, particularly in features of the corm and flower size. These are most clearly explained in *Flora of the USSR* (Komarov 1935) under its invalid name of *C. liparochiadys*. Further information can be found in Oganezova (2011).

Although a somewhat obscure species, *Colchicum woronowii* is in cultivation.

Colchicum × agrippinum

Colchicum × agrippinum hort. Angl. ex Baker, *J. Linn. Soc., Bot.* 17: 425 **(1879)**
Type: 'In hortis Anglis', not located
Syns *C. tessulatum* Mill., *C. tessellatum* hort. Angl. in Baker (1879) pro syn., non *C. tessellatum* Salisb. (1796)

Corm: Ovoid to broad ovoid, 5.5–6.5 × 4.9–6cm, often as broad as long, foot, 0.5–1.8cm long, neck slender, 3.5–6.2cm long
Leaves: 3–4, appearing in early to mid-winter, erect to ascending, often incurving, lowermost leaf elliptic-lanceolate, 26–29 × 3.5–4.7cm, uppermost leaf narrow elliptic-lanceolate, 19–22 × 1.8–2.1cm, dull bluish green, undulate
Flowers: Numerous, open, starry chalices, whitish with bold pinkish purple tessellation and flushing; perianth tubes white with some pinkish purple flushing towards the top, 3.5–6cm long
Tepals: Spreading widely apart, all about the same length, often slightly twisted, the outer oblong-elliptic to lanceolate-elliptic, 3.5–5.5cm long, the inner narrow elliptic-oblong
Stamens: Filaments pinkish purple, darker towards the base; anthers dull purplish brown
Styles: Long and slender, deep purple with hooked tips; stigmas shortly decurrent
Fruit capsules: Not produced
Flowering period: Late August–September
Origin: Not known from the wild but in cultivation for more than 150 years
Cytology: 2n=*c*.45

This is a fine garden plant with posies of small, tessellated flowers produced in great quantity. It is reported to be a hybrid, with a chromosome number midway between that of the putative parents (*C. autumnale* 2n=36, 38 and *C. variegatum* 2n=54). It is distinguished from the former by its earlier flowering time and tessellation, and from the latter by its erect and less undulate leaves (Mathew & Starling 1980). This hybrid appears to be sterile and is only known as a single clone.

Unlike *C. variegatum*, this hybrid is an easy garden plant, multiplying freely in a range of soils, provided they are not waterlogged. It is one of the first autumn colchicums to flower and deserves a place at the front of flower borders, but is especially effective on rock gardens or raised beds. It holds the RHS Award of Garden Merit and is rated at H5 (hardy in an average UK winter, to -15°C). It has been suggested by Mathew (1982) that the unusual name derives from the fact it might have originated in Cologne, known in Roman times as Colonia Agrippinensis.

Colchicum × agrippinum is an early-flowering hybrid inheriting its tessellation from C. variegatum.

Colchicum × alberti

Colchicum × alberti Regel, *Trudy Imp. S.-Peterburgsk Bot. Sada.* 8:647 **(1884)**
Type: Kyrgyzstan: "Pass Jassy zw. Urgent u. der Alabuga, alt. 9,000-11,000 ped.", 1 Jun. 1880, *A. Regel s.n.*, (lectotype LE)

Corm: Oblong-ovoid, 2.5–5cm long, with a membranous or somewhat coriaceous brown tunic
Cataphylls: White, to 17cm long, sometimes flushed purplish at the top
Leaves: Generally 3–5, partly developed at flowering but then shorter than the flowers, at maturity elliptic-oblong to elliptic-oblovate, to 20cm long, occasionally longer
Flowers: Pale yellow, cream or white, often with purplish flushing or lines on the exterior, funnel-shaped, opening widely to star-shaped in bright light
Tepals: Narrow-oblong to linear-lanceolate, 1.5–3.5cm long, ridged at the base within and with short lamellae along the ridges
Stamens: Filaments pale yellow, greenish or whitish; anthers basifixed, yellow, equaling the filaments
Styles: Greenish or yellowish, equaling or slightly overtopping the stamens; stigma punctiform
Fruit capsules: Not recorded
Habitat: As for the parent species
Flowering period: April–June, February–March in cultivation
Distribution: Likely to occur where the two species grow in close proximity in the wild in Central Asia
Cytology: Not recorded

Although first described as a species, *Colchicum alberti*, this plant is now recognised to be a natural hybrid between *C. kesselringii* and *C. luteum*. Natural hybrids between these two species have been recorded from the wild on a number of occasions and hybrids have also been created in cultivation. On the whole they more closely resemble *C. luteum*, but usually have pale yellow or cream-coloured, occasionally white, flowers, and with or without purple at the base of the tepals or along the perianth tube which is a character of *C. kesselringii*. Some plants in cultivation appear to be backcrosses with one or either parent and these look very similar to that parent. Introgression between the parent species and *C.* × *alberti* cannot be ruled out in some populations found in the wild..

Cultivars of this hybrid include 'Jānis', 'Jeanne', 'Lucky Selfmade' and 'Moonlight'.

Colchicum × alberti tends towards *C. luteum* in appearance but has paler flowers.

Colchicum × ambiguum

Colchicum* × *ambiguum Grey-Wilson, *Colchicum: The complete guide*: 557 (2020)
Type: Cultivated plant (accession no. W853856A) collected at RHS Garden Wisley, Surrey, UK, 2 Sep.. 1997 (holotype WSY)
Syn. *C. tenorei* (as *C. tenorii*) sensu J.G. Baker (1879) & E.A. Bowles (1952), non Parl.

Corm: Ovoid to oblong, 3.5–5cm long, with a chestnut-brown subcoriaceous tunic
Cataphylls: White, 4–9cm long, just reaching the soil surface
Leaves: (3–)4(-5), lowermost leaf narrow-elliptic to lanceolate-elliptic, 32–36.5 × 5.5–6cm, apex acute, often somewhat acuminate, uppermost leaf narrow lanceolate-elliptic, 28–38 × 3.4–3.6cm, apex acuminate, mid-blue-green, moderately shiny, with several longitudinal ridges
Flowers: 1–4, funnel-shaped, mid-rose-purple, faintly tessellated and with a median white stripe on the inner surface in the lower third to a half of each tepal; perianth tube white, 6.5–10.2cm long
Tepals: Elliptic-oblanceolate, 3.9–6.6 × 1.1–1.8cm, inner tepals slightly shorter than the outer, apex acute or subacute, sometimes shortly apiculate
Stamens: About one-third to a half the length of the tepals; filaments whitish, tinged purple towards the top, greenish and scarcely swollen at the base; anthers creamy-yellow to mid-yellow, 7–9mm long
Styles: Reddish purple, very slender, somewhat to markedly overtopping the stamens, slightly curved or hooked at the tip; stigmas punctiform
Fruit capsules: Not produced
Flowering period: August–early September
Origin: Not of wild origin
Cytology: 2n=72

The earliest material under the name *C. tenorei* (or *C. tenorii*) is found in the herbarium of Parlatore (FI) and consists of two Italian collections. The first is by Michele Tenore from 1858 (Italy, Basilicata, 'In pascuis Samnii Lucaniae, Marzo 1858') and collected as *C. bivonae*, and the second is by Giovanni Gussone from 1856 ('Napolia Castel di Sangro, Nov. 1856') and collected as *C. byzantinum*. Persson (2007) states quite categorically that both specimens are *C. cilicicum* and she selects the Tenore collection as a lectotype of *C. tenorei* Parl., although she regarded the name as a synonym of *C. cilicicum*, pointing out that 'What Tenore included as *C. bivonae* (in *Fl. Neapol. Prodr. App.* 5: 11, 1826), later taken

Although one of the smaller autumn-flowering hybrids, Colchicum × ambiguum is a prolific increaser.

up by Parlatore (1860) as a synonym under *C. tenorei* seems to be at least partly *C. lusitanum* Brot. as judged from description and localities'.

The origin of the plant currently grown in gardens as *C. tenorei* (*C. tenorei* hort.) is a bit of a mystery. There is clear evidence that it has been in cultivation since at least the end of the 19th century (Baker 1879). Bowles (1952) stated that he had '*C. tenorei*' in his garden that matched Baker's description, although there is no mention of anther colour. Baker's description (1879), under the name *C. tenorii*, was based on material from Henry Elwes's garden at Colesbourne, Gloucestershire, but no material relating to this has been located in the herbarium at Royal Botanic Gardens, Kew (K), where it might be expected to reside. Both Baker (1879) and Bowles (1952) stated that it originated in Italy, which is linked to the fact that they had identified it as *C. tenorei* Parl., but it is clear that such material, as already observed, is probably referable to *C. cilicicum*. That species was known to be in cultivation in Italy at the time and D'Amato had indicated (1957b) in his revision of Italian species of *Colchicum* that *C. cilicicum* had 'escaped from cultivation and perhaps naturalised'.

Persson (2007) found that material of *C. tenorei* hort. has a chromosome number of 2n=72. No other *Colchicum* species to date has been found to have 2n=72, so this suggests a hybrid origin for the cultivated plant. It appears to be a hybrid with *C. cilicicum* (2n=54) as one possible parent and a species with a higher chromosome number as the other. Persson (2007) suggested that the other parent might be *C. lusitanum*, which is native to the western and central Mediterranean region including Italy, and was certainly cultivated at the time. It has a chromosome count of 2n=108, and in the genus as a whole only it and *C. balansae* have this chromosome count, and both species are autumn-flowering. There is no indication that *C. balansae*, native to Greece and Turkey, was in cultivation in Italy in the 19th century or was naturalised. *Colchicum lusitanum* would have been present, so it is possible that it is the other parent. However, the numbers do not add up: *C. cilicicum* (2n=54) × *C. lusitanum* (2n=108) would give a chromosome count of 2n=81, if no chromosome material is lost in the process. There is another contender, *C. neapolitanum*, which is also native to parts of northern Italy and elsewhere. This has a chromosome count of 2n=90 and, crossed with *C. cilicicum*, would give a count of 2n=72, a perfect match for *C. tenorei* hort. At the same time, the characteristics of the cultivated plant do, in many respects, fall midway between these two possible parent species. It is also sterile and not known to produce fruit, a feature indicative of its hybrid origin.

The name *C. tenorei* as described by Parlatore cannot be used for this likely hybrid in cultivation under that name, as it is clear that the Tenore specimens upon which the Parlatore name is based are actually *C. cilicicum*. There is no indication from Italian authors, as far as can be ascertained, that there might be a hybrid in the wild. It therefore stands to reason that a new name has to be applied to the hybrid (*C. tenorei* hort.) and *C.* × *ambiguum* is here designated.

RIGHT **The faintly tessellated flowers of *Colchicum* × *ambiguum* have a white median stripe to each tepal.**

BOTTOM RIGHT ***Colchicum* × *ambiguum* is vigorous enough to naturalise in short grass.**

The prime characters of *C.* × *ambiguum* are the yellow anthers and the reddish purple, hooked styles, along with the faintly tessellated tepals. Baker (1879) treated *C. tenorei* (as *C. tenorii*) as a distinct species, with *C. bivonae* sensu Tenore and *C. byzantinum* sensu Tenore as synonyms. The description is ambiguous, stating quite clearly 'antheris purpurascentibus' whereas they are actually yellow, but is correct with 'styli purpurascentes' and flowers as 'lilacino-purpureus obscure tessellatus'. Bowles (1952) also stated that the plant is distinctive for the 'crimson colour of the stigmatic crook'. Persson (2007) notes that Baker's description of *C. tenorei* 'is at variance with Parlatore's original description and herbarium material. Again there seems to be confusion with *C. lusitanum* which generally has purplish or fuscous anthers'. Further discussion can be found in Govaerts & Persson (2008).

In the UK at least, there only seems to be one clone of *C.* × *ambiguum* in cultivation. Although relatively small-flowered and compact compared to the other autumn-flowering colchicums, *C.* × *ambiguum* is a prolific increaser and quite widely cultivated. It flowers at around the same time as *C. cilicicum*. Until recently there was a fine stand of of this hybrid, under the name *C.* × *tenorei*, at the National Trust's Felbrigg Hall garden in Norfolk. The planting was mainly concentrated as a band, many metres long, following a box hedge in the walled garden. Unfortunately, the *Colchicum* collection at Felbrigg Hall is now much diminished. Currently, under the name *C.* × *tenorei*, it holds the RHS Award of Garden Merit and is rated at H5 (hardy in an average UK winter, to -15°C). This award now applies to *C.* × *ambiguum*.

Colchicum × *ambiguum* is sometimes confused with the cultivar 'Pink Star', but for the differences between them see the Cultivars chapter.

RIGHT **Although forming substantial clumps, *Colchicum* × *ambiguum* is quite a compact plant.**

BOTTOM RIGHT **Yellow anthers and faint tessellation are characters of *Colchicum* × *ambiguum*.**

Colchicum × byzantinum

Colchicum × byzantinum Ker Gawl., *Bot. Mag.* 26: t. 1028 **(1807)**
Type: Clusius, *Rar. Pl. Hist.*, upper left illustration on p. 200
Syns *C. autumnale* var. *major* hort., *C. autumnale* var. *patens* (Schultz) Rouy, *C. patens* Schultz, *C. veratrifolium* S. Arn.

Corm: Ovoid to subglobose, 5–7.5 × 7–11.5cm, broader than long, foot short and broad, 0.5–2.2cm long, neck stout, 8.4–11.2cm long
Cataphylls:?
Leaves: 4–5(–6), appearing in late winter and spring, ascending, lowermost leaf elliptic-oblanceolate, 36–44.5 × 10.8–14cm, uppermost leaf narrow elliptic, 38–41.5 × 3.5–4.8cm, deep, slightly blue, green, moderately shiny, generally with several longitudinal folds
Flowers: (1–)2(–many), wide goblet-shaped, pale to mid-pink, unscented, each tepal with a white, often rather vague stripe from the base reaching two-thirds to the top, each tepal tipped with a tiny purple dot; perianth tube white, 5–9.5cm long (longer in semi-shaded positions in the garden)
Tepals: Outer tepals elliptic-obovate, 5–7 × 1.5–3.5cm, apex obtuse, inner tepals slightly shorter than the outer, oblong-obovate, apex obtuse
Stamens: Filaments white; anthers yellow
Styles: Overtopping the stamens, white with a purple tip; stigmas shortly decurrent
Fruit capsules: Not or rarely produced
Flowering period: Late August–September
Origin: Not known from the wild but in cultivation for more than 400 years
Cytology: 2n=45

This is a reliable and floriferous hybrid, with attractive, though rather pale, flowers, that tolerates a wide range of garden conditions, including sun or dappled shade. The handsome leaves are some of the largest of the autumn-flowering colchicums.

Colchicum × byzantinum has long been known in cultivation. It was first named in 1601 by Carl Clusius after the ancient city of Byzantinum that was founded by the Greeks and is now Istanbul in Turkey. Clusius's name was formally established as a binomial in 1807. Although the name suggests a Turkish origin it has never been found in the wild. Past records of *C. × byzantinum* from Turkey can mostly, if not all, be refered to *C. cilicicum*, with which it has been much confused.

It is presumed to be of hybrid origin, with *C. cilicicum* as one of the parents. The fact that it is thought to have arisen in gardens means that another species must have been cultivated with *C. cilicicum* all those years ago. Karin Persson

Colchicum × byzantinum is an ancient hybrid known in cultivation for more than 400 years.

(2007) has suggested that if *C. cilicicum* (2n=54) is one of the parents then the other parent must have a chromosome number of 2n=36, as the hybrid has 2n=45. This revelation narrows the possibilities, and as only *C. autumnale* was widely cultivated at the time that would seem to be the likely contender.

Colchicum × *byzantinum* is excellent for naturalising between shrubs or for grouping at the front of a flower border, but it is also well suited to naturalising in grass. Mature corms, which can be the size of a clenched fist, produce a large number of flowers, making it especially appealing.

It was at one time widely sold as a windowsill novelty. The large corms would be placed in a dish on a sunny windowsill where they would flower happily and then could subsequently be planted in the garden.

This hybrid formerly bore an AGM but this was rescinded at the recent RHS trial (2014–2017) held at Hyde Hall in Essex. This was primarily because the flower colour was considered to be rather insipid and it can not compete with more modern, better coloured selections.

Reginald Farrer was reasonably complimentary of it in *The English Rock-Garden* (1928), saying: '*Colchicum byzantinum*: …may also be known by its knobbly irregular great corm, the size and shape of a closed fist. Its flowers are after the giant style of *C. Bornmülleri*, but rather smaller and paler, with the segments shorter, broader and blunter than in *C. speciosum* and *C. latifolium*. The leaves are big and corrugated, following in spring after the blossoms, which have been abundantly and unanimously produced in autumn'.

Colchicum × *byzantinum* is only known as a pale pink-flowered clone and the white-flowered cultivar 'Innocence'. The latter has purple tepal tips and purple style tips with occasional aberrant pink tepals. Plants offered as 'Album' with supposedly pure white flowers are likely to be 'Innocence', as the purple tipping is sometimes barely visible and varies according to plant and season.

RIGHT **Mature corms of *Colchicum* × *byzantinum* can produce an impressive number of flowers.**

FAR RIGHT **The purple tips of the tepals and styles of *Colchicum* × *byzantinum* are obvious in the white-flowered 'Innocence', which can also have odd pink tepals.**

BOTTOM RIGHT **The front of a border makes a good location for *Colchicum* × *byzantinum*.**

Cultivars

It is the cultivars of *Colchicum* that provide the most rewarding garden plants. This is because they are generally easier to grow than the species, are more vigorous and have larger flowers. The majority of cultivars are autumn-flowering hybrids of unknown parentage, but with primarily *C. autumnale*, *C. bivonae* and *C. speciosum* in their ancestry, with also *C. cilicicum* and *C. turcicum* playing a part in some. A smaller number of cultivars, some of which are spring-flowering, are direct selections of species, or belong to one of the few named hybrids such as *C.* × *alberti* or *C.* × *byzantinum*. These ones are listed first in this chapter, alphabetically by species or hybrid name. These are followed by the hybrid cultivars in alphabetical order by cultivar name.

DISCOVERY

For each cultivar an indication of its origin is given by citing the name of the nursery or individual responsible for its introduction. For those where this information is not known, the country in which the cultivar first arose is provided. More detailed information, such as the names of individuals involved in the discovery, as well as the location of relevant nurseries and the route of the cultivar into horticulture, is given in the Checklist.

DATE

The date attributed to the cultivar is the earliest confirmed date for the cultivar being in existence. In the case of mature plants that have later been recognised as cultivars, the date attributed is when the name was given to that cultivar.

COMPARABLE CHARACTERS

Characters that are helpful for comparing the traits of each cultivar are given in a standardised form for most entries. The abbreviations used are:

HL = Height in mature leaf in cm
HF = Height in flower in cm
F = Flowering period by month 1–12, e = early, m = mid, l = late

A selection of *Colchicum* cultivars, hybrids and species photographed on 13 September 2018 at the trial at RHS Garden Hyde Hall.

TOP ROW
LEFT to RIGHT
C. 'Benton End'
C. speciosum
C. × *agrippinum*
C. × *byzantinum* 'Innocence'

SECOND ROW
LEFT to RIGHT
C. 'Felbrigg'
C. 'Little Woods'
C. 'Waterlily'
C. 'Rosy Dawn'

THIRD ROW
LEFT to RIGHT
C. autumnale
C. × *ambiguum*
C. 'Glory of Heemestede'
C. 'Pink Star'

BOTTOM ROW
LEFT to RIGHT
C. 'Autumn Queen'
C. × *byzantinum*
C. autumnale 'Nancy Lindsay'
C. cilicicum 'Purpureum'

RHS AWARD OF GARDEN MERIT

The RHS Award of Garden Merit (AGM) is given to plants considered excellent for ordinary use in appropriate garden conditions. The award is currently held by 20 colchicums; 18 cultivars, and two that are simply known under their hybrid binomials and covered in the Species chapter, *C.* × *agrippinum* and *C.* × *ambiguum*. Cultivars with an AGM are indicated in the headings, alongside an RHS hardiness rating. Hardiness ratings in the RHS system range from H1 to H7 and all colchicums with an AGM are classed as H5 (hardy in a cold UK winter, to -15°C).

IDENTIFICATION NOTES

In the cultivar descriptions, the following should be noted:

1. Leaf measurements are given for fully grown leaves of mature plants. In some cultivars the leaves are all very similar in size and shape, but in others the uppermost leaf can be significantly smaller and narrower than the lowermost. Leaves of most of the large-flowered cultivars are not normally fully developed until mid-spring. The time of their development varies considerably from one cultivar to another. For example, those of 'Poseidon' start to develop shortly after flowering and are semi-mature by late winter, while those of 'Glory of Threave' do not appear above ground until late winter and are not mature until mid-spring.
2. Corm details are for mature flowering corms. Foot (hypopodium) and neck details can vary considerably within the same group of plants.
3. Flower shape can vary, dependent on whether the bloom is newly opened or mature. Unless otherwise stated, flower shape is for fully mature blooms, as are tepal measurements.
4. Flower colour can vary, not only according to whether plants are in sun or shade, but also on flower maturity and time of day.
5. Perianth tube length is taken from ground level to the base of the tepals. The perianth tube extends considerably further underground through the centre of the cataphyll.
6. Style length often increases as flowers mature. The style often starts shorter than the stamens but ends up being equal or longer, and sometimes considerably longer than the stamens.
7. Tepal tessellations often become more marked as the flowers mature.
8. Anther colour refers to the colour before dehiscence. Once dehisced, anthers are often masked by the pollen which is nearly always yellow, orange-yellow or occasionally cream.
9. Some cultivars, such as 'Daendels', 'Glory of Threave' and 'Rosy Dawn', regularly set fruit. Others scarcely ever do, or not at all, but this can vary to some extent from year to year.

Cultivars considered excellent for the garden, such as *Colchicum autumnale* 'Nancy Lindsay', have the RHS Award of Garden Merit.

BOTANICAL DESCRIPTIONS

We have been able to provide full botanical descriptions for the majority of hybrid cultivars. These descriptions are based on verified plants that were cultivated in the RHS trial of large-flowered *Colchicum* at RHS Garden Hyde Hall in Essex during the period of 2014 to 2018 (Grey-Wilson 2019) and on plants from private collections with reliably identified material.

TABLE OF CHARACTERS

The table on the following two pages is intended as a quick guide to locating cultivars with particular characteristics for the garden. As well as cultivar name and species attribution it gives flowering period and details of the flowers such as whether double-flowered or marked with tessellations. No attempt has been made to distinguish between the different shades of pink to purple in the table, but if significantly bicoloured with a white centre to the flower or with white tips to the tepals, this is indicated. Those with white or yellow flowers are indicated.

Characters of *Colchicum* cultivars, including AGM

Cultivar	AGM	Species	Autumn-flowering Aug	Sept	Oct	Nov	Spring-flowering	Double-flowered	White	White-centred	White-tipped	Yellow	Tessellated
'Alboplenum'		*C. autumnale*		●	●				●	●			
'Album'	H5	*C. autumnale*		●	●				●				
'Album'	H5	*C. speciosum*		●	●				●				
'Anopolis'		*C. macrophyllum*	●	●							●		●
'Antares'		*Colchicum*		●	●						●		●
'Apollo'		*C. bivonae*	●	●							●		●
'Artur Klark'		*Colchicum*		●	●						●		
'Atrorubens'	H5	*C. speciosum*		●	●						●		
'Autumn Herald'		*Colchicum*		●	●						●		●
'Autumn Queen'	H5	*Colchicum*		●	●						●		●
'Beaconsfield'		*Colchicum*		●	●						●		
'Benton End'	H5	*Colchicum*		●	●						●		●
Bornmuelleri Group		*C. speciosum*		●	●						●		
'Boxford'		*Colchicum*		●	●						●		●
'Chequers'	H5	*C. speciosum*		●	●						●		●
'Constable'		*Colchicum*		●	●						●		●
'Cretan White'		*C. macrophyllum*	●	●					●		●		
'Daendels'		*Colchicum*		●	●						●		●
'Darwin'		*Colchicum*		●	●	●					●		
'Dick Trotter'		*Colchicum*		●	●						●		
'Disraeli'		*Colchicum*		●	●						●		●
'Dombai'		*C. speciosum*		●	●						●		
'Dorothee Kersen'		*C. autumnale*	●	●	●						●		
'E.A. Bowles'		*Colchicum*		●	●						●		
'E.K. Balls'		*Colchicum*	●	●							●		
'Emerald Town'		*Colchicum*		●	●						●		●
'Enigma'		*Colchicum*		●	●						●		
'Fabergé's Silver'		*Colchicum*		●	●						●	●	
'Felbrigg'	H5	*Colchicum*		●	●						●		
'Felbrigg Violet'		*Colchicum*		●	●						●		
'Flamenco Dance'		*Colchicum*		●	●						●		
'Fuller's Mill'		*Colchicum*		●	●						●		
'Giant'		*Colchicum*		●	●						●		●
Giganteum Group		*C. speciosum*		●	●						●		
'Glory of Heemstede'		*Colchicum*		●	●						●		
'Glory of Threave'	H5	*Colchicum*		●	●						●		
'Golden Baby'		*C. luteum*					●					●	
'Gothic Style'		*Colchicum*		●	●						●		
'Gracia'		*Colchicum*		●	●						●		
'Harlekijn'		*Colchicum*		●	●					●	●		
'Herbstkugel'		*Colchicum*		●	●						●		
'Hora Sfakion'		*C. macrophyllum*	●	●							●		
'Huxley'	H5	*Colchicum*		●	●					●	●		
'Innocence'	H5	*C. × byzantinum*		●	●				●		●		

Cultivar	AGM	Species	Aug	Sept	Oct	Nov	Spring-flowering	Double-flowered	White	White-centred	White-tipped	Yellow	Tessellated
'Jānis'		C. × alberti						●				●	
'Jarka'		Colchicum		●							●		●
'Jenny Robinson'		Colchicum		●							●		●
'Joseph'		Colchicum		●							●		
'Kiss Me Quick'		Colchicum		●							●		
'Larisa'		Colchicum		●							●		●
'Lilac Bedder'		Colchicum	●	●									
'Lilac Wonder'		Colchicum		●									●
'Little Woods'	H5	Colchicum		●							●		
'Lucky Selfmade'		C. × alberti						●				●	
'Lysimachus'		Colchicum		●							●		
'Maximum'		C. speciosum		●							●		
'Mells Park'		Colchicum		●							●		
'Mount Etna'		C. bivonae	●	●									
'Nancy Lindsay'	H5	C. autumnale		●									
'Neptun'		Colchicum		●							●		
'Norman Barratt'		C. montanum											
'Oktoberfest'		Colchicum			●						●		●
'Papa Rema'	H5	C. bivonae	●	●									
'Paul Furse'		C. speciosum		●							●		
'Paul Furse Early'		C. speciosum		●							●		
'Pink Goblet'	H5	Colchicum		●							●		
'Pink Star'	H5	Colchicum		●							●		
'Pleniflorum'		C. autumnale		●				●					
'Poseidon'		Colchicum		●							●		
'Pride of Holland'		Colchicum		●							●		
'Purpureum'	H5	C. cilicicum		●									●
'Redgrave'		Colchicum		●							●		
'Revelation'		C. speciosum		●									
'Rosy Dawn'	H5	Colchicum		●							●		●
'Rubrum'		C. speciosum											
'Snow of Highland'		C. kesselringii					●		●				
'Snowwhite'		C. szovitsii					●		●				
'Spartacus'		Colchicum		●							●		●
'Tivi'		C. szovitsii					●						
'Vakhsh'		C. luteum					●					●	
'Valentine'		C. doerfleri					●						
'Vardahovit'		C. szovitsii					●						
'Waterlily'	H5	Colchicum		●				●		●			
'Whitton Globe'		Colchicum		●							●		
'William Dykes'		Colchicum		●							●		●
'World Champion's Cup'		Colchicum		●							●	●	
'Yeti'		C. kesselringii					●		●				
'Zephyr'		Colchicum		●							●		

Colchicum × agrippinum and Colchicum × ambiguum also hold the AGM and are H5

Colchicum × alberti

'Jānis'

This natural hybrid between *C. kesselringii* and *C. luteum* has pale yellow flowers. It was discovered in 1989 by Arnis Seisums, along with a similar hybrid later called 'Jeanne', growing wild near the airfield of Tovil'-Dora in Tajikistan. When first exhibited in the UK in 2001 it was listed as 'clone no. 1' (Rolfe 2001).
Seisums, 1989

'Lucky Selfmade'

This hybrid has pale yellow flowers that have some greenish spotting at the base of the tepals. The leaves are relatively broad and can exceed 30cm in length. A cross between *C. kesselringii* 'Snow of Highland' and *C. luteum*, it was raised by Leonid Bondarenko of Lithuanian Rare Bulb Garden nursery. A similar hybrid using *C. kesselringii* 'Yeti' instead is called 'Moonlight' and has darker yellow flowers.
Bondarenko, early 21st century

Colchicum autumnale

'Alboplenum'

The double, white flowers of this cultivar are larger than those of the typical species and with inner tepals that are shorter than the outer ones and of uneven length. The flowers are quite heavy and tend to fall over in time, or after heavy rain. In most fully double flowers the stamens are replaced by extra perianth parts. In 'Waterlily' no trace of stamens can be found, but in this cultivar several of the inner tepals are accompanied by a perfectly formed stamen, these about a quarter the length of the tepals and with a yellow anther, 6-8mm long. See the entry for 'Waterlily' for a fuller discussion of flower doubling in *Colchicum*.

There are suspicions that there are at least two, perhaps more, clones of double white *C. autumnale* in circulation, as not all plants have stamens and the flower shape and size is quite variable. Despite this, they are all sold as 'Alboplenum'. The double pink cultivar of *C. autumnale*, 'Pleniflorum', is far more uniform in its tepal arrangement.

Flowers 1-5, occasionally more, mostly 9-11cm across with numerous tepals forming a rather uneven white cluster; tube (4–)6-10cm long, somewhat twisted; tepals markedly uneven, the outer 6-9 the largest, elliptic-oblanceolate, 6.8-7.4 x 1.2-1.4 cm, subobtuse, spreading widely apart, the inner numerous, ascending to upright, very uneven, the larger elliptic-oblanceolate to linear-elliptic, 2.5-4.2 x 0.7-1 cm, the smallest 2.4-3.5 x 0.2-0.4cm, acute, often with an anther, 6-8mm long, or part anther, attached.
Origin unknown, pre-1872
HL: 25-35 HF: 9-15 F: m9-m10

'Album' AGM H5

Widely grown in gardens, this is a proven cultivar of long-standing that multiplies well. The creamy-white flowers are at their best when they form tight posies. They are borne on perianth tubes that are twice as long as the tepals, sometimes longer, and with a slight tinge of green towards the base. From three to 12 flowers can be produced per corm. As they age, the outer flowers fall prostrate.

The flowers can look slightly greyish in some light and this has given rise to the cultivar name 'Old Bones', generally now considered to be a synonym of 'Album'. They contrast with those of *C. speciosum* 'Album' which are both larger and purer white. Having said that, *C. autumnale* 'Album' as sold today is not uniform and it is wise to select out the better forms when they are in flower. The best have flowers with larger and broader tepals, giving them a more rounded appearance. White-flowered variants of *C. autumnale* can be seen in the wild occasionally and certainly pop up in naturalised colonies in Britain and elsewhere, so it is hardly surprising that they reveal quite a lot of variation.

Plants to 35cm tall in leaf. Leaves 3–4, ascending, deep somewhat dull green with a slight sheen, somewhat fluted and twisted, the largest narrowly oblong-elliptic, 27–30.8 × 4–5.8cm. Flowers 3–12, narrow goblet-shaped at first, opening widely at maturity, creamy white at first but ageing to greyish white; tepals elliptic to elliptic-oblong or elliptic-obovate, mostly 3–4 × 1–1.4cm, outer ones longer than inner ones; tube white with a slight tinge of green towards the base, 11–16.5cm long; stamens with white filaments tinged green towards the base, anthers pale creamy yellow; styles white, hooked at the tip. Fruit capsules produced occasionally. Origin and date unknown
HL: 35 HF: 9–14 F: 9–m10

'Dorothee Kersen'

Very similar to 'Nancy Lindsay' in flower, this variegated cultivar has pink flowers with darker perianth tubes but with bicoloured, creamy yellow and green leaves. The variegation is a narrow or broad marginal band, affecting every leaf to some degree. Occasionally the central green zone may also reveal some pale yellowish lines. It belongs to *C. autumnale* subsp. *pannonicum*.

Kersen, 2019
HL: 22–28 HF: 8–13 F: l8–9

'Nancy Lindsay' AGM H5

The flowers of this cultivar are a good, attractive pink with deep pink perianth tubes and stand up well to the vagaries of the weather. It is a fine and reliable cultivar for a variety of positions in the garden, including naturalising in grass, and bulks up well.

Nancy Lindsay, daughter of socialite garden designer Norah Lindsay, is presumed to have collected it in 1936 in Romania, while travelling back from a plant-collecting expedition to Iran with Alice Fullerton. It belongs to *C. autumnale* subsp. *pannonicum*.

Leaves 5–6(–7), elliptic, the largest to 32 x 6.9cm, mid-yellow-green to somewhat glaucous green. Flowers 5, narrowly chalice-shaped, deep lilac-pink; tepals 5.4–6cm long, faintly tessellated on the outside, inside with a white stripe for up the half the length; anthers pale yellow; styles white with a hooked tip; perianth tube 8.8–10.6cm long, pink, paling towards base.

Lindsay, 1936
HL: 32–45 HF: 12–17 F: I8–9

'Pleniflorum'

This double-flowered cultivar is similar to white-flowered 'Alboplenum', but the flowers are a soft, mid-pink and generally rather more even. There are traces of anthers attached to some of the innermost tepals but this is less obvious in this cultivar than in 'Alboplenum'. See the entry for 'Waterlily' for a fuller discussion of flower doubling in *Colchicum*.
Origin unknown, pre-1874
HL: 35-45 HF: 8-12 F: 9-e10

Colchicum bivonae

'Apollo'
This fine cultivar reflects the prime characteristics of the species, with lightly scented flowers, up to six per corm. They are funnel-shaped, the tepals more or less equal, 6–8cm long, and whitish with bright violet-rose or purple-violet flushing and tessellations. Each tepal has a slight central fold running lengthwise. Within, the flowers have a white, star-shaped centre with the star points stretching two-thirds the length of the tepals as a thin white band. The anthers are purple-brown and the styles are thickened and very curved at the tip, overtopping the stamens. The perianth tubes are greenish white.

It was originally grown as an unnamed selection by Van Tubergen, but named by Antoine Hoog in 1995 when introduced by Hoog & Dix Export, the Netherlands.

Van Tubergen, 1995
HL: 38–45 HF: 12–18 F: 18–9

'Mount Etna'

As its name implies, this is a cultivar derived from an Italian collection of *C. bivonae*. The flowers are rather paler than those of most selections of *C. bivonae* grown in gardens and lightly tessellated, but pretty none-the-less.
Flowers usually 2 to 5 per corm, goblet-shaped, bright rose-purple and moderately tessellated towards the margins of the tepals, basal one-third of tepals white on both sides; outer tepals elliptic-obovate, 6.2–7.5 × 2.8–3.6cm, apex obtuse; inner tepals elliptic-obovate, 5.9–6.2 × 2.7–3.1cm, apex obtuse; perianth tube greenish white, 6–7.8cm long, about half the length of the tepals; stamens with greenish white filaments and purple-brown anthers; overtopped by white styles.
Origin and date unknown
HL: 35–48 HF: 12–15 F: 18–9

'Papa Rema' AGM H5

While many of the cultivars of *C. bivonae* have not thrived in gardens or perform poorly, 'Papa Rema' is an exception and is the finest so far. It is vigorous, early and free-flowering, soon multiplying to substantial clumps in the open garden.

It originates from a collection made by Peter and Penny Watts in the Papa-rema Gorge on the slopes of Mount Olympus in Greece at an altitude of about 850m. It has performed exceptionally well in the garden of Robin White, who used to run Blackthorn Nursery, in Hampshire.

Corms large, 5–6.4 × 5.2–7.2cm, asymmetrical, the foot indistinct, occasionally to 1cm long, the neck relatively stout, 3.5–6.5cm long. Leaves 12–30cm long. Flowers 2 to 6 per corm, often appearing more numerous due to clumping of the corms, relatively large, goblet- to funnel-shaped, deep rose-purple with strong, chequered tessellation and a white centre reaching about one-third the distance up each tepal; tepals narrowly to broadly elliptical, occasionally somewhat obovate, 6–8cm long, usually with one or two longitudinal folds; anthers deep brownish purple; stamens much exceeded by the very slender, slightly curved styles that are slightly swollen at the apex and with the stigma decurrent for 2–4mm.
Watts, 2018
HL: 45 HF: 12–30 F: m8–9

Colchicum × byzantinum

'Innocence' AGM H5

The white flowers have small purple tips to each tepal and the tip of the style, a feature that is more pronounced in some plants than others. This variation may reflect growing conditions, as it varies from year to year in the same plants. In addition, plants may produce partly pinkish purple flowers. This chimeralism can affect the whole flower, or just appear as a stripe down a tepal. The purple tips to tepals and style suggest it belongs to *C. × byzantinum*, with which it agrees In other respects, although its leaves can be more yellowish green.

'Innocence' was named by horticulturist and *Colchicum* expert Chris Brickell after an RHS assessment of colchicums at Felbrigg Hall, Norfolk, in 1996 (Anon. 1998, Rolfe 2000). The original plants had probably come from Van Tubergen nursery in the Netherlands. Plants with supposedly pure white flowers are offered under the unpublished names of f. *album*, var. *album* and 'Album'. They might all be referrable to 'Innocence' as its purple tipping is sometimes barely visible, and varies according to plant and season.

Leaves 5–6, elliptic, broad, distinctly ridged, yellow-green, the largest 32.5 × 10.4cm. Flowers 10–13cm in height, narrowly to broadly goblet-shaped, tepals 5–6.5cm long, white, tipped purple, occasionally with a purple stripe or two on tepals; perianth tube white; filaments white; styles white with strong purple tips.
Brickell, 1996
HL: 40–45 HF: 10–13 F: l8–m9

Colchicum cilicicum

'Purpureum' AGM H5

A delightful and highly floriferous cultivar, this has distinctive mid-reddish purple flowers that form tight posies. The tepals are noticeably darker at the tip and contrast with the white perianth tubes. In shape, the flowers start off as narrow chalices but soon open to broad funnels.

In the garden 'Purpureum' will thrive equally well in full sun or part shade, although the colour is more vibrant in brighter aspects. The corms multiply freely and require dividing every three years, otherwise they become over-congested and start to lose vigour.

Corms, asymmetric, 6.5–9.5 × 7.4–12cm, with a foot 2–2.6cm long and a neck 4.5–8.5cm long. Leaves 4–5, more or less erect, narrowly elliptic to narrowly elliptic-lanceolate, 30cm or more in length, dark to mid-green and glossy. Flowers 5–25; tepals to 7.5cm long; inner tepals pale purple, darker at apex, lightly tessellated, central vein mid-yellow-green becoming white on lower two-thirds and at base, outer tepals with paler reverse; also darket at the apex; filaments white, yellow-green at base; styles straight or slightly curved at the apex, white, with purple punctiform stigmas decurrent for not more than 0.5mm; perianth tube white.

Origin unknown, 1928

HL: 20–40 HF: 10–20 F: 9–10

Colchicum doerfleri

'Valentine'

A floriferous cultivar, this has pale rose pink, weather-resistant flowers produced for several weeks over the period encompassing Saint Valentine's Day (14 February) in an average year.

Originally an unnamed clone grown in the Netherlands, it was named by Antoine Hoog Authentic Plants in 2007 when supplied to overseas nurseries. It is often incorrectly assigned to *C. hungaricum*.

Origin unknown, 2007
HL: 20–25 HF: 6–10 F: 2

Colchicum kesselringii

'Snow of Highland'
This selection has relatively large, pure white flowers that open to wide funnels. The tepals have a slightly more acute apex than that of 'Yeti'. The perianth tubes retain some of the purple colouring usually seen in the species.

It was found by Leonid Bondarenko of Lithuanian Rare Bulb Garden nursery in a population of *C. kesselringii* in Central Asia.

Bondarenko, early 21st century

'Yeti'
An attractive selection, this has larger flowers than the species that open to broad funnels. They are pure white and lack the maroon-purple stripe on the exterior of the tepals normally seen on the species. The tepals have a more rounded apex than that of 'Snow of Hghland'.

Like 'Snow of Highland', it was found by Leonid Bondarenko in a population of *C. kesselringii* in Central Asia.

Bondarenko, early 21st century

Colchicum luteum

'Golden Baby'

This cultivar was one of the first selections of C. *luteum* made by Leonid Bondarenko of Lithuanian Rare Bulb Garden nursery. It has dark green leaves and anthers that are purplish brown prior to dehiscense. It has also been offered under the name 'Golden Elf'.
Bondarenko, 2009

'Vakhsh'

Collected by Leonid Bondarenko from near Vakhsh reservoir in Tajikistan, this cultivar has large, rich yellow flowers with a hint of orange.
Bondarenko, early 21st century

Colchicum macrophyllum

'Anopolis'

With its prettily patterned, relatively pale flowers, this was selected because the tepals were broader than other plants in the same population. Like most selections of C. macrophyllum it is rather slow to increase in the open garden. However, it will thrive in a bulb frame.

The original corms were collected on Crete by author and photographer John Fielding from near Anopolis. This town is at the foot of the White Mountains at the western end of the island where most of the populations of C. macrophyllum are located.

Flowers 3–5 per corm, opening in full sun from funnel-shaped to almost star-like; tepals 6.4–7 × 1.7–2.7cm, white with pale pink tessellations, flushed deeper pink towards the top, fading to white in the throat; stamens about two-thirds the length of the tepals, filaments white anthers purple-brown, 13–14mm long; styles white, pink-flushed towards the hooked tip; perianth tube 8–9.8cm long, white.
Fielding, 2000
HL: 35–45 HF: 8–13 F: m8–m9

'Cretan White'

This is an attractive, pure white cultivar of the normally lilac-purple to purple-violet species. It was collected on Crete by UK plant breeder Peter Moore near Anopolis and named by him.

Flowers 1–6 per corm, broadly funnel-shaped, pure white; tepals elliptic to narrow-elliptic, slightly keeled on the reverse, the outer 6–6.2 × 1.7–1.8cm, apex subobtuse, the inner slightly shorter than the outer, 5.6–5.9 × 1.5–1.7cm; stamens about half the length of the tepals, filaments white, anthers creamy yellow, 11–12mm long; styles are white, somewhat longer than the stamens and hooked at the tip; perianth tube 4.5–7cm long.
Moore, late 20th century
HL: 34–43 HF: 8–12 F: l8–m9

'Hora Sfakion'

The flowers of this cultivar are of average size for *C. macrophyllum* and quite pale but with pleasing light tessellations, which contrast with purple-brown anthers. As with all *C. macrophyllum* cultivars, the handsome, pleated leaves appear long after flowering, generally in late winter and early spring. They are arguably the most handsome leaves of any *Colchicum* species and are sometimes mistaken for those of a *Veratrum*.

The original corms were collected by author and photographer John Fielding on Crete in the vicinity of Hora Sfakion, a village near Anopolis.
Fielding, 2000
HL: 35–45 HF: 8–12 F: m8–m9

Colchicum montanum

'Norman Barratt'

The corms that were eventually given this cultivar name were collected by Norman Barratt in the French Pyrenees in the 1950s (Barratt 1951). They persisted in cultivation and in September 2002 a pan was shown by Bob and Rannveig Wallis at an Alpine Garden Society show. It was then given the cultivar name 'Norman Barratt', under the species *Merendera montana* (Wallis 2003).

Bob & Rannveig Wallis, 2003

Colchicum speciosum

'Album' AGM H5

This white-flowered cultivar of *C. speciosum* is without question the finest white autumn crocus grown in gardens. It is a prolific grower that multiplies freely, producing stout, long-lasting goblets that stand up well to the vagaries of the weather. It has been a popular garden plant since it was raised in about 1901. Although pale-flowered variants of *C. speciosum* are found in the wild, no plant matching 'Album' has been seen.

Corms large, ovoid, the largest measuring 12.5 × 10.2cm, often with a triangular foot up to 3cm long and a long neck, usually 8–13cm long. Leaves 3–6, more or less erect, elliptic to oblong-lanceolate, to 32 x 9cm, glossy, slightly yellowish green, plain or somewhat fluted. Flowers 1–3 per corm, goblet-shaped opening to broadly bowl-shaped, pure white; tepals 6.2–7.6 × 2.3–3.2cm, the outer somewhat longer than the inner; stamens consisting of green filaments and yellow anthers; styles white, generally overtopping the stamens, stigmas decurrent for 2–4mm; perianth tube lime-green, 13–16cm long. Fruit capsules occasionally produced.
J. Backhouse & Son of York, 1901
HL: 50 HF: 18–24 F: 9–e10

'Atrorubens' AGM H5

Of all the cultivars of *C. speciosum*, 'Atrorubens' has the darkest flowers, making a fine contrast with some of the others, especially 'Album'. This is a good garden plant, multiplying well, but perhaps not as freely as 'Album', and it is readily available in the horticultural trade.

Leaves 5, elliptic to oblong-lanceolate, to 21 x 10.5cm, bright yellowish green and somewhat glossy, with a few folds lengthways. Flowers goblet-shaped, rich purple-crimson or purple-red with faint tessellations close to the margins of the tepals in the upper half, and white in the lower third of the tepals within, and with a thin median white stripe extending up to two-thirds of each tepal; tepals 7–7.8cm long, the outer rather longer than the inner; stamens with whitish filaments and yellow anthers; styles white, overtopping the stamens, hooked at the tip; perianth tube 14–17cm long, rich purple, greenish towards base. Fruit capsules occasionally produced.
J. Backhouse & Son of York, pre-1914
HL: 36 HF: 21–25 F: m9–10

Bornmuelleri Group

This cultivar group name is established in this book for plants primarily from the western end of the range of *C. speciosum* in Turkey. They are distinguished primarily by their purple-brown or purple anthers, as opposed to the yellow anthers of *C. speciosum*. The group name is based on *C. bornmuelleri* Freyn, a species named from a collection made in 1889 by Joseph Bornmüller from western Turkey at 1,800m in Ak-Dagh. Although *C. bornmuelleri* is regarded as a synonym of *C. speciosum* by Persson (2007) and others, plants in cultivation, some of known wild origin, require some form of recognition. At this stage in our understanding of the genus we prefer to establish a cultivar group, *C. speciosum* Bornmuelleri Group. It differs from *C. speciosum* primarily in its paler, more funnel-shaped flowers, the anther colour as mentioned above, and in the length of the decurrent stigmas which are 0.5–1.5mm as opposed to 2–4mm in *C. speciosum*. See the Species chapter under *C. speciosum* for a fuller discussion.

Colchicum speciosum Bornmuelleri Group is rare in cultivation, but it can be seen in the collections at Royal Botanic Garden Edinburgh, and was included in the 2014–2018 trial at RHS Garden Hyde Hall. A different plant, often sold as *C. bornmuelleri*, which has yellow anthers and lacks pubescent filament channels, is now known as 'Joseph'.

Wild origin, 1889/2020
HL 38–45 HF: 6–22 F: I8–9

'Dombai'

A distinctive cultivar, this has narrow flowers shaped rather like champagne flutes and of a rather vivid, purplish pink colour. There is some faint chequering near the margins, and white in the throat.

Unfortunately, it is rarely offered for sale, although it is present in a number of private collections. It was introduced and named by Jānis Rukšāns of Rare Bulb Nursery in Latvia who obtained it from Tallinn Botanic Garden in Estonia. The name refers to a ski resort in the Russian Caucasus, and it is likely that it originated from a collection of *C. speciosum* in that area.

Flowers narrow, champagne-flute-shaped, vivid, purplish pink, with faint chequering near the margins, and white in the throat extending one-third of the way up the tepals; tepals elliptic, 5.5–7.5 × 2–2.2cm, narrowed and rather claw-like towards the base; stamens half the length of the tepals, filaments greenish, anthers yellow, styles white, much exceeding the stamens; perianth tube 8–11cm long, greenish white.

Rukšāns, pre-1989
HL: 28–36 HF: 14–18 F: 9–e10

Giganteum Group

This cultivar group name is established in this book for plants, presumably of Turkish origin, characterised by having pale flowers that are larger, broader and more funnel-shaped than *C. speciosum* and earlier into bloom. Plants showing characteristics of *C. giganteum* appear in populations of *C. speciosum* in eastern and north-eastern Turkey and more commonly in Iranian populations. The group name is based on *C. giganteum* hort. ex Stef, a species named in 1926 from plants cultivated by Scottish horticulturist Sam Arnott and presumably of Turkish origin. Although *C. giganteum* is regarded as a synonym of *C. speciosum* by Persson (2007) and others, it deserves some sort of recognition horticulturally, so a cultivar group, *C. speciosum* 'Gianteum Group', seems most appropriate. It differs from *C. speciosum* primarily in having larger, broader, more funnel-shaped, pale flowers with a white centre, white perianth tube, pale yellow anthers and being earlier into bloom. See the Species chapter under *C. speciosum* for a fuller discussion.

Colchicum speciosum Giganteum Group is still in cultivation, having been grown for many years, and was included in the 2014–2018 trial at RHS Garden Hyde Hall. It is sometimes sold under the name *C. illyricum*.

Presumably wild origin, 1926/2020
HL 35–45 HF: 18–22 F: 8–10

Giganteum Group 'Chequers' AGM H5

A large, vigorous, weather-resistant, early-flowering cultivar, this comes into bloom at the first hint of autumn. The flowers are quite pale in colour but are nonetheless pleasing, opening to a wide funnel shape when fully developed. It is a selection of *C. speciosum* Giganteum Group and originated at Chequers, the garden of the late Jenny Robinson in Boxford, Suffolk.

It is the only cultivar to possess superfluous curled and reduced tepals in the centre of what looks like a perfectly normal flower. This is a very useful diagnostic feature that distinguishes it from other early-flowering cultivars derived from Giganteum Group. These extra tepals replace one or several of the stamens and sometimes bear a portion of an anther, but not all the flowers possess this feature. It does not appear to produce fruits.

Corm narrow ovoid, 4.8–6 × 3.8–4cm, foot oblong, 1.2–2.5cm long, neck slender, 5.5–8.6cm long. Leaves 4–5, ascending, spreading and curving, elliptic to oblanceolate or linear-elliptic, 22.5–30 × 4.8–7.5cm, bright mid-green, somewhat glossy, fluted lengthways. Flowers 2–7(–9), funnel-shaped, pale mauve-pink, untessellated, with a prominent white throat extending halfway up each tepal, green at the very base within, opening from white buds, somewhat twisted at maturity; tepals elliptic-oval to oval-obovate, the outer 9.5–10.2 × 3.2–3.7cm, the inner somewhat shorter, 9.1–9.4 × 2.9–3.2cm; filament channels finely pubescent; flowers often with 1–3(4) smaller superfluous tepals, twisted and curled in the centre of the flower; stamens with pale green filaments, and brownish purple anthers with yellow pollen; styles slender, pale greenish white, 6.8–7.5cm long, exceeding the stamens, slightly curved at the tip, stigmas yellowish and shortly decurrent; perianth tube stout, greenish white with a hint of grey, 14–21cm long.

Grey-Wilson, 2014
HL: 42–50 HF: 26–32 F: m8–m9

Giganteum Group 'Revelation'

A very floriferous and elegant cultivar, this stands well in the garden, even in inclement weather. The pale, pastel-coloured flowers contrast well with some of the deeper coloured cultivars. In some ways it is reminiscent of the earlier flowering 'Chequers', but it has narrower flowers without the extra contorted tepals within.

Named for the first time here, this distinctive and pretty colchicum is reputed to have originated as a wild-collected plant, presumably from Turkey, that was growing in the garden of E.A. Bowles at Myddelton House, Enfield, in Middlesex. It was entered into the RHS assessment of colchicums at Felbrigg Hall, Norfolk, in 1996 as number 16. It was named there as *C. bornmuelleri* but it is neither that species nor the plant widely circulating under that name in the horticultural trade. 'Revelation' clearly belongs to *C. speciosum* Giganteum Group, based on its large, pale, funnel-shaped flowers.

Corm ovoid, 7–11cm long; tunic mid-brown. Leaves 4–5, bright and somewhat glossy green, 27–30 × 5–5.8cm, the lowermost elliptic, the uppermost linear-elliptic. Flowers 1–3, funnel-shaped, white in bud but gradually gaining colour, at maturity pale to mid-lilac-pink, with a bold white centre reaching just over halfway up the tepals and with a ragged margin as it diffuses into the lilac; outer tepals lilac-pink overall except paling towards the margins in the lower half, oval to oval-obovate, 70–78 × 26–31mm, apex obtuse; inner tepals slightly shorter, elliptic-obovate, 65–75 × 23–27mm, apex subobtuse to slightly retuse; filament channels minutely puberulous; stamens two-fifths the length of the tepals, with green filaments and purple-brown anthers; styles white, slightly curved at the tip and just overtopping the stamens, stigmas very shortly decurrent for not more than 0.5mm; perianth tube pale greenish white, with a slight hint of lilac at the top, 11–13(–17)cm long. Faintly scented.

Bowles, 2020
HL: 34–40 HF: 16–19 F: m9–e10

'Maximum'

This late-flowering, old Irish cultivar is noted particularly for its profuse, medium-sized, concolorous flowers which show good weather resistance and are borne on relatively short perianth tubes. Plants produce a great tuft of flowers and, as the corms clump up well, substantial posies of 40 or 50 can be borne close to the ground. The flowers are rather smaller than those of most cultivars of *C. speciosum*.

'Maximum' is a selection from Tom Smith of Daisy Hill Nursery in Newry, Northern Ireland. Smith established the nursery in 1887. There is some doubt that the plant which goes under the name 'Maximum' today is the same as that originally described, and there are discrepancies in the description given by E.A. Bowles (1924). Bowles found it a slow increaser, which is certainly not the experience of many who grow it today.

Leaves 4(–5), the largest elliptic-oblanceolate, to 35 × 5.2cm, dark green, with several longitudinal ridges, unusually neat for an autumn colchicum. Flowers up to 15 per corm, broadly goblet-shaped at maturity, sometimes almost funnel-shaped, 6 × 2.5cm, concolorous in rich rose-purple or dark pink, with a whitish throat; tepals with a thin, central, white stripe extending to half the length and with a hint of tessellation towards the margins; perianth tube relatively short, 5.5–9cm long, and flushed with pink. Fruit capsules rarely produced.
Smith, pre-1924
HL: 35–40 HF: 15 F: m9–10

'Paul Furse'

This cultivar produces few, rather small, chalice-shaped flowers of rich deep purple, darkest towards the tips but with prominent white centres. The corm has a very pronounced, oblong foot.

A number of collections of *C. speciosum* from Iran and Turkey by Paul and Polly Furse from the 1950s and 1960s have survived in gardens to this day. Indeed, several entries at the trial of autumn-flowering *Colchicum* at RHS Garden Hyde Hall were traceable to Furse collections. This is the only cultivar named 'Paul Furse', but other Furse collections have been dubbed inadvertently with the same cultivar name, or simply listed as 'ex Paul Furse'.

Corm ovoid, up to 6.5cm long, with a very pronounced oblong foot, 2.5–5.8cm long. Leaves 4–5, the largest oblong-elliptic, to 32 × 5.6cm, deep green, glossy, scarcely ridged. Flowers 1–3 per corm, rather small, chalice-shaped, rich deep purple and darkest towards the tips, with prominent white centres up to half the length of the tepals; perianth tube relatively stout, greenish white.

Furse, 1950s–1960s

HL: 30–38 HF: 9–15 F: m8–e9

'Paul Furse Early'

This is a fine, early-flowering selection that was named by Michael Salmon and first listed in his monocot catalogue of 1990. The plant is based on a Paul and Polly Furse collection, presumably from north-eastern Turkey, or perhaps north-western Iran.

Plants come into flower in August, the mid-purple flower reminiscent of those of some forms of *C. autumnale*, although this is a selection from *C. speciosum*. The flowers are shaped like narrow and rather elegant chalices and are twice the size of those of 'Paul Furse'. The leaves that follow in the spring are quite large and ascending, generally 4 or 5 in number, elliptic-oblong to elliptic-oblanceolate, the largest some 30 × 5cm, mid-green, smooth and with a sheen.

Salmon, 1990
HL: 32–42 HF: 8–12 F: 8

'Rubrum'

A fine cultivar of *C. speciosum*, this produces narrow, chalice-shaped flowers of a similar hue to 'Atrorubens' but rather paler. The flowers tend to remain quite narrow throughout their life and are devoid of any suggestion of tessellation. Other differences from 'Atrorubens' include the silky appearance of the flowers, especially effective in bright sunshine, the prominent white, star-like throat, and the paler, pinker perianth tubes. Fruit capsules are not normally produced. Plants of similar colour can often be found in the wild, especially in north-eastern Turkey.

Origin unknown, pre-1900
HL: 35–50 HF: 14–18 F: m9–10

Colchicum szovitsii

'Snowwhite'
A cultivar with pure white, slender flowers that differs from other cultivars of this species in having pure white perianth tubes, rather than cream or greenish, and flowers that open one to two weeks later.
Seisums, pre-1997
HL: 28–35 HF: 8–14 F: 2–3

'Tivi'
This white-flowered cultivar with broad tepals has large, star-shaped flowers rather than the more usual bowl shape of the species. It was found by Arnis Seisums near the village of Tivi in Armenia.
Seisums, late 20th century
HL: 28–35 HF: 8–14 F: 2–3

'Vardahovit'
A fertile and frost-resistant cultivar, this has large, rounded, glistening white flowers that have a pink tinge on opening. It was found by Arnis Seisums near the town of Vardahovit in Armenia.
Seisums, late 20th century
HL: 20–26 HF: 6–9 F: 2–3

Hybrids

'Antares'

Charming and relatively small-flowered, the flowers of this cultivar are quite distinctive. They are generally rather pale, with a prominent white centre occupying most of the middle on the inside, except for the upper third of each tepal. The tepals are attractively parallel-veined and bear a prominent midrib that runs two-thirds of the distance up each one. This colchicum is occasionally available in the horticultural trade.

Corm narrow-ovoid, 4.7–5.3 × 2.5–2.8cm, foot absent or short and rounded, to 0.8cm long, neck slender, 3.5–6.4cm long. Leaves 5 or 6, ascending to somewhat spreading, particularly in the upper half, 23.5–33 × 5.8–6.5cm, deep green, somewhat glossy, with several ridges lengthways, lowermost leaf elliptic-oblong, uppermost leaf linear-elliptic. Flowers pale pink-lilac, whitish within with a greenish base, but purple towards the tip; filaments pale green and anthers yellow; styles whitish, equalling the stamens.
Visser, 1977
HL: 43–49 HF: 22–26 F: 9–e10

'Artur Klark'

This late-flowering cultivar is readily identified by the thin white edge to its tepals, especially the outer three. Raised by Leonid Bondarenko of Lithuanian Rare Bulb Garden nursery, it was selected as a brightly coloured seedling obtained from *C. bornmuelleri* hort., now known as 'Joseph'.

Corm oblong-ovoid, 6.5–9cm long, with a mid-brown tunic. Leaves not recorded. Flowers 1–3, goblet-shaped, rich rose-magenta with a bold white centre, the white diffusing up to half the length of the tepals, the outer three with a thin, central white line reaching close to the apex, tepals all with a thin white margin; outer tepals elliptic-obovate to oblanceolate, 65–70 × 23–35mm, apex obtuse to slightly retuse; inner tepals somewhat shorter, apex subobtuse to subacute; filament channels minutely puberulous; stamens one-third the length of the tepals, filaments greenish white, anthers pale yellow; styles white, filiform, long overtopping the stamens, with a punctiform stigma; perianth tube pale greenish white, 11.3–12.2cm long. Not scented.
Bondarenko, 2010s
HL: 28–35 HF: 18–22 F: m9–e10

'Autumn Herald'

This selection is a mid-season cultivar with good weather resistance. There is a particularly good stand of it at East Ruston Old Vicarage garden in Norfolk which houses a National Plant Collection of *Colchicum*. 'Autumn Herald' is occasionally available from nurseries.

Corms ovate-oblong, 7–9.5cm long, with a dark brown tunic. Leaves unrecorded. Flowers 2–4, goblet-shaped, mid-pink merging into creamy-white outside in the lower third and the lower two-thirds inside, greenish at the base of the throat; perianth tube greenish white, 9.8–11cm long; outer tepals elliptic-obovate, 70–72 × 34–35mm, apex obtuse; inner tepals elliptic-oblanceolate, 66–67 × 25–26mm, apex subobtuse; stamens one-third the length of the tepals, with greenish white filaments and yellow anthers, outer stamens clearly shorter than the inner; filament channels sparsely and minutely puberulous; styles white, equalling the stamens, the stigmas very shortly decurrent for 0.5–1mm.

W. E. Th. Ingwersen, mid-20th century

HL: 38–46 HF: 17–22 F: 18–9

'Autumn Queen' AGM H5

The finest of the early-flowering autumn cultivars, this multiplies freely in the garden and is as good in the border as it is for naturalising in grass. Plants occasionally produce fruit and these contain good, viable seed. It has therefore been known to self-seed in the garden and seedlings can vary in the substance and depth of tessellation of the flowers. 'Autumn Queen' has a strong influence of *C. bivonae* in its parentage.

A derivative of 'Autumn Queen' has been distinguished as 'Nutt's Green Star'. This apparent selection by plantsman Richard Nutt has a green, star-like marking at the base within the flowers. Close inspection by the authors, however, has shown that this green marking, a result of swollen filament bases, is often present in the young flowers of 'Autumn Queen' and fades as the flowers age. It is therefore regarded as a synonym of 'Autumn Queen'.

Corm ovoid, 5–8.3 × 4.2–7.2cm, foot narrow triangular, 0.8–2.6cm, apex obtuse, neck relatively stout, 6–8.5cm long. Leaves (3)4–5, ascending to spreading, slightly yellow-green, moderately glossy, with several ridges lengthways, the margin somewhat undulate; lowermost leaf elliptic-oblanceolate, 29–33 × 5.6–6.2 cm; uppermost leaf elliptic oblanceolate, to 28 × 5.4cm. Flowers 1–5, narrow goblet-shaped at first, opening to wide bowls at maturity, mid to deep purple-pink to rosy violet, moderately tessellated in the upper half, with a prominent white throat reaching one-third the length of the tepals; tepals elliptic to elliptic-oblanceolate, the outer 8–8.2 × 3.3–3.5cm, the inner 6.9–7.2 × 2.3–2.7cm; stamens with pale green filaments and purplish brown anthers, swollen and greenish at the base; styles pale to deep purple, 4.2–4.5cm long, slightly curved at the tip; perianth tube moderately stout, 12–15cm long, greenish white. Mildly fragrant of primroses.

Zocher & Co., 1900–1905

HL: 38–43 HF: 19–24 F: e8–e9

'Beaconsfield'

A vigorous and reliable cultivar, this is both floriferous and weather resistant. At maturity the flowers are characteristically Y-shaped and an attractive shade of mid-purple. The faint tessellations that are visible on the tepals probably indicate *C. bivonae* in its parentage, although the general demeanour of the flowers and foliage also suggest some influence of *C. autumnale*.

Corm oblong or oblong-ovoid, 6–9cm long. Leaves 4 or 5, yellowish green, glossy, 25.5–28 × 3.2–3.6cm, somewhat twisted, moderately ridged longitudinally; lowermost leaf elliptic, apex obtuse, the uppermost leaf narrow-elliptic, apex subacaute. Flowers 1–5, narrow, Y-shaped chalices, mid-purple, faintly tessellated, with a prominent white centre for about half the length of the tepals within and extending as a central white line for two-thirds the length; tepals appearing relatively narrow because of a tendency to incurve along the edges, the outer elliptic-oblanceolate, 8–8.5 × 3–3.2cm, apex subobtuse, inner tepals elliptic, 8.1–8.3 × 2.2–2.5cm, apex subobtuse; stamens with whitish filaments and purple-brown anthers; styles purplish or lilac-pink, slightly exceeding the stamens and with a cream, slightly curved and expanded tip; perianth tube 9–11.5cm long, greenish white, purplish towards the top.

Origin unknown, pre-1924

HL: 38–42 HF: 17–19 F: 9–m10

'Benton End' AGM H5

One of the finest autumn-flowering cultivars, this has bold, intensely coloured flowers of excellent substance that stand up well to the weather. Planted alongside cultivars such as 'Pink Goblet' and *C. speciosum* 'Album', which flower at the same time, it can make a striking statement. It multiplies freely in the open garden.

This cultivar is readily confused with 'E.A. Bowles' and is likely to have a similar parentage involving *C. speciosum* and possibly *C. turcicum*. 'E.A. Bowles' has slightly smaller, even more richly coloured flowers, and a deeper coloured perianth tube. In 'Benton End' the flower buds are quite slender and the central white line inside each tepal reaches almost to the top, particularly on the inner tepals. However, in 'E.A. Bowles' the flower buds are more rounded and the white line extends only halfway up from the base of the tepal. 'Benton End' produces fruit capsules in some seasons, but not consistently so.

The cultivar was named in 1998 by Chris Brickell following an RHS assessment of colchicums at Felbrigg Hall, Norfolk. Benton End was the home at Hadleigh in Suffolk of the artist and gardener Sir Cedric Morris (1889–1982) from where this cultivar presumably originated. There is some question over the cultivar named 'Cedric Morris' and whether it is the same as 'Benton End', but it is treated as a synonym in this book, as is 'Cedric's Drake' (Anon. 1998) and 'Cedric's Darkest of All'.

Corm ovoid to ovoid-oblong, 4–5.5 × 2.8–4.2cm; foot absent or short, 0.2–0.5cm long; neck stout, 3.5–7.2cm long. Leaves 4–5, ascending to somewhat spreading, 24.5–28.5 × 5.2–8.5cm, bright glossy green, lightly ridged lengthways; lowermost leaf elliptic- to oval-oblong, uppermost leaf narrow elliptic-oblanceolate. Flowers 1–5, goblet-shaped, deep purple-pink, somewhat shiny on the exterior, with a white centre within extending to about one-third up the tepals, each tepal also with a thin, white, median stripe to the top; tepals concave, oval-elliptic to -obovate, the outer 7.8–8.7 × 2.7–3.2cm; stamens with greenish filaments and pale yellow anthers; styles slender, white, sometimes flushed purple at the slightly curved or twisted tip; perianth tube stout, white, flushed greyish purple towards the top, darkening with age, 14–16.5cm long. Not scented.
Benton End, 1998
HL: 42–50 HF: 21.5–24 F: 9–e10

'Boxford'

With pretty yet quite pale flowers, 'Boxford' is an early-flowering, small cultivar of great charm that performs well in the garden.

It originated at Chequers, the garden of the late Jenny Robinson in Boxford, Suffolk, who died in 2010. Two related collections, among others, were gathered from the garden by two of the authors (CG-W and RL) in 2010 and numbered JR1 and JR2. JR1 was considered the better of the two and subsequently named 'Boxford' by Grey-Wilson in 2014, although its first formal description is presented here. JR2, which has not been formally named, is distinguished by having slightly larger, lilac-pink flowers with a white, star-shaped centre, the white extending up the tepals for only one third. Both are vigorous and multiply freely in the garden. They clearly have *C. autumnale* in their parentage. Both occasionally produce fruit capsules.

Corm ovoid-oblong, 3.5–7.5cm long. Leaves 4–5, stiffly ascending to erect, 35–36.5 × 6.4–6.7, deep green, moderately glossy, somewhat twisted, shallowly ridged longitudinally; lowermost leaf oblong-elliptic, slightly hooded at the rounded apex, the uppermost leaf linear-lanceolate, apex subacute. Flowers 2–5, narrowly chalice-shaped, mid-pink with a white centre extending to halfway up the tepals, greenish at the base; outer tepals obovate-oval, 5.7–6.2 × 2.3–2.5cm, apex subobtuse; inner tepals elliptic-obovate, 5.4–5.6 × 1.7–1.8cm, apex subobtuse; stamens with greenish filaments and yellow anthers, browning with age; style white, flushed pink in the upper half, exceeding the stamens at maturity, slightly hooked at the tip; perianth tube slender, 14–19cm long, greenish white, flushed pink in the upper half.

Grey-Wilson, 2014
HL: 36–40 HF: 20–25 F: l8–9

'Constable'

This cultivar is similar in some respects to 'Lilac Wonder', especially in its floriferousness and flower colour. The flowers of 'Constable', however, stand up better in the garden on stouter perianth tubes. Those of 'Lilac Wonder' tend to fall over and quickly become disarrayed, making it a less satisfactory plant in most gardens, although it fares better when planted in rough sward.

Despite its name, the origin of which is unclear and is not a reference to landscape artist John Constable, this floriferous cultivar originated in the Netherlands.

Corm narrow-ovoid, 5.5–8 × 3.5–7cm. Leaves 4–6, ascending, spreading in the upper half, 25–34.5 × 6–7, somewhat bluish mid-green, moderately glossy, unridged or slightly ridged lengthways; lowermost leaf elliptic-oblong, apex obtuse, the uppermost leaf narrow elliptic-oblanceolate to linear-elliptic, apex acute. Flowers 3–9, chalice-shaped to begin with but opening to more or less funnel-shaped, deep pink, each tepal with a central white line in the lower half, faintly tessellated; outer tepals elliptic-oblanceolate, 7–7.5 × 2.1–2.3cm; inner tepals elliptic, 6–6.4 × 1.5–1.7cm; stamens with white filaments and pale creamy yellow anthers; style very slender, white, slightly exceeding the stamens, with a hooked tip, the stigmas long-decurrent; perianth tube white, 11–15cm long.
Origin and date unknown
HL: 46–55 HF: 18–22 F: m9–e10

'Daendels'

Although rather undistinguished in appearance, this is is a floriferous and reliable cultivar that is well-suited to cultivation and the flowers are quite sturdy and weather-resistant. It has obvious affinities with 'Joseph' and 'Redgrave' which have rather similar colouring. The flowers of all three have the curious habit of emerging white in bud, then gradually assuming their pink coloration.

Corm narrow ovoid, 5–6.2 × 2.5–2.8cm;, foot absent to just present, to 0.5cm long; neck relatively stout, 3.5–6.8cm long. Leaves 4–5, ascending, somewhat curved, 26–28.3 × 7–7.3cm, deep green, moderately glossy, slightly fluted lengthways; lowermost leaf elliptic-oblanceolate, apex subobtuse, the uppermost leaf elliptic-oblong, apex subacute. Flowers 2–7, broadly chalice-shaped, rose-pink with a white throat, yellowish at the base, the tepals with a faint white stripe for two-thirds from the base, not tessellated; outer tepals elliptic-obovate, 6.5–7.5 × 1.9–2.3cm, apex obtuse; inner tepals elliptic-oblanceolate, 6.3–7.2 × 1.7–2.1cm, apex subobtuse; stamens with white filaments, yellowish towards the base, and anthers yellow, browning with age; styles straight, white, equalling the stamens; perianth tube creamy or greenish yellow, 11–17cm long. Fruit capsules often produced.
Zocher & Co., 1900–1905
HL: 40–46 HF: 17–22 F: 9–m10

'Darwin'

This is a fine plant of excellent colour, much sought-after by colchicum connoisseurs. It is prized for its deep pink, globose chalices which have a defining sheen, as well as its late-flowering habit.

   'Darwin' is rare in cultivation and scarcely ever available. This is a pity because it is the last of the autumn-flowering cultivars to bloom, mainly in November, filling the gap between late autumn and winter. The prime reason for its scarcity is that plants increase very slowly, often only replacing themselves each season. Fruit capsules are occasionally produced but appear to be sterile, and if seedlings were produced they would not necessarily come true to type.

Corm oblong-ellipsoid, 12–15cm long at maturity, with a dark brown tunic and with a distinct triangular foot. Leaves dull, mid-green, slightly fluted; lowermost leaf 32–35 × 5.8–8cm. Flowers globular chalice-shaped, deep lustrous pink, the tepals with faint parallel lines; outer tepals oval-obovate, 5.3–5.5 × 3.8–4cm, inner tepals obovate, 4.8–5 × 2.8–3cm, all tepals generally with inrolled margins, especially towards the top; stamens with white filaments and cream-coloured anthers; styles whitish, equalling or slightly longer than the stamens, strongly curved at the tip; perianth tube deep pinkish purple, 8–11cm long.

R.O. Backhouse of Herefordshire, 1930s

HL: 35–40 HF: 15–18 L: F: m10–11

'Dick Trotter'

A late-flowering cultivar with large, almost straight-sided tepals, this increases well from corms, quickly forming substantial clumps and it often sets fruit. It is similar to two other first-class cultivars, 'Huxley' and 'Pink Goblet'. For differences between them, see 'Huxley'.

The original plant of 'Dick Trotter' was found in the orchard at Brin, the garden of Dick Trotter, near Inverness, although it was probably named later by his daughter Elizabeth Parker-Jervis of P-J Nursery, Longworth, Oxfordshire (Rolfe 2009). Interestingly, both this cultivar and 'Pink Goblet' arose from a batch of seedlings from fruits set on plants of *C. speciosum* 'Album'. It is perhaps strange that two pink plants should arise from a pure white cultivar.

Corm ovoid, 4.8–5 × 4–5.2cm; foot rather obscure, occasionally more obvious, generally 0.8–3.2cm long; neck stout, 7–16cm long. Leaves (3–)4(–5), ascending to curving, 29–30.3 × 6.8–7.2cm, yellowish green, glossy, with several longitudinal pleats; lowermost leaf elliptic-oblanceolate, apex obtuse; uppermost leaf elliptic-oblanceolate, apex subacute. Flowers 2–8, narrowly chalice-shaped, opening more broadly with age, purple-pink, with a pronounced, almost metallic sheen on the exterior, with a greenish throat extending into a white centre and with a central white line extending two-thirds the way up each tepal, to give a star-shaped pattern; tepals more or less parallel-sided, the outer tepals elliptic- or oval-obovate, 6.6–7.8 × 2.4–3cm, apex obtuse, the inner tepals elliptic, 5.8–7.2 × 2.3–2.6cm, apex obtuse; stamens with greenish white filaments and yellow anthers; style white, very slender, eventually greatly exceeding the stamens, slightly curved at the tip; perianth tube relatively stout, 12–16cm long, greenish lined and flushed purple, especially towards the top.

Trotter, 1930s
HL: 40–52 HF: 17–23 F: m9–m10

'Disraeli'

A beautifully tessellated cultivar, this has large flowers that have rather wavy tepal margins and is certainly one of the best. 'Disraeli' is rare in cultivation, but is occasionally offered by nurseries.

'Disraeli' and 'Glory of Heemstede' are rather similar in size and could be confused, flowering at more or less the same time. Both owe much of their character to the parent species *C. bivonae*. 'Glory of Heemstede' has rather darker flowers overall and the white throat extends up the tepals as a central band while the tepals are flat and unruffled. 'Disraeli', on the other hand, has a more prominent white throat with a more distinct demarcation between the white throat and the violet-mauve of the upper half of the tepals, which are somewhat waved at the margin. Both cultivars have purple-brown anthers, also inherited from *C. bivonae*. Fruit capsules are occasionally produced but they are small and contain few viable seeds.

Corm broad-ovoid, 10.2–11.8 × 8.2–9.5cm; foot obscure or up to 10mm long; neck 7.5–13cm long. Leaves usually 5, ascending, 25–26.5 × 6.6–7.2cm, the lowermost separated from the upper leaves by a pronounced gap, bright glossy green, with several ridges lengthways; lowermost leaf elliptic, apex obtuse, the uppermost leaf oblanceolate, apex subobtuse. Flowers deep violet-mauve with prominent tessellations, tepals slightly undulate at the margin; outer tepals oval-elliptic, 6.2–7.5 × 2.3–2.8cm; inner tepals elliptic-obovate, 5.9–6.8 × 2.1–2.5cm; stamens about one-third the length of the tepals, with greenish filaments and purple-brown anthers; styles purplish, slightly overtopping the stamens; perianth tube white, pink-flushed towards the top, 10–16cm long.

Zocher & Co., 1900–1905
HL: 55–60 HF: 18–25: F: m9–m10

'E.A. Bowles'

A fine cultivar, this has rich, glossy, purple, goblet-shaped flowers of good substance, with a greenish white star at the base within, made primarily by the swollen bases of the filaments. Unfortunately, it did not perform well at the RHS trial at Hyde Hall and therefore failed to achieve an AGM. However it is a superb garden plant and readily available in the horticultural trade. It is very similar to, and easily mistaken for, 'Benton End', and the differences are discussed under that cultivar. It is likely to have a similar parentage involving *C. speciosum* and possibly *C. turcicum*. Fruit capsules are often produced.

This charming cultivar has a good pedigree, having originated in the garden of Edward A. Bowles at Myddelton House, Enfield, in Middlesex, and found there after his death. It was named and distributed by Dick Trotter and the name is likely to have been first published by Elizabeth Parker-Jervis in a list of P-J Nursery, Longworth, Oxfordshire, in the 1980s.

Corm broad-ovoid to narrow ovoid, 4.2–7.6 × 3.8–5.5cm, foot oblong, 0.5–3.5 × 1.2–2cm, apex obtuse, often prominent, neck stout, 4.5–10.5cm long. Leaves 4–5, stiff and ascending to upright, 23–26 × 8.8–9.4cm, moderately glossy, mid to deep green, fluted; lowermost leaf oval-oblong, apex obtuse; the uppermost leaf elliptic-oblanceolate, apex subacute. Flowers goblet-shaped, rich pink-purple, shiny on the exterior, greenish star-shaped right at the base within, otherwise with a white centre extending one third the way up the tepals, with the median white line extending to two-thirds of each tepal; tepals concave, oval-obovate to almost oblong, the outer 6.8–7.1 × 2.5–2.7cm, the inner 5.9–6.2 × 2.1–2.3cm; perianth tube dark pinkish purple; stamens with green filaments and pale yellow anthers; styles white, sometimes tipped pink, shorter than or slightly exceeding the stamens;
perianth tube dark pink, 13–14.8cm long. Not scented.
Bowles, 1980s
HL: 43–47 HF: 19–22 F: m9–e10

'E.K. Balls'

The rather pale flowers of this floriferous cultivar are highly attractive en masse and make a fine contrast to the majority of more vibrant cultivars. In colour it comes quite close to another pale cultivar, 'Gracia', although the flowers of 'E.K. Balls' are more lilac and the tepals are obtuse-tipped rather than subobtuse or even subacute. Both cultivars have a prominent white centre which merges with the colour in the upper half of the tepals. Another difference can be seen in the anthers which are purple-brown in 'E.K. Balls' and yellow in 'Gracia'.

Corm narrow-ovoid, 5-7.5cm long. Leaves usually 4, elliptic to elliptic-oblanceolate, 25-29.5 × 7-7.4cm, mid-green and somewhat shiny. Flowers goblet-shaped, pale lilac-pink, untessellated, with an irregular white centre extending for three-fifths up the inside of the tepals, the exterior white towards the base of the tepals with lilac-pink coloration extending down the keel to the top of the perianth tube, outer tepals often curving outwards at the top at full maturity; outer tepals elliptic-oval to elliptic-oblanceolate, relatively narrow, 6.8-7 × 2.4-2.5cm, apex obtuse to retuse; inner tepals shorter than the outer, elliptic-oblanceolate, 6.3-6.5 × 1.9-2.1cm, apex obtuse, often retuse; filament channels thinly and finely pubescent; stamens one third the length of the tepals, with pale green filaments and purple-brown anthers, 7-8mm long; styles white, slightly overtopping the stamens; perianth tube greenish white, 9.2-11cm long.
Origin and date unknown
HL: 43-43 HF: 11-14 F: m8-9

'Emerald Town'

The large, goblet-shaped flowers of this cultivar are the same size of those of *C. speciosum*. They are an intense deep purple and contrast strikingly with the bright green, relatively short and stout perianth tubes. The tepals bear a slight hint of tessellation but otherwise they are uniformly coloured.

This is a recent introduction from Leonid Bondarenko of Lithuanian Rare Bulb Garden nursery and perhaps one of his best for garden use. However, it has not yet been widely assessed in UK gardens.

Bondarenko, 21st century
HL: 38–45 HF: 17–23 F: 9–e10

'Enigma'

At the RHS trial in 2015 it was noticed that plants from suppliers in the Netherlands under the name 'Spartacus' were different from those sourced from the UK under this name and also at the National Plant Collection at East Ruston Old Vicarage in Norfolk. Interestingly, plants sourced from one company in the Netherlands, Dix Export BV, turned out to be two different selections. We think the trialled plant should retain the name 'Spartacus' and the Netherlands cultivar is here named 'Enigma'. However, plants from the Netherlands will still be sold as 'Spartacus' and will probably cause confusion for years to come.

'Enigma' is a taller cultivar in bloom with narrowly goblet-shaped, rather than funnel-shaped, flowers. The tepals are more acute at the apex and there is no hint of chequering as in 'Spartacus'. In addition, the white flower centre is more clearly defined in 'Spartacus', while in 'Enigma' there is a hint of green at the base. Perhaps the most useful defining character is in anther colour; pale yellow in 'Enigma' and brownish purple in 'Spartacus'. 'Enigma' is almost the first colchicum to come into bloom in late summer. 'Spartacus', on the other hand, is a mid-season cultivar, flowering primarily in mid-September.

Corm ovoid, 7.5–8.5 × 7–9.2cm, with a mid-brown tunic. Leaves 4–6, erect to ascending, 23–28 × 6.5–7.5cm, bright and moderately glossy green, the lowermost elliptic, the uppermost linear-elliptic, all leaves with a few, shallow, longitudinal ridges. Flowers 2–5, narrowly chalice-shaped, pale pink merging into white in the lower third on the outside, untessellated; outer tepals oblanceolate, 5.4–5.8 × 1.4–1.5cm, apex subobtuse to subacute; inner tepals slightly shorter; all tepals finely puberulous along the filament channels; stamens one third the length of the tepals with white filaments, greenish at the base but scarcely swollen, bearing pale yellow anthers; styles very slender, white, two-thirds the length of the tepals, hooked at the tip, stigmas decurrent for about 1mm; perianth tube white with a slight hint of green, 9.5–11.5cm long.
The Netherlands, 2020
HL: 28–38 HF: 14–17 F: 8–e9

'Fabergé's Silver'

Named because of its beautifully formed, egg-shaped flowers, these keep their shape throughout the flowering period. It is related to 'Joseph' and 'Redgrave' and all three have lilac-pink flowers with white throats and yellow anthers, with some white on the exterior. 'Fabergé's Silver' has a silvery-lilac wash, with white confined to the apex of the rounded tepals.

This is a recently named cultivar from Leonid Bondarenko of Lithuanian Rare Bulb Garden nursery and is gradually becoming more widely available. Unfortunately, some corms being sold by others under the name are a rather poor, unidentified colchicum which is not 'Fabergé's Silver', so it is wise to check the source before ordering plants.

Corm ovoid to ovoid-oblong, 4.3–9 × 3.4–5cm; tunic pale brown; foot triangular, 1.2–1.4 × 1.8–2.1cm. Leaves 5–6, ascending, 22–30 × 6.5–8cm, mid-green with a silvery sheen, the lowermost elliptic-lanceolate, the uppermost narrow-elliptic. Flowers rounded goblet-shaped, mid-pink with a hint of tessellation, white in the lower third within, green at the very base, the outer tepals with a creamy white patch in the upper part, the outside of the flowers with a distinctive silvery-lilac wash or patina; outer tepals ovate-oval, 5.6–6.2 × 4.8–5cm, the inner oval-elliptic, rather shorter and not more than 3cm wide; stamens with greenish white filaments and yellow anthers; styles white, greatly exceeding the stamens. Fruit capsules not produced as far as is known.

Bondarenko, 2017

HL: 30–40 HF: 18–25 F: m9–m10

'Felbrigg' AGM H5

One of the finest cultivars, this is noted for its vigorous habit, early-flowering and sturdy, long-lasting, weather-resistant and rather squat blooms. While 'Felbrigg' is often referred to as a large selection of *C. cilicicum*, it is clearly of hybrid origin. The flowers do not have the long stamens and styles of that species and the tepals are distinctly hooded at the top. In addition, there is no hint of tessellation, a feature of most forms of *C. cilicicum*.

It was first noted following an RHS assessment of colchicums at Felbrigg Hall, Norfolk, in 1996. At the time it was referred to informally as "Myddelton" (Anon. 1998) and when many of the colchicums were sold off this one was listed as number 20. It was named 'Felbrigg' by Christopher Grey-Wilson after receiving an RHS Preliminary Commendation on 25 September 2010 (Rolfe 2013).

Corm broad-ovoid, 4.2–6.3 × 6.8–7.5cm, broader than long, foot narrow-triangular, neck stout, 5–7.8cm long. Leaves 4–5, spreading, 23–35 × 8.8–10cm, bright glossy green, somewhat ridged lengthways; lowermost leaf oblong-elliptic, the uppermost leaf elliptic-oblanceolate, apex subacute. Flowers broadly goblet- to bowl-shaped, with the tips curved inwards at the top, pink-purple but paler in the shade, with a greenish base within and a white star shape with the narrow arms of the star reaching halfway up each tepal, closely veined; outer tepals broad oval-obovate, apex obtuse, 4.8–5.4 × 2.2–2.5cm; stamens with whitish filaments and brownish or brownish purple anthers; styles white, flushed purple at the slightly curved tip, exceeding the stamens; perianth tube stout, white, 7–10cm long. Faintly scented. Fruit capsules not produced.
Origin unknown, 2010
HL: 48–55 HF: 12–16 F: 9–e10

'Felbrigg Violet'

This cultivar was seen by the authors in the garden of John and Diana Morley at Stoven in Suffolk and is named for the first time here. It had originally been found at Felbrigg Hall, Norfolk, by Richard Hobbs and subsequently some corms were passed on to John Morley. It is a handsome cultivar with distinctive, deep violet-purple flowers. The plant was labelled 'Violet Queen', an old cultivar that may no longer be in existence.

The original 'Violet Queen' was described as rich violet-purple, but with moderate or strong tessellations and a bold white centre. The cultivar in question has neither of these last two features; the tepals are uniformly coloured overall and are very lightly tessellated. Despite this, 'Felbrigg Violet' is a fine plant and a good addition to any collection of autumn-flowering colchicums.

Leaves 25–35 × 6.5–8.4cm. Flowers narrowly funnel-shaped, deep violet-purple, lightly tessellated, with a median white stripe in the lower half of each tepal within, the tepals often a little pinched and twisted; outer tepals narrow-elliptic to elliptic-oblanceolate, 6.7–7.3 × 2.1–2.4cm, apex subobtuse; inner tepals somewhat shorter than the outer, narrow-elliptic, 6.4–6.7 × 2–2.2cm; filament channels finely pubescent along the ridges; stamens half the length of the tepals, filaments greenish at the base, violet-flushed in the upper part, apex subacute, anthers yellow, 8–9mm long; styles whitish, flushed violet towards the tip, slightly longer than the stamens; perianth tube greenish white, suffused violet-purple in the upper part, 10–12.5cm long. Fruit capsules occasionally produced but generally rather small and containing few viable seeds.

Grey-Wilson & Leeds, 2020
HL: 38–44 HF: 17–19 F: m9–m10

'Flamenco Dance'

This cultivar, raised by Leonid Bondarenko of Lithuanian Rare Bulb Garden nursery from the same seed pod as 'Larisa', is a deliberate cross between 'Dick Trotter' and *C. speciosum* 'Ordu'. The result is an elegant, narrow, funnel-shaped flower with its rather narrow tepals characteristically furrowed longitudinally in the centre, and set on vivid, greenish yellow perianth tubes.

Corm ovoid, 9 × 6cm with an obvious foot; tunic pale orange-brown. Flowers 2–5, rather slim, funnel-shaped, dark magenta-purple with a whitish centre occupying the lower third of the chalice; outer tepals elliptic-oblanceolate, 4.8–7 × 20–30mm, apex subacute; inner tepals more elliptic and slightly shorter, apex subacute; stamens one third the length of the tepals with greenish filaments and yellow anthers; styles white, very slender and straight, greatly exceeding the stamens; perianth tube vivid greenish yellow. Fruit capsules not recorded.
Bondarenko, 2010
HL: 25–35 HF: 16–20 F: l9–m10

'Fuller's Mill'

During the trial of *Colchicum* at RHS Garden Hyde Hall the judging panel visited several gardens in East Anglia noted for their *Colchicum* collections. These included Fuller's Mill at West Stow in Suffolk, the home of the late Bernard Tickner, a former head brewer at Greene King and a keen horticulturist. Among the many interesting colchicums seen there in 2018, one was singled out by two of the authors (CG-W and RL) as being a fine and distinctive plant and is named here as 'Fuller's Mill'. Its origin is not recorded and, as Tickner had recently died, the judges were unable to elicit any further information.

'Fuller's Mill' has obvious affinities with 'Benton End', 'E.A. Bowles' and 'Glory of Threave', especially the last with which it shares the elegant goblet-shaped flowers and very long perianth tubes. Despite their height, these two cultivars stand up surprisingly well to inclement weather. 'Fuller's Mill' can be distinguished from 'Glory of Threave' by its colour, a rich rose-purple, whereas the latter is deep purple. In 'Fuller's Mill' the thin central white central stripe reaches the tepal tip while in 'Glory of Threave' it stops short. Differences can also be seen in the shade of yellow of the anthers. Like 'Glory of Threave', fruit capsules are reliably borne most years.

Corm subglobose to ovoid, 4.5–6 × 5.4–6.4cm, foot triangular, often very distinct, 0.8–2.5cm long, neck stout, 5–11cm long. Leaves 3–4, ascending, mid-green, somewhat glossy, scarcely ridged lengthways; lowermost leaf narrow-elliptic, 17.5–22 × 5.8–7cm, apex subobtuse, uppermost leaf linear-elliptic, 13.5–15.5 × 3–4cm, apex acute. Flowers 1–3, goblet-shaped, borne on very long perianth tubes, with incurved tepals, deep rose-purple, somewhat glossy, with a white throat within reaching to halfway and a thin white line extending three-quarters or more to the top; outer tepals broad oval to oblong-obovate, 6.8–7.2 × 2.5–2.7cm, apex obtuse and shortly apiculate; inner tepals noticeably shorter; stamens with green filaments and orange-yellow anthers 11–12mm long; styles white, somewhat longer than the anthers, slightly curved at the shortly deccurrent stigma; perianth tube deep reddish purple, greenish towards the base, 11–14.5cm long, the cataphyll just showing above ground. Not scented.
Origin unknown, 2020
HL: 38–45 HF: 17–19 F: l9–m10

'Giant'

One of the largest-flowered cultivars, this has rather pale, coarse blooms. From established clumps they are produced en masse but have a tendency to fall over after a short time in bloom. This cultivar is, however, extremely vigorous in the garden and is quick to multiply. For this reason it is quite often offered for sale, sometimes unidentified and without a name.

'Giant', often incorrectly listed as 'The Giant', is apparently the result of a cross between *C. speciosum* Bornmuelleri Group and *C. speciosum* Giganteum Group. There is a suspicion that much of the stock of 'Giant' is virused, which often manifests as streakiness in the flowers.

Corm broad-ovoid, 5–8.2 × 7.2–10.2cm, broader than long; foot often pronounced, 0.7–2.8cm long, occasionally rather obscure; neck stout, 6.5–12.5cm long. Leaves 5–6, ascending, 27–28.5 × 4.9–45.2cm, mid-green, moderately glossy, fluted; lowermost leaf elliptic, apex subobtuse, uppermost leaf linear-elliptic to elliptic-oblanceolate, apex subacute. Flowers 5–many, coarse, large, rosy-lilac with a prominent white throat extending one-third up the tepals, very faintly tessellated within towards the margins, the tepals somewhat fluted; outer tepals elliptic, elliptic-oblanceolate, 8.3–8.9 × 3–3.4cm, apex subobtuse; inner tepals elliptic, 7.7–8.1 × 3–3.3 cm, apex subacute; stamens with white filaments and purple-brown anthers; styles pink, exceeding the stamens, curved at the tip, the stigmas shortly decurrent; perianth tube creamy-white, 15–20cm long. Not scented.

Zocher & Co., 1900–1905

HL: 44–50 HF: 19–23 F: 9–m10

'Glory of Heemstede'

The deep violet-mauve, tessellated flowers of this bold cultivar appear towards the end of the autumn flowering season. It owes much of its genetic make up to *C. bivonae* and it has much in common with 'Autumn Queen'. The flowers are more richly coloured than the latter, with more even tepals, obtuse at the apex, and it flowers later. Although a good and easy garden plant, 'Glory of Heemstede' did not perform well at the RHS trial in 2014-18 but it was one of the best at the RHS assessment of colchicums at Felbrigg Hall in 1996.

This cultivar has been listed as 'Conquest' by several English nurseries, including P-J Nursery in the 1980s, but 'Glory of Heemstede' was published first. Also, take care when buying as some plants offered as 'Glory of Heemstede' are 'Autumn Queen' and others are inferior seedlings.

Corm broad-ovoid to subglobose, 3.5–5.2 × 3–5.4cm; foot squarish, 0.5–3.6cm long, sometimes very pronounced; neck relatively stout, (3–)4.2–6.5cm long. Leaves (4–)5(–6), ascending to curving outwards, slightly yellowish, green or bright green, moderately glossy, slightly ridged lengthways; lowermost leaf elliptic to elliptic-oblong, 22–25.5 × 5–5.6cm, apex subobtuse; uppermost leaf linear-oblanceolate to lanceolate-elliptic, 23.7–26.8 × 1.7–2.8cm, apex subacute. Flowers 2–6, narrowly chalice-shaped at first, opening to broadly bowl-shaped at maturity, deep purple-pink or rich purple, strongly tessellated, with a narrow purple edge to the tepals and a white throat in the lower third, each tepal with a central white stripe in the lower half; stamens with pale, greenish white filaments and purple anthers; styles pale pink, white, flushed pinkish purple in the upper half, curved at the tip, just overtopping the stamens; perianth tube white with a faint greenish tinge, (7.5–)9–15cm long.
Zocher & Co., 1900–1905
HL: 38–42 HF: 18–23 F: 9–e10

'Glory of Threave' AGM H5

This is a very elegant cultivar with relatively small, narrowly wine-glass-shaped flowers borne on exceptionally long perianth tubes. The dark flowers bear a metallic sheen matched by few other cultivars, this and the rich purple colour make it a distinctive plant. It multiplies well vegetatively and often produces fruit with viable seed. 'Glory of Threave' is a fine garden plant that is a delight to see and well worth while seeking out.

'Glory of Threave' has been confused with both 'E.A. Bowles' and 'Benton End' and is likely to have a similar parentage involving *C. speciosum* and possibly *C. turcicum*. Included in an RHS assessment of colchicums at Felbrigg Hall, Norfolk, in 1996, it had been supplied by the National Trust for Scotland's Threave Garden and Estate labelled as 'E.A. Bowles' (Anon. 1998). It proved to be a very vibrant cultivar of exceptional quality flowering in mid to late season, but it was not 'E.A. Bowles'. The name 'Glory of Threave' was proposed by Chris Brickell and agreed by the donors (Rolfe 2006). It entered wider cultivation around 1996 when corms were sold from Felbrigg Hall in their annual sales.

Corm broadly oval, 4–4.7 × 3.8–4.7cm; foot triangular, 1–3.2cm long; neck relatively stout, cylindrical, tunicated, dark brown, 3.8–10cm long; tunic dark chestnut-brown. Leaves (3–4)5, ascending, curving outwards, 25–27.2 × 9.7–10.5cm, bright deep green, shiny, somewhat fluted; lowermost leaf elliptic, the uppermost narrow elliptic-oblanceolate. Flowers narrowly wine-glass-shaped, deep purple with a pronounced sheen on the outside, paler towards the base, with a white centre within forming a star shape, the white extending halfway up the tepals from the base; tepals oblong-elliptic, the outer 5.8–6.5 × 2.7–2.9cm, the inner 5.5–5.8 × 2.4–2.5cm; stamens with pale greenish white filaments and deep yellow anthers; styles very slender, white, 3.8–4cm long, exceeding the stamens, slightly hooked at the tip; perianth tube slender, 17–20cm long, greenish at first but darkening to match the tepals as the flowers mature. Faintly scented

Origin unknown, 1996
HL: 44–50 HF: 22–26 F: m9–m10

'Gothic Style'

The magenta-pink flowers of this cultivar with its pronouncedly ridged tepals and bright green perianth tube make it readily recognisable. Raised by Leonid Bondarenko of Lithuanian Rare Bulb Garden nursery, its parents are likely to have been 'Dick Trotter' and 'Huxley'. The raiser has subsequently renamed it as 'Looking Up' but it is included here under its original name.

Leaves 4, spreading widely, 15–20 × 5–7cm, mid-green, moderately ridged and with a sheen, the lowermost elliptic-oval, the uppermost lanceolate. Flowers 2–4, narrowly funnel-shaped, opening more widely at maturity, pink-magenta, faintly mottled, with a prominent white centre occupying half the length of the tepals; tepals elliptic to elliptic-oblanceolate, relatively narrow and prominently parallel-ridged, the largest 7–8 × 2.5–3.5cm, the inner somewhat shorter than the outer; stamens about one third the length of the tepals, with yellow anthers; styles slender, white, slightly hooked at the tip, greatly exceeding the stamens, sometimes almost as long as the tepals; perianth tube substantial, 8–11cm long, lime-green, the colour bleeding into the base of the tepals. Bondarenko, early 21st century
HL: 20–26 HF: 14–16 F: l9–10

'Gracia'

A neat, pretty cultivar that greatly enhances any collection of autumn-flowering colchicums. Unfortunately, it was only fleetingly available in the horticultural trade. It is similar to 'E.K. Balls' and the differences are outlined under that cultivar.

Leaves 15-20 × 5-7cm. Flowers goblet-shaped, pale rosy pink, finely tessellated, white in the throat and extending one third the way up the tepals, to two-thirds as a narrow median stripe, the base greenish; outer tepals elliptic-oval to obovate, 6.9-7.3 × 2.7-3cm, apex subobtuse, keeled in the lower third on the outside; inner tepals elliptic-obovate, 6.8-7.5 × 2.6-2.9cm, apex subobtuse to obtuse; stamens one-third the length of the tepals with very slender green filaments and yellow anthers turning brown on ageing, 10-11mm long; styles white, much exceeding the anthers, two-thirds the length of the tepals; perianth tube white with a hint of green, 8.3-10.7cm long.

Visser, 1974

HL: 32-40 HF: 14-17 F: 9-e10

'Harlekijn'

This unusual cultivar has the dubious honour of being the first bicolour to enter the horticultural trade. It is of little appeal except to those keen to amass a full collection of cultivars. The flowers generally arise white and gradually gain purple-pink coloration in the lower part, although this never reaches far up the tepals. In addition, the tepals are somewhat inrolled along the margins, giving the flowers a distinctly pinched appearance.

'Harlekijn' is clearly related to 'Jarka', 'Joseph' and the recently recognised 'Redgrave' and 'World Champion's Cup', all derived from what used to be grown as *C. bornmuelleri* (now 'Joseph'), but the last three have regular-looking flowers without the pinched and slightly contorted look of 'Harlekijn'. All four cultivars are of a similar colour, basically purplish pink with white towards the top, the white appearing as a distinctive patch in 'Redgrave'.

Corm ovoid, 6.5–7 × 5.4–5.8cm; foot obscure, when present rounded, up to 1 cm long; neck 3.8–6.7cm long. Leaves usually 4, ascending, mid-green, somewhat glossy, with a few slight ridges lengthways; lowermost leaf narrow-elliptic to elliptic-oblanceolate, 23.8–24.5 × 4–4.2cm, apex subacute; uppermost leaf narrow-elliptic, 22–22.8 × 2.6–3.6cm, apex acute. Flowers bicoloured, purple-pink in the lower half, creamy-white above, the tepals often inrolled in the upper half; tepals elliptic to ovate-elliptic, 5.2–6.5 × 2.4–2.8cm, apex subobtuse; perianth tube greenish white, 7–13cm long.
Visser, 1988
HL: 39–43 HF: 15–25 F: m9–m10

'Herbstkugel'

An attractive cultivar, this bears stout, relatively thick-tubed flowers that are a characteristic broad, goblet shape with relatively short tepals when fully developed. Flowering in mid-season, it has excellent weather resistance.

This selection was originally bought in the early 1980s as 'Huxley', along with some other misnamed colchicums, from a flower shop in Prague by landscaper and snowdrop enthusiast Hagen Engelmann. After 30 years of trying to establish its true identity he concluded it did not match any existing cultivars, so he named it 'Herbstkugel'.

Corm almost globose, 5–7.5cm; foot very small; tunic mid-brown. Flowers 1–5, broad, rounded goblet-shaped, mid-pinkish purple, untessellated, the tepals whitish towards the base and inside with a median white stripe extending about half way from the base; outer tepals ovate elliptic to oval, 5.8–6.2 × 3.5–3.7cm, apex obtuse, keeled on the exterior in the lower third; inner tepals slightly shorter than the outer, elliptic-obovate, 5.4–6 × 3–3.2cm, apex obtuse to subobtuse; stamens two-fifths the length of the tepals, with white filaments that are greenish towards the base, and yellow anthers, 7–8mm long; styles white, purple at the tip, straight, equalling or slightly longer than the stamens; perianth tube off-white, stout, 6–9cm long. Fruit capsules unknown.

Origin unknown, 2010s
HL 30–45 HF: 12–15 F: e9–m10

'Huxley' AGM H5

This is a very beautiful cultivar with goblets of pastel lilac-pink or rose-lilac. It is late-flowering with a distinct U-shaped flower profile when fully open. It is readily identified by its colour and the rather gappy flowers with concave tepals, especially in the upper half. The tepals bear the faintest trace of tessellations towards the top.

'Huxley' belongs to a small group of cultivars that includes 'Dick Trotter' and 'Pink Goblet'. All three have flowers with a silky sheen and remarkably smooth tepals (characters derived from *C. speciosum*) , a whitish throat and yellow anthers. Early in the flowering period, before the tepals spread widely apart, it can be mistaken for 'Pink Goblet'. 'Pink Goblet' has more goblet-shaped flowers, with markedly concave tepals without a trace of tessellation, that do not become wide and gappy as in 'Huxley', while the central white stripe fails to reach the top of each tepal. 'Dick Trotter' has a rather deeper colour overall, the perianth tube is deeper pink and there is a greenish centre to the flowers, mostly provided by the base of the stamens. The white base to the tepals continues up for about two-thirds as a thin white central stripe, this failing to reach the top. Other small differences can be observed, such as in 'Dick Trotter' and 'Huxley' the thin wispy styles overtop the stamens while in 'Pink Goblet' they generally fall short of the stamens or just equal them, and the tepals of 'Pink Goblet' are more strongly incurved giving their profile a more rounded, goblet shape.

It was considered worthy of an AGM at the recent trial at RHS Garden Hyde Hall, but this was only witheld because of its limited availability. It is sometimes offered, mainly via Dutch wholesalers who market it as 'Rosy Wonder', a superfluous later name.

Corm narrow-ovoid, 7.2–8 × 4.8–5.4cm; foot often rather obscure but in some specimens to 2.5cm long; neck relatively stout, 8.5–14.5cm long. Leaves 4–5, ascending, spreading in the upper half, 25–34 × 7–10cm, bright green or slightly yellowish green, somewhat glossy, slightly fluted lengthways; lowermost leaf oblong-oblanceolate, the uppermost leaf linear-elliptic. Flowers large, openly goblet-shaped, with a gappy appearance when full open and markedly concave tepals, rosy lilac or lilac-pink, faintly tessellated in the upper part, with a white throat reaching one-third the way from the base, gradually diffusing into pink with a slender white central stripe reaching the top of each tepal; tepals elliptic-obovate, the outer 6.6–7.8 × 2.4–3cm, the inner 5.8–7.2 × 2.3–2.6cm; stamens with yellowish white filaments and yellow anthers; styles straight, white, slightly exceeding the stamens, stigmas punctiform; perianth tube greenish white, flushed purple in the upper part, 13–16cm long.
R.O. Backhouse of Herefordshire, 1930s
HL: 38–50 HF: 19–23 F: m9–e10

'Jarka'

This is another bicolour, like 'Harlekijn', both having purple flowers with white tips. Harlekijn' has the upper third or half of all six tepals white or cream, while in 'Jarka' only the upper third of the outer three tepals is white. There is a marked tendency in 'Harlekijn' for all the tepals to roll inwards, giving the flowers a rather curious, not particularly attractive, pinched appearance. In 'Jarka' this tendency is less marked or scarcely present, making it a far more attractive and satisfactory selection.

'Jarka' and 'Harlekijn' are probably derived from 'Joseph' which has fuller, more goblet-shaped flowers which also reveal some white on the outer tepals. Along with 'Redgrave', in these four cultivars the flowers start off white overall and gradually assume the pink or pinkish purple colour. 'Jarka' was introduced by Jānis Rukšāns of Rare Bulb Nursery in Latvia who obtained it from a contact in the Czech Republic.

Rukšāns, 1992
HL: 38–45 HF: 15–22 F: 9–m10

'Jenny Robinson'

This is a distinct and easily recognised cultivar that flowers mid-season. The richly coloured flowers with bold, irregularly defined, white centres flecked with colour and white marks on the margins of the lower half of the tepal exteriors are useful diagnostic features.

This excellent cultivar originated at Chequers, the garden of Jenny Robinson in Boxford, Suffolk, and was named by horticulturist Matt Bishop in 2011. There were many clumps of *Colchicum* in her garden, with well-known cultivars such as 'Rosy Dawn' seeding around, probably giving rise to the novel plants now included in this book, such as 'Boxford', 'Chequers' and this one. 'Jenny Robinson' was not included in the trial of autumn-flowering *Colchicum* at RHS Garden Hyde Hall, otherwise it might have gained an AGM.

Corm oblong to narrow-ovoid, 4–7cm long; tunic mid-brown. Leaves 4–6, elliptic to elliptic-lanceolate, 22–27 × 5.8–6.5cm, deep green, moderately shiny, with a few, shallow, longitudinal ridges. Flowers quite stout, goblet-shaped at first but funnel-shaped at maturity, deep magenta-purple, throat white (with some flecking) extending half way up the tepals, faintly tessellated in the upper marginal areas, white along the margin outside in the lower half, giving a slightly stripy effect that extends down onto the top of the perianth tube on either side of the magenta-purple keel; outer tepals oval-obovate, 6.4–6.8 × 3–3.2cm, apex obtuse and often slightly retuse; inner tepals somewhat shorter than the outer; filament channels finely pubescent along the ridges; stamens one-third to two-fifths the length of the tepals, with stout, green filaments and yellow anthers, 8–9mm long; styles white, overtopping the stamens; perianth tube greenish, striped purple-magenta towards the top, stout, 11.5–13cm long.
Robinson, 2011
HL: 33–38 HF: 18–20 F: 9

'Joseph'

This cultivar, which appears to be of hybrid origin between *C. speciosum* and another species or cultivar, is here given the name 'Joseph', after Joseph Bornmüller. The name is intended to replace that of *C. bornmuelleri* hort., a name used in horticulture for plants that are clearly not *C. bornmuelleri* Freyn. The latter name was coined for a species from the central Anatolian region of Turkey. This Turkish plant is robust with flowers about the size of those of *C. cilicicum* but which are more funnel-shaped, rosy purple with a well-marked, white throat, and with fine hairs along the ridges of the filament channels. The most obvious character is its purple or purple-brown anthers. In the wild *C. bornmuelleri* Freyn and the related *C. giganteum* appear to form a cline with the more familiar and widespread *C. speciosum*. In this book they are treated as a single species but separated horticulturally as *C. speciosum* Bornmuelleri Group and *C. speciosum* Giganteum Group. Most plants in cultivation as *C. bornmuelleri*, now called 'Joseph', do not comply with the description of the wild species because they have yellow anthers and lack the pubescent filament channels.

The recently named 'Redgrave' is similar in many respects to 'Joseph'. The flowers of 'Joseph' are more funnel-shaped while those of 'Redgrave' are openly bowl-shaped. The white on the outside of the tepals of 'Joseph' is rather ill-defined and ragged, while that of 'Redgrave' is a well-marked patch in the upper half of the outer tepals. Cultivars that appear to be derived from 'Joseph' include 'Artur Klark', 'Fabergé's Silver', 'Harlekijn', 'Jarka', 'Redgrave' and 'World Champion's Cup'.

Corm large, ovoid, 8–12cm; foot short, blunt or absent; tunic dark brown, tough. Leaves 5–6, at first ascending, but later spreading in the upper half, 22–27 × 4–4.8cm, mid-green and somewhat shiny, the lowermost leaf generally rather distant from the upper which are closely clustered at the top of a pseudostem; lowermost leaf elliptic-oblong, apex obtuse to subobtuse; uppermost leaves linear-elliptic, apex subobtuse to subacute. Flowers (1–)2–5(–7), large, broadly goblet-shaped with incurving tepals, white in bud, then lilac-pink or rose-pink with a prominent, rather uneven white throat extending to halfway up the tepals, the outer tepals with a central white patch on the outside extending to the apex; outer tepals oval-obovate to ovate-elliptic, (5.7–)7.4–7.9 × (3.4–)3.7–3.9cm, apex obtuse-apiculate; inner tepals slightly shorter than the outer; stamens with greenish filaments and yellow anthers, browning on ageing, one third the length of the tepals; styles white, straight, more or less equalling or overtopping the stamens, slightly curved at the top, shortly decurrent; perianth tube stout, greenish, 9.5–15cm long, grooved. Scented faintly of primroses. Fruit capsules occasionally formed.
Origin unknown, 2020
HL: 40–45 HF: 17–23 F: 9–m10

'Kiss Me Quick'

This is an attractive free-flowering cultivar that is occasionally offered for sale, although it can be difficult to obtain. The flowers are of medium size and a pleasing clear rosy lilac. One of its outstanding features is its ability to produce a large cluster of flowers, up to nine, from a single corm. These are quite stocky and stand up to the weather well. 'Kiss Me Quick' should certainly be more widely grown if the name does not put you off!

Corm ovoid to oblong-ovoid, 4.8–5.6 × 3.2–3.6cm; foot prominent, triangular, to 2.2 × 1.5cm; neck to 7cm long, occasionally longer. Flowers 4–7(–9), broadly goblet-shaped, mid to quite deep rose-lilac, with a white centre extending about half way up the tepals, outer tepals oval-obovate, 6–6.4 × 2.2–2.4cm, apex obtuse; inner tepals elliptical, very slightly shorter than the outer; stamens with greenish white filaments and yellow anthers; styles white, straight, about half the length of the tepals, equalling or slightly longer than the stamens, sometimes curving towards at the tip, stigmas punctiform; perianth tube greenish white, flushed purple, especially towards the top, 6.5–9cm long.
Origin and date unknown
HL: 25–23 HF: 10–15 F: 9–e10

'Larisa'

A hybrid between 'Dick Trotter' and *C. speciosum* 'Ordu' by Leonid Bondarenko of Lithuanian Rare Bulb Garden nursery, this was raised from the same seed pod as 'Flamenco Dance'. It has elegant, slim, pale flowers that stand out from the majority of deeper coloured cultivars. It is well worth seeking out but not often available.

Corm narrow-ovoid, relatively small, 5.5–6 × 2–2.5cm; tunic mid-brown. Flowers narrowly funnel-shaped, pale mid-pink with a whitish centre; tepals narrow oval to oval oblanceolate, 6.6–7 × 1.3–1.5cm, apex subobtuse to subacute, the inner very slightly shorter than the outer, all with a hint of tessellations towards the margins; stamens about half the length of the tepals, with greenish white filaments and pale yellow anthers; styles slender, white, slightly exceeding the stamens; perianth tube pale green, the green running up the midrib in the lower part of each tepal. Fruit capsules unknown.

Bondarenko, 2010
HL: 38–45 HF: 10–12 F: 9–e10

'Lilac Bedder'

This little-known cultivar has flowers that stand up to the weather reasonably well. It has relatively large, lilac to light violet-purple flowers with slightly darker veins, the tepals amethyst-violet inside and with a prominent white centre. At one time it was sold by Ingwersen's Birch Farm Nursery in Sussex but is no longer readily available. It is not particularly distinguished and certainly inferior to many other more recently named selections.

Corm narrow-ovoid, 5.5–8cm long; tunic pale to mid-brown tunic. Flowers 2–3, goblet-shaped, mid-lilac-pink with a white throat extending up the lower third of the tepals, the junction between the white and lilac pronouncedly ragged, tepals tending to incurve in the upper part; outer tepals narrow-obovate, 6.8–7 × 2.5–2.8cm, apex obtuse to more-or-less truncated at the apex; inner tepals elliptic-obovate, 6.1–6.4 × 2.5–2.8cm, apex obtuse to more-or-less truncated at the apex; filament channels minutely puberulous; stamens two-fifths the length of the tepals, with greenish filaments and pale purplish brown anthers; styles white, extending slightly beyond the stamens at maturity, slightly curved at the tip, the stigmas punctiform to slightly decurrent (for no more than 0.5mm), purple-tipped; perianth tube pale greenish white with a pinkish flush towards the top, 10.2–11.5cm long.

Visser, 1974
HL: 26–35 HF: 22–26 F: m8–l9

'Lilac Wonder'

A widely available, floriferous cultivar with large flowers of excellent colour, especially if planted in full sun. Unfortunately, the flowers tend to flop over in the slightest breeze or after rain, as the perianth tubes are quite slender. This feature aside stopped, what is in all other respects a fine cultivar, from attaining an AGM at the trial at RHS Garden Hyde Hall. The flowers stand up better when the corms are naturalised in rough grass.

This was at one time, along with what was then sold as *C. bornmuelleri* (now 'Joseph'), marketed as a windowsill 'bulb'. These were purchased in late summer in the dormant stage, placed dry in a bowl situated on a windowsill where they would eventually flower, then afterwards planted in the garden. Although widely available, it is often misnamed and listed under the names 'Premier' or 'The Premier'.

The lost cultivar 'Mr Kerbert' was said by Bowles (1924) to be very like 'Lilac Wonder' but with a significant white throat to the flowers, along with white channels extending two-thirds up the length up the tepals.

Corm ovoid, 5.5–7 × 2.8–7cm, with a deep brown rather tough tunic, foot triangular, to 2.5cm long, apex obtuse, sometimes indistinct, neck relatively stout, 4.6–8.2cm long. Leaves 4–5, ascending, spreading in the upper half, 31–34.5 × 6–7.2cm, bright mid-green, somewhat glossy, scarcely fluted if at all; lowermost leaf oblanceolate, with the margin often rolled back towards the top; uppermost leaf narrow elliptic. Flowers freely produced, up to 12, occasionally more, narrow goblet-shaped at first, opening widely and more starry at maturity, rose-purple or rose-violet (lilac-pink in shade), with very faint tessellations, with a thin white stripe along the centre of each tepal for two-thirds from the base, tepals characteristically creased lengthways, narrow oblong to elliptic oblanceolate; outer tepals 8–8.4 × 2.1–2.3cm; inner tepals 7–7.3 × 1.8–2cm; stamens with white filaments, stained green near the base, and pale yellow anthers; styles white, slightly exceeding the stamens, hooked at the tip, stigmas shortly decurrent; perianth tube white flushed rose-lilac or rose-purple towards the top, rather slender, 12–18cm long.

Zocher & Co., 1900–1905

HL: 40–47 HF: 19–26 F: m9–m10

'Little Woods' AGM H5

A charming, relatively small-flowered cultivar with neat, goblet-shaped flowers on sturdy perianth tubes, giving it good weather resistance. It is best grown in full sun, otherwise it can become leggy. The flowers last rather longer than those of many other cultivars and keep their shape. Although the flowers are not the largest of the AGM selections, they are produced in quantity and the foliage that follows in the spring is substantial. Plants occasionally produce fruit capsules containing viable seed.

'Little Woods' has affinities with *C. autumnale*, although the flowers of this hybrid are more broadly chalice-shaped and the tepals are rather wider and of firmer texture. Although not readily available, 'Little Woods' is a fine garden plant, multiplying freely in most well-drained soils. The original stock was distributed by Elizabeth Parker-Jervis of P-J Nursery, Longworth, Oxfordshire, in the 1980s.

Corm subglobose-ovoid, 3.2–3.5 × 3–4.2cm, generally as wide, or wider than long; foot triangular, generally 0.5–1.8cm long, sometimes obscure; neck slender, 3.5–7.6cm long. Leaves 4–5, ascending, spreading in the upper half, 26–35.5 × 7–8.5cm, mid to deep green, moderately glossy, lightly ridged lengthways; lowermost leaf lanceolate-elliptic to elliptic, apex subobtuse, uppermost leaves narrow lanceolate-elliptic, apex subacute. Flowers 3–9, goblet-shaped, but with wider tepals, mid-pink, tepals with a median white stripe for half to two-thirds the length from the base, finely veined; outer tepals elliptic obovate, 4.5–7 × 1.8–2.5cm; inner tepals elliptic-oblanceolate, 4.2–5.5 × 1.6–2.2cm; stamens with white filaments and pale yellow anthers; styles white, greatly exceeding the stamens, with a deep purple, hooked tip, stigmas shortly decurrent; perianth tube pure white, 11–14cm long.
Parker-Jervis, 1980s
HL: 45–54 HF: 10–15 F: m9–10

'Lysimachus'

Generally performing well in the open garden, this cultivar is rather slow to increase compared with more vigorous and better-known ones. Raised in 2006 by nurseryman Antoine Hoog, it is intermediate in most characters between its two parents, *C. autumnale* 'Drama Bunch' and probably *C. haynaldii*. It is named after Lysimachus (360–281BC), a Thessalian general under Alexander the Great who later became king of Thrace, Asia Minor and Macedon.

Flowers 2–5, funnel-shaped at maturity, rose-purple with a white star shape in the centre, the star arms reach about halfway up the centre of each tepal; tepals oval, 5.5–8cm long, apex obtuse, the margins characteristically curving inwards; stamens quite short, about a quarter the length of the tepals, with white filaments flushed with purple, styles thread-like, just overtopping the stamens.

Hoog, 2012
HL: 28–36 HF: 10–14 F: 9

'Mells Park'

This is a relatively undistinguished cultivar that has been grown in gardens for some years. It has no special features that distinguish it, although the corms multiply well in the garden. It is one of a number of mid-pink colchicums that are fine for naturalising in grass as it has the general appearance of *C. autumnale*.

It is named after the country estate of Mells Park near Frome in Somerset, which has a house designed by Edwin Lutyens and a garden planned by Gertrude Jekyll. It is sometimes listed incorrectly as 'Mellis Park'

Corm oblong-ovoid; tunic mid-brown. Leaves 4–5, mid-green and somewhat shiny, 24–25 × 6.2–6.6cm, ridged longitudinally; lowermost leaf elliptic; uppermost leaves elliptic to elliptic-lanceolate. Flowers mid-pink, very similar in shape and size to those of larger-flowered forms of *C. autumnale*; tepals 4–6cm long. Fruit capsules often produced, these more or less ovoid and to almost 6cm in length.
Originated from Mells Park, mid 20th century
HL: 38–44 HF: 12–16.5 F: 9

'Neptun'

This is a recent cultivar with relatively narrow flowers bunched together in small posies. It has not yet been grown widely in the UK so little is known of its performance.

Corm narrow-ovoid, 5.5–9cm long; tunic dark brown. Leaves usually 4, ascending, 15–20 × 6–8cm, mid-green, heavily ridged lengthways; lowermost leaves oblong-elliptic; uppermost leaves elliptic. Flowers 2–6, narrowly funnel-shaped, violet-purple, finely tessellated with a white star in the throat extending one-third the way up the tepals; tepals narrow oval to oval-elliptic, 6.8–8 × 2.5–3.5cm; stamens about half the length of the tepals, with whitish filaments and yellow anthers; styles white, sometimes pink-flushed towards the tip, very slender, much exceeding the stamens at maturity; perianth tube creamy white, flushed with pink, 6.5–10cm long.

Origin unknown, 2017
HL: 20–25 HF: 14–18 F: 9

'Oktoberfest'

A late-flowering cultivar with sumptuous deep rose-purple flowers and a bold white centre, 'Oktoberfest' can be likened to a late-flowering 'Rosy Dawn'. It is an equally fine cultivar although it did not perform particularly well in the trial at RHS Garden Hyde Hall. It has grown well in the authors' gardens in average garden soils in full sun where it multiplies relatively quickly. The flowers are more richly coloured than 'Rosy Dawn' and have a rather less prominent white throat.

Corm ovoid to ovoid-oblong, 5–8cm long. Leaves 4(5), elliptic-oblong to oblong-oblanceolate, 21–25 × 5.5–7cm, mid to rather bright green with a moderate shine. Flowers 1–5, goblet-shaped at first but more funnel-shaped at maturity, mid-rose-purple with some faint tessellations in the upper part, particularly towards the margins of the tepals, with a white centre reaching for just over a third from the base of the tepals; outer tepals broad oval-elliptic, 7.9–8.4 × 3.6–4cm, apex obtuse; inner tepals somewhat smaller, obovate, 7.2–7.5 × 3.4–3.5cm, apex obtuse; stamens with greenish filaments and pale to mid-yellow anthers, browning on ageing; styles white, sometimes with a faint pinkish flush towards the top, equalling or overtopping the stamens, stigmas shortly decurrent at the curved tip; perianth tube greenish, purple-flushed in the upper part, 9–15cm long. Fruit capsules unknown.

Rollich, date unknown
HL: 35–40 HF: 16–22 F: l9–e11

'Pink Goblet' AGM H5

One of most distinctive autumn colchicums, this has beautiful, rounded goblet-shaped flowers of soft lilac-pink, enhanced by a silky patina on the exterior. It flowers at the same time as 'Benton End' and the two can be very effective planted next to one another. The flowers are held on stout, pale greenish, pink-flushed perianth tubes, making them capable of withstanding the vagaries of the autumn weather. The only cultivars with which it might be confused are 'Dick Trotter' and 'Huxley'. For differences between them, see 'Huxley'. The styles of 'Pink Goblet' are often strangely kinked at the apex.

This charming cultivar, along with 'Dick Trotter', arose from a batch of seedlings sown from *C. speciosum* 'Album'. Dick Trotter reportedly broadcast the seed on his compost heap at Brin, his garden near Inverness in Scotland. 'Pink Goblet', a truly pink colchicum, was perhaps the finest plant to come out of these random sowings.

Corm broadly ovoid, 3.2–5.2 × 2.8–5cm, abruptly extended above into a slender, dark brown neck, 3.8–10cm long; foot pronounced, triangular to oblong-triangular, 1–4.5cm long when mature; tunic dark chestnut-brown. Leaves 4–5, ascending, curving in the upper half, elliptic-oblanceolate, 17–30 × 7.4–8.8cm, mid-green, moderately shiny, poorly pleated. Flowers 1–4(–6), broadly goblet-shaped, bright, glossy, rather pale pink, often with a hint of lilac, deeper on the outside, with a rather ill-defined, whitish centre within, occupying the lower one-third to half of the tepals, and with a central white stripe extending for two-thirds of the outer tepals and three-fifths of the inner, tepals oblanceolate, pronouncedly concave; outer tepals 7.2–8.2 × 2.7–3.2cm; inner tepals somewhat shorter; stamens yellow; styles white, 2.4–2.9cm, shorter than or about the same length as the stamens, slightly curved and sometimes kinked at the tip; perianth tube greenish below, pale pink above, 15–17cm long. Scented of honey. Fruit capsules occasionally produced.
Trotter, 1930s
HL: 35–48 HF: 22–25 F: 9–m10

'Pink Star' AGM H5

This cultivar is thought to be a hybrid involving *C. cilicicum* or *C. × byzantinum* and appears to be sterile, or at least does not appear to produce any fruits. However, it is a prolific multiplier in the garden and, being rather smaller than the majority of the autumn-flowering cultivars, makes a good feature at the front of a border. Its generous production of flowers is probably owed to *C. × byzantinum*.

It is unfortunate that this excellent, floriferous plant has been much confused in gardens with *C. laetum*, a southern Russian and Caucasian species that is only occasionally seen in cultivation. 'Pink Star', a name suggested by Chris Brickell after an RHS assessment of colchicums at Felbrigg Hall, Norfolk (Anon. 1998), is sometimes listed as *C. laetum* hort. It is also confused with *C. × ambiguum* whose flowers are a similar size. However, the flowers of *C. × ambiguum* have broader, more substantial tepals of a deeper purple-pink or rosy purple and are faintly tessellated, with anthers that are creamy yellow rather than yellow and styles that are reddish purple throughout.

Corm ovoid, 3.5–6.2 × 4.8–7.5cm, generally broader than long; foot 1–2.8cm long; neck slender, 4.5–8.6cm long. Leaves 3–4, ascending, 33–40.5 × 8–9.5cm, deep and slightly bluish green, glossy, pleated longitudinally; lowermost leaf elliptic-oblong, apex subobtuse, the uppermost leaf narrow lanceolate-elliptic, apex subacute. Flowers borne in profusion, chalice-shaped at first but opening to wide stars, pale purple-pink, tepals rather narrow and blunt with a median whitish stripe for half to two-thirds from the base, sometimes with a faint hint of chequering; outer tepals narrow-oblanceolate, 3.8–5.8 × 0.9–1.4cm, apex obtuse; inner tepals narrow-elliptic-oblanceolate, 3.5–5.5 × 0.7–1.2cm, apex subobtuse; stamens with white filaments and yellow anthers; styles very slender, greatly overtopping the stamens, white, tipped pale purple; perianth tube white, 4–8.5cm long.
Origin unknown, 1998
HL: 42–50 HF: 8–14cm F: l8–9

'Poseidon'

An excellent, robust cultivar, this flowers early and has quite squat, substantial, weather-resistant flowers. The large and luxuriant leaves begin to appear shortly after flowering and are part-developed by winter, leaving them prone to some frost damage during severe weather but this, in the UK at least, is not enough to cause any lasting harm, although the leaf tips can sometimes be spoilt. This is in contrast to some cultivars, 'Glory of Threave' for example, in which the leaves only begin to appear in late winter.

This cultivar has been offered under names such as 'Jarkoslavan', 'Jaroslavna', 'Jaroslawna', 'Jochem Hof' and 'Jochum Hof' but 'Poseidon' is the name registered by its breeder.

Corm subglobose to broad-ovoid, 3.8–5.2 × 4–5.4cm; foot sometimes rather obscure but usually obvious, 5.5–8.5cm long; neck relatively stout, 3.8–6.5cm long. Leaves 4–6, ascending, curving outwards in the upper half, 18–31 × 9.2–11.2cm, deep green, glossy, stiff and fleshy, moderately fluted; lowermost leaf elliptic to ovate-elliptic; uppermost leaf narrow elliptic-lanceolate. Flowers 2–7, goblet- to funnel-shaped, rather stocky, deep purple, somewhat paler towards the base of the tepals, whitish within at the base, extending up the midvein of each tepal to about a quarter the length, tepals elliptic to elliptic-oval; outer tepals 6.5–6.8 × 2.4–2.6cm; inner tepals 6.3–6.5 × 2–2.2cm; stamens with purple filaments and deep yellow anthers; styles purple, long and slender, straight, 6.4–6.7cm long, greatly exceeding the stamens at maturity, stigmas punctiform; perianth tube stout, 12–15cm long, whitish, flushed purple, deeper towards the top. Faintly fragrant. Fruit capsules not produced.
de Groot & de Zilk, 2007
HL: 38–47 HF: 17–21 F: 18–9

'Pride of Holland'

This name, coined here by two of the authors (CG-W and RL), is for plants produced in the Netherlands and sold as 'Violet Queen'. They have untessellated, pink flowers without a hint of violet and a white throat. The original 'Violet Queen', an old cultivar from the early 20th century, had larger, violet, well-tessellated flowers with a bold white throat. It appears to be no longer in cultivation, although there is always a chance it might survive in an old garden somewhere. A different plant, sometimes found in private collections as 'Violet Queen', is also named in this book, as 'Felbrigg Violet'.

Corm ovoid, somewhat asymmetrical, with a short neck. Leaves mid-green, somewhat lustrous, with some longitudinal ridging; lowermost leaf oblong-elliptic, to 55 x 23cm, apex subobtuse; uppermost leaf much reduced usually, linear-oblanceolate, to 20 x 1.5cm, acute. Flowers 1–3, funnel-shaped, pink with a hint of lilac, untessellated and with a white throat occupying the lower third of the flower; outer tepals elliptic-obovate, 7–8.5 x 3.0–3.4cm; inner tepals elliptic-oblanceolate, slightly shorter, 6.7–8.2 x 2.4–2.9cm; stamens with white filaments with a slightly swollen yellowish base, and yellow anthers; styles white, slender, straight, exceeding the stamens, slightly curved at the tip; perianth tube greenish white, relatively short and about the same length as the tepals.
Grey-Wilson & Leeds, 2020
HL: 34–40 HF: 13–16 F: 9–e10

'Redgrave'

This is a robust and distinctive plant, easily recognisable by the prominent, pale whitish patch on the outside and towards the tip of each tepal. In this respect it comes closest to 'Joseph' and is almost certainly derived from that cultivar, which used to be known as *C. bornmuelleri* hort.

This interesting colchicum was found in a garden in Redgrave, Suffolk, by Christopher Grey-Wilson and named by him in 2017. Its origin is unclear although the owner of the garden thought that her late husband acquired them on a trip to the Continent.

Corm narrow ovoid, 5–6.2 × 3.8–4.8cm; foot short, triangular, 1–1.3cm long, sometimes rather obscure; neck stout, 4.8–7.2cm long. Leaves (3)4, ascending, 26–29.5 × 5.2–6.6cm, mid-green, somewhat glossy, slightly ridged lengthways; lowermost leaf elliptic to elliptic-oblong, apex obtuse; uppermost leaf narrow oblong, apex subobtuse. Flowers mid-rose-pink, paler or whitish towards the margins, with a prominent white central patch towards the top, particularly noticeable on the exterior; tepals oval-obovate, 6.5–7.8 × 3.5–4.2cm, the inner slightly shorter than the outer; stamens with greenish filaments and yellow anthers; styles white, overtopping the stamens, with a shortly decurrent stigma; perianth tube greenish white, 9–14cm long. Fruit capsules not normally formed.

Origin unknown, 2017
HL: 46–58 HF: 10–16 F: m9–e10

'Rosy Dawn' AGM H5

One of the very best, most floriferous and dependable cultivars, this has lightly tessellated flowers with bold white centres. The best identification feature is the arrangement of tepals which, when viewed from above, form distinct triangles, especially the inner ones. 'Rosy Dawn' is excellent for naturalising in grass, where it will come into flower several days after those in a flower border, nicely extending the season. Corms can multiply three- or even fourfold annually.

'Rosy Dawn' will often produce fruit and seed around in the garden. The seedlings appear to come true to type. There is, however, a suspicion that two clones are available. Both are identical in most respects except that one comes into flower eight to ten days later than the other, but this requires further investigation. Much of the stock in gardens came directly or indirectly from the Suffolk gardens of either Cedric Morris or Jenny Robinson.

Corm ovoid, 4–5 × 2.8–3.4cm, extending at the top into a long, relatively stout, cylindrical, dark brown neck, 6–10.5cm long; foot absent or very short, broad triangular; tunic dark chestnut-brown. Leaves 4–6, ascending to erect, curving outwards gradually, 22–25 × 7.5–8.2cm, deep green, moderately shiny, generally with 4–6, shallow, fluted pleats; lowermost leaf elliptic-oblong; uppermost leaf narrow lanceolate-elliptic. Flowers 1–3, funnel-shaped with flattish to slightly concave tepals spreading at about 70°, more widely apart as the flowers fade, the inner and outer tepals, when looked at from above, forming rough, interlocked triangles, bright rose-pink, tessellated in the upper half, with a white centre extending to the middle of each tepal, tepals elliptic-obovate; outer tepals 7.2–8.5 × 3.3–3.8cm; inner tepals slightly shorter, 6.8–7.8 × 2.6–3.1cm; stamens yellow at first, soon browning; style whitish, very slender, just exceeding the stamens, very slightly curved at the tip; perianth tube greenish 16.5–19cm long, relatively stout. Fragrant of primrose. Fruit capsules often produced.
Barr & Sons, 1948
HL: 38–49 HF: 24–27.5 F: 9–e10

'Spartacus'

This cultivar produces compact bunches of short, lilac-pink flowers with white centres, faint tessellation and a slight perfume.

Raised in 2007 by nurseryman Antoine Hoog, it is a cross between *C. autumnale* 'Drama Bunch' and *C. bivonae*. It is sometimes listed as *C. autumnale* 'Spartacus' or *C. bivonae* 'Spartacus'. See 'Enigma' for details of a cultivar wrongly grown as 'Spartacus'.

Corm broadly ovoid, 8.5–8.8 × 8.5–9.2cm, generally as wide as long; foot obscure, to 5mm long at the most if present. Leaves 5–7, erect to ascending and somewhat spreading, bright mid-green with a slight sheen, with several shallow ridges lengthways; lowermost leaf elliptic, 23–30 × 5.5–7.5cm, apex subobtuse; uppermost leaf linear-elliptic, 18.5–20 × 2.5–3cm, apex subacute. Flowers 3–8, chalice- to somewhat funnel-shaped, particularly crocus-like, pale lilac-pink fading to white in the lower half, each tepal with a rather indistinct central white stripe reaching two-thirds from the base, occasionally with some faint tessellation towards the basal margin; outer tepals elliptic-oblong to elliptic-oblanceolate, (4.3–)5–6.3 × (1.2–)1.5–2cm, apex obtuse; inner tepals narrow-elliptic, (3.9–)4.5–5.4 × (1–)1.4–1.6cm, apex subobtuse; stamens one-third to half the length of the tepals, filaments white, anthers yellow and around 7mm long; styles 3.1–3.3cm long, overtopping the stamens, curved at the tip which often has a faint pink tinge; perianth tube greenish white, sometimes slightly pink-flushed at the top, 7–8.4cm long. Fruit capsules not produced.

Hoog, 2012
HL: 43-47 HF: 12-15 F: 9-m10

'Waterlily' AGM H5

The finest fully double-flowered colchicum, 'Waterlily' is well known and widely available. Aptly named, the flowers are large, full, multi-tepalled and long-lasting, although they tend to flop over in inclement weather. They will support each other when closely planted. Nonetheless, it is one of the most distinctive and eagerly sought garden cultivars.

There are three double-flowered colchicums at present in the horticultural trade, the other two being *C. autumnale* 'Alboplenum' and *C. autumnale* 'Pleniflorum', the former being the only white-flowered one. In many double-flowered plants it is usually flower parts such as sepals, stamens, or sometimes even the styles and ovary, that become petal-like. It is frequently the stamens that develop into petal-like structures and on occasions this transition is imperfect so that a stamen may only be part-converted to a petal. This can be seen clearly in 'Chequers', where most flowers bear one, sometimes as many as six, curled and often contorted extra tepals inside the six normal ones, although there are not enough extra tepals to regard it as a double-flowered cultivar. These inner tepals are much smaller than the outer and often bear parts of the filament or anther in their structure.

This phenomenon of developing extra tepals from other flower parts does not appear to be what has happened in the conventional double-flowered colchicums. In these there are many more flower parts, manifested as tepals, than can be accounted for in a normal *Colchicum* bloom, which has six tepals, six stamens and three styles. In 'Waterlily' there are more than 50 tepals, which present a gradual transition with the outer the largest and the inner the smallest, and there is usually no trace of stamens. In the two double-flowered cultivars of *C. autumnale* there are fewer and a rather more uneven number of tepals, but certainly in excess of 30. Curiously, in both these cultivars stamens are often, but not always, present and are always associated with the outermost tepals and join with the accompanying tepal in a similar manner to those in an ordinary colchicum flower. There do not appear to be any transitional stamens or staminodes but the stamen number is very many more than six, as might be expected. In both these cultivars the central tepals are generally short and upright, forming a brush-like structure in the centre of the flower. It is difficult to explain this proliferation in flower parts, particularly in the numbers of tepals, and the impression given is that several flowers have somehow merged together. In many plants that do this, some fasciation is generally visible, but in these double colchicums no fasciation is obvious.

'Waterlily' reportedly arose from a cross between *C. autumnale* 'Alboplenum' and *C. speciosum* 'Album' (Bowles 1952, Mathew 2000), strangely both of which are white-flowered selections.

Corm broad-ovoid, 8.6–9.8 × 9–10.2cm, broader than long, foot broad and rounded, 1.2–1.8cm long, neck stout, 7–12cm long. Leaves (3–)4(–5), ascending, spreading in the upper half, 21–35 × 6–7cm, often with an extra leaf curled at ground level, bright mid-green, moderately glossy, with several folds lengthways; basal leaf, when present to 18.5 × 5.2cm, contorted and curled; lowermost leaf narrow elliptic to elliptic-oblanceolate; uppermost leaf narrow elliptic to elliptic-oblanceolate. Flowers 2–6, rosy-lilac, fully double and 'waterlily-like', to 12cm across, with 40 or more tepals that decrease in size towards the centre of the flower; tepals elliptic to elliptic-oblanceolate, the innermost linear-elliptic, all spreading widely apart, the largest to 7.5 × 1.8cm; stamens usually absent, occasionally 1–3, part-staminode with a whole or part anther attached; styles absent; perianth tube 7–11cm long, white, flushed rosy-lilac towards the top. Fruit capsules not produced.
Zocher & Co., 1900–1905
HL: 36–42 HF: 14.5–18 F: m9–m10

'Whitton Globe'

A little-known cultivar, this has attractive, rounded, goblet-shaped flowers that slowly open to a generous bowl shape. The rosy purple colour is smooth and even, paling to white at the base of the tepals, while the exterior has a slight sheen. Although slow to increase, 'Whitton Globe' is a sturdy plant in flower, standing up well to the vagaries of autumn weather. The authors have only ever seen it in the garden of John and Diana Morley at Stoven in Suffolk but it was selected by bulb enthusiast Richard Hobbs.

Flowers relatively small yet sturdy, goblet-shaped, mid-rose-purple, untessellated, whitish at the very base of the tepals and with a median white stripe extending up from the base in the lower two-thirds, with a slight sheen on the exterior; outer tepals elliptic-oval to elliptic-obovate, 5.2–5.4 × 2.3–2.5cm, apex subobtuse; inner tepals slightly shorter than the outer, elliptic-obovate, 4.8–5 × 1.9–2.1cm, apex subacute; stamens about half the length of the tepals, anthers mid-yellow, 6–7mm long; styles whitish, flushed with purple towards the top, slightly exceeding the stamens; perianth tube stout, 6–7mm in diameter, green, flushed rose-purple, especially towards the top, 7–9cm long. Not scented.

Hobbs, late 20th century
HL: 35–40 HF: 12–15 F: 9

'William Dykes'

This is a fine and reliable cultivar from the early 20th century. Elizabeth Parker-Jervis, in a catalogue of 1983, stated it was from the same seed bed as 'Lilac Wonder' and that it was also known as 'Intermediate Dykes'. William Rickatson Dykes (1877–1925) was an authority on *Iris* in particular, both as a collector, breeder and writer, and it was named after his death.

Corm narrowly ovoid, 6.3–8 × 4–6.2cm; foot obscure or rounded and up to 0.8cm long; neck 10–13cm long. Leaves 4–5, ascending, 34.5–43.5 × 7.6–11.2cm, deep green, glossy, with several ridges lengthways; lowermost leaf elliptic, apex subobtuse; uppermost leaf linear-elliptic, apex acute. Flowers freely borne, goblet-shaped, particularly crocus-like, pale lilac, sometimes very faintly tessellated, paler, often whitish towards the base within, with a median white stripe within extending about halfway up each tepal, tepals incurving in the upper half; outer tepals oval-obovate, 6–7.1 × 2.6–3cm; inner tepals similarly shaped, slightly smaller, 5.7–6.7 × 1.8–2.5cm; stamens with greenish filaments and yellow anthers; style white, 3.8–4.1cm long; perianth tube greenish white, flushed lilac towards the top, 10.5–12.7cm long.

Zocher & Co., 1900–1905

HL: 50–54 HF: 16–19 F: 9–m10

'World Champion's Cup'

A newish cultivar, raised by Leonid Bondarenko of Lithuanian Rare Bulb Garden nursery, this is distinguished primarily by its large, goblet-shaped flowers. It is similar in most respects to 'Redgrave' but with rather larger, taller chalices with broader outer tepals that give the flowers a fuller, more rounded appearance. In addition, the outer tepals are white only at the tip, whereas in 'Redgrave' the white extends to half the length of the outer tepals from the tip, sometimes more.

This cultivar, along with 'Harlekijn', 'Jarka' and the recently recognised 'Redgrave' are all derived from what used to be grown as *C. bornmuelleri* hort., now known as 'Joseph'. In all these, the flowers commence white in bud, gaining colour after a day or so. There is a tendency in some of them, especially 'Harlekijn', for the tepals to become inrolled or even contorted.

Corm large, ovoid, 8–12cm long; tunic mid-brown tunic. Flowers large, goblet-shaped, white in bud but gradually gaining colour to maturity, pink with white tips to the tepals, the throat white, reaching half the length of the tepals; outer tepals obovate, 7–7.8 × 4–4.3cm, apex obtuse; inner tepals slightly shorter than the outer; stamens half the length of the tepals, with greenish white filaments and cream to pale yellow anthers; styles white, slender, straight, equalling or slightly shorter than the stamens, slightly curved at the tip, stigmas shortly decurrent; perianth tube stout greenish, 11–14cm long.
Bondarenko, 2018
HL: 35–42 HF: 18–22 F: m9–e10

'Zephyr'

This bold, rather stout cultivar has beautifully formed, deeply coloured, goblet-shaped flowers with stout perianth tubes that stand up well to the vagaries of autumn weather. The large flowers are lilac-purple with a broad, white star in the throat where the white radiates up the centre of each tepal as a thin stripe.

Although the precise origin of 'Zephyr' is unclear, the cultivar reveals some of the features of *C. bivonae*, especially in the flower colour and tessellations which, in this instance, are not well marked. It probably owes its origin to a hybrid involving one of the cultivars of *C. bivonae* and another species or cultivar.

Corm ovoid, 7–10cm long; tunic mid-chestnut-brown. Flowers goblet-shaped, deep purple-pink with faint tessellation towards the tepal margins, greenish towards the base outside, with a bold white centre inside diffusing half way up the tepals; outer tepals oval-obovate, 78–80 × 35–37mm, apex obtuse; inner tepals slightly shorter than the outer, elliptic-obovate, 71–73 × 30–32mm, apex obtuse; filament channels puberulous; stamens half the length of the tepals, with stout, green filaments and mid-yellow anthers; styles whitish, straight, slightly longer than the stamens, stigmas punctiform to slightly decurrent for not more than 0.5mm; perianth tube stout, 7–8mm diameter, 10.5–13cm long. Faintly scented.

Visser, 1985
HL: 35–42 HF: 18–22 F: l9–m10

Checklist of epithets

This checklist provides botanical and cultivar epithets that have been applied in *Colchicum*. It is not exhaustive, but lists names that are most likely to be encountered in the literature.

The names are arranged alphabetically with cross-references between synonyms and accepted names. For each cultivar its species or hybrid attribution is provided in brackets. Excluded species (those with a name in *Colchicum* but considered here as belonging to *Androcymbium*) and synonymy for botanical epithets under other genera are listed at the end.

Descriptions of the cultivars are also given where possible, although some are brief due to lack of information. More detailed descriptions of the most commonly grown cultivars can be found in the Cultivars chapter and accounts of the species are in the Species chapter. Where known, information on the discovery, introduction and earliest known publication details of the cultivar is given. Information has been cross-checked against the cultivar register maintained by the Koninklijke Algemeene Vereeniging voor Bloembollencultuur as International Cultivar Registration Authority for bulbous, cormous and tuberous-rooted plants (www.kavb.nl/english/registration), originally published as van Scheepen (1991).

This checklist should help to ensure the correct application of cultivar names in this genus, preventing misnaming and also the repetition of cultivar names in the future.

Name in **bold:** Currently accepted name, with the species it is attributed to in brackets.
Name in roman: Synonym, with the species it is attributed to (or was attributed to when it was first published) in brackets, followed by an equals (=) sign and the accepted name.
E Earliest identified publication of the name with a description.
R The name of the person or company who raised, discovered, selected or introduced the cultivar.
I Year when first offered for sale.
RHS Details of RHS awards: AGM = Award of Garden Merit; [AGM] = historic AGM, now rescinded; AM = Award of Merit; BC = Botanical Certificate; FCC = First Class Certificate; PC = Preliminary Commendation.
D Description. For species, descriptions can be found in the Species chapter, and descriptions of widely grown cultivars can be found in the Cultivars chapter.
N Notes providing additional information about the cultivar, including references to the name where the earliest publication has not been located, confused applications of the name, parentage and whether still in cultivation..

Colchicum L., *Sp. Pl.* 1: 341 (1753) **Type:** *Colchicum autumnale* L., designated by Hitchcock, *Prop. Brit. Bot.* 148 (1929)

actupii Fridl.   = *C. lusitanum*
aegyptiacum Boiss.   = *C. ritchii*
afghanistanicum Mischtsch. ex Czerniak.
  = *C. robustum*
'**Afrodite**' (*C. bivonae*)   **R:** Raised by Antoine Hoog from seed of *C. bivonae* from Mount Giona and introduced by Antoine Hoog Authentic Plants. **I:** 2018. **D:** Reverse bicolour, lacks vigour and best grown under cold glass.
agrippinum auct., non hort. Angl. ex Baker
  = *C. variegatum*
× **agrippinum hort. Angl. ex Baker** (*C. autumnale*

× *C. variegatum*)   **E:** *J. Linn. Soc., Bot.* 17: 425 (1879). **RHS:** AM 1974 (Mathew & Starling 1980), AGM (H5) 1993.

'Alba Plenum'   = 'Alboplenum'

× **alberti** Regel (*C. kesselringii* × *C. luteum*)   **E:** *Trudy Imp. S.-Peterburgsk Bot. Sada.* 8: 647 (1884). **RHS:** PC 2001 (as 'clone 1', later given cultivar name of 'Jānis')

'Alboplenum' (*C. autumnale*)   **R:** Unknown. **I:** Pre-1872. **RHS:** FCC 1872. **D:** See Cultivars chapter.

'Album' (*C. autumnale*)   **RHS:** AGM (H5) 2017. **R:** Unknown. **I:** Unknown. **D:** See Cultivars chapter.

var. *album* Gray (*C. autumnale*)   = 'Album'

'Album' (*C.* × *byzantinum*)   = 'Innocence'

f. *album* (*C.* × *byzantinum*)   = 'Innocence'

var. *album* (*C.* × *byzantinum*)   = 'Innocence'

'Album' (*C. speciosum*)   **R:** Registered by J. Backhouse & Son of York. **I:** 1901. **RHS:** [AGM] 1924, AGM (H5) 1993. **N:** Chittenden (1927). **D:** See Cultivars chapter.

algeriense Batt.   = *C. lusitanum*

alpinum DC.   **E:** *Fl. Franc. [de Candolle & Lamarck]*, ed. 3. 3: 195 (1805).

alpinum var. *parvulum* (Ten.) Baker   = *C. alpinum*

amabile Heldr.   = *C. bivonae*

× **ambiguum** Grey-Wilson (*C. cilicicum* × ?*C. neapolitanum*)   **E:** *Colchicum: The Complete Guide* (2020): 557. **RHS:** AGM (H5) 1997.

'Anopolis' (*C. macrophyllum*)   **R:** John Fielding. **I:** 2000. **D:** See Cultivars chapter.

ancyrense B.L. Burtt   = *C. triphyllum*

andrium Rech. f. & P.H. Davis   = *C. pusillum*

androcymbioides (Valdés) K. Perss.   **E:** *Bot. Jahrb. Syst.* 127(2): 169 (2007).

'Annecy' (*C. autumnale*)   **R:** Cotswold Garden Flowers. **I:** 2009. **N:** Named by John Morley from a collection made in the French Alps near Annecy. **D:** Early-flowering (August to early September) and a good pink colour; height in leaf 35–42cm; height in flower 10–15cm. Scarcely distinguishable from many wild plants of this extremely variable species.

'Antares' (*Colchicum*)   **R:** Registered by P. Visser Czn. in 1977. **I:** 1977. **D:** See Cultivars chapter.

antepense K. Perss. **E:** *Bot. Jahrb. Syst.* 127: 169 (2007).

antilibanoticum Gomb.   **E:** *Bull. Soc. Bot. France* 104: 286 (1957).

'Apollo' (*C. bivonae*)   **R:** Originally cultivated by Van Tubergen and named by Antoine Hoog in 1995 when introduced by Hoog & Dix Export. Registered by Kwekerij Huisman in 2004. **I:** 1995. **D:** See Cultivars chapter.

'Arak' (*C. persicum*)   **R:** Selected and named by Antoine Hoog who received material in 2007 from Arnis Seisums who collected it near Arak in Iran. Introduced by Antoine Hoog Authentic Plants. **I:** 2015. **D:** A uniform, seed-raised selection distinguished by wide open, almost star-shaped flowers, thick tepals and widely spreading leaves. Hardy and flowering starts in September, peaking in mid-October.

arenarium Waldst. & Kit.   **E:** *Descr. Icon. Pl. Hung.* 2: 195, t. 179 (1803–1805).

arenarium var. *umbrosum* Ker Gawl.   = *C. umbrosum*

arenasii Fridl.   **E:** *Acta Bot. Gallica* 146(2): 158, fig. 1 (1999).

armenum B. Fedtsch.   = *C. szovitsii*

'Artur Klark' (*Colchicum*)   **R:** Leonid Bondarenko, Lithuanian Rare Bulb Garden. **I:** 2010s. **D:** See Cultivars chapter.

asteranthum Vassiliad. & K. Perss.   **E:** *Preslia* 74(1): 57 (2002).

atropurpureum Stapf ex Stearn   = *C. autumnale* 'Atropurpureum'

'Atropurpureum' (*Colchicum*) **N:** See discussion under *C. autumnale* in Species chapter.

'Atrorubens' (*C. speciosum*)   **R:** Registered by J. Backhouse & Son of York. **I:** Pre-1914. **RHS:** [AGM] 1952, AGM (H5) 2012. **D:** See Cultivars chapter.

atticum Spruner ex Tommas.   **E:** *Flora* 23: 730 (1840).

'Attlee' (*Colchicum*)   **R:** Registered by P. Visser Czn. in 1984. **I:** 1984.

'Autumn Herald' (*Colchicum*)   **R:** W.E. Th. Ingwersen, West Sussex, and registered by Ingwersen. **I:** mid 20th century. **D:** See Cultivars chapter.

'Autumn Queen' (*Colchicum*)   **R:** Registered by Zocher & Co. **I:** 1900–1905. **RHS:** [AGM] 1952, AGM (H5) 2012. **D:** See Cultivars chapter

autumnale L.   **E:** *Sp. Pl.* 1: 341 (1753).

autumnale var. *album* Gray   = 'Album' (*C. autumnale*).

autumnale var. *algeriense* (Batt.) Batt. & Trab.   = *C. lusitanum*

autumnale var. *atropurpureum*   = *C. autumnale* 'Atropurpureum'

autumnale var. *corsicum* (Baker) Firoi   = *C. corsicum*
autumnale var. *gibraltaricum* Kelaart   = *C. lusitanum*
autumnale var. *major* Tubergen   = *C.* × *byzantinum*
autumnale var. *minor* hort.   = *C. autumnale*
autumnale subsp. pannonicum (Griseb. & Schenk) Nyman   **E:** *Consp. Fl. Eur.*: 743 (1882).
autumnale var. *pannonicum* (Griseb. & Schenk.) Baker.   = *C. autumnale* subsp. *pannonicum*
autumnale var. *patens* (Schultz) Rouy   = *C.* × *byzantinum*
autumnale var. *tenorei* (Parl.) Fiori   = *C.* × *tenorei*
autumnale var. *vernum* L. nom. nud.   = *C. autumnale* subsp. *vernum*
autumnale subsp. vernum (Reichard) K. Richt.   **E:** *Pl. Eur.* 1: 190 (1890).
'Baker' (?*C. persicum*) **R:** Unknown. **I:** Unknown. **RHS:** BC 2015.
bakhtiaricum Matin & Iranshahr   = *C. varians*
balansae Planch.   **E:** *Ann. Sci. Nat., Bot.* sér. 4, 4: 145 (1855).
balansae var. *macrophyllum* Siehe ex Hayek   = *C. cilicicum*
'Balbithan'   = 'Glory of Threave'. **N:** 'Balbithan' is a later name for 'Glory of Threave'. The name has its origins at Balbithan House, Aberdeenshire, where Mary McMurtrie ran a small nursery and painted watercolours; she died in 2003, aged 101. McMurtrie presumably obtained corms of the plant from Threave Garden, Dumfries and Galloway, where it was labelled 'E.A. Bowles'.
baytopiorum C.D. Brickell   **E:** *Notes Roy. Bot. Gard. Edinburgh* 41(1): 49 (1983).
'Beaconsfield' (*Colchicum*)   **R:** Unknown. **I:** Pre-1924 **D:** See Cultivars chapter.
'Benton End' (*Colchicum*)   **E:** *Extracts from the Proceedings* (*RHS*) 122 (1998). **RHS:** AGM (H5) 1997 as *C. speciosum* 'Cedric Morris'. **D:** see Cultivars chapter.
bertolonii Steven   = *C. cupanii*
biebersteinii Rouy, nom illeg.   = *C. triphyllum*
biflorum hort.   = *C. szovitsii*
bifolium Freyn & Sint.   = *C. szovitsii*
bifolium var. *pleiophyllum* Bornm.   = *C. varians*
bivonae Guss.   **E:** *Cat. Pl. Hort. Boccadifalco* 1821: 4 (1821).
bivonae subsp. *euboeum* (Boiss.) Nyman   = *C. euboeum*

boissieri Orph.   **E:** *Atti Congr. Int. Bot. Firenze 1874* 1876: 31 (1876).
'Boldness' (*C. kesselringii*) **R:** Arnis Seisums. **I:** Offered by Jānis Rukšāns, Rare Bulb Nursery, Latvia, in 2005. **D:** White flowers with a narrow, but very bright, purple-violet stripe on the outside of the tepals (Rukšāns 2007).
borisii Stef.   = *C. autumnale*
bornmuelleri Freyn   = *C. speciosum* Bornmuelleri Group
Bornmuelleri Group (*C. speciosum*) **E:** *Colchicum: The Complete Guide* (2020): 469. **N:** A cultivar group name established in this book for plants primarily from the western end of the range of *C. speciosum* in Turkey which have distinguishing purple-brown or purple rather than yellow anthers. The group name is based on *C. bornmuelleri* Freyn (*Ber. Deutsch. Bot. Ges.* 7: 319 (1889); type – Turkey, Amasya, Ak Dağ, 1,800m, *Bornmüller s.n.*; holotype – BRNM). **D:** See Species chapter and Cultivars chapter, both under *C. speciosum*.
bornmuelleri hort.   = 'Joseph'
'Bowles' Form' (*C. bivonae*)   = *C. bivonae* **N:** Plants are occasionally seen in gardens under this name, which equates with *C. bowlesianum*, itself a synonym of *C. bivonae*. It is not the same as 'E.A. Bowles', which is derived from *C. speciosum*.
bowlesianum B.L. Burtt   = *C. bivonae*
bowlesianum sensu Feinbrun, non. B.L. Burtt   = *C. feinbruniae*
'Boxford' (*Colchicum*)   **E:** *Colchicum: The Complete Guide* (2020): 485. **I:** 2014. **D:** See Cultivars chapter.
'Bozkir'   = *C. boissieri*. Not a cultivar, simply a high-altitude form of the species offered by Jānis Rukšāns, Rare Bulb Nursery, Latvia.
brachyphyllum Boiss. & Hausskn.   = *C. szovitsii* subsp. *brachyphyllum*
brevistylum Feinbrun   = *C. decaisnei*
bulbocodioides Brot.   = *C. montanum*
bulbocodioides M. Bieb., non Brot.   = *C. triphyllum*
bulbocodium Ker Gawl   **E:** *Bot. Mag.* tab. 1028 (1807).
bulbocodium subsp. versicolor (Ker Gawl.) K. Perss.   **E:** *Bot. Jahrb. Syst.* 127(2): 178 (2007).
bulbocodium subsp. versicolor var. edentatum (Schur) K. Perss.   **E:** *Bot. Jahrb. Syst.* 127(2): 179 (2007).
bulgaricum Velen.   = *C. autumnale*

burttii Meikle   **E:** *Bot. Mag.* 181(3): 134 (1977).

× **byzantinum** Ker Gawl. (*C. cilicicum* × ?*C. autumnale*)   **E:** *Bot. Mag.* 26: t. 1028 (1807).

byzantinum var. *cilicicum* Boiss.   = *C. cilicicum*

callicymbium Stearn & Stef.   = *C. haynaldii*

candidum Schott & Kotschy ex Baker   = *C. balansae*

candidum var. *hirtiflorum* Boiss.   = *C. kotschyi*

'Carrot Line' (*C. luteum*)   **R:** Leonid Bondarenko, Lithuanian Rare Bulb Garden. **I:** 21st century. **D:** Derived from repeated selection of seedlings for larger and more colourful flowers. Flowers large, orange-yellow.

castrense De Laramb.   = *C. longifolium*

catacuzenium Heldr. ex Stef.   = *C. triphyllum*

caucasicum (M. Bieb.) Spreng.   = *C. trigynum*

caucasicum sensu Stefanov, non. (M. Bieb.) Spreng.   = *C. wendelboi*

caucasicum var. *raddeanum* (Regel) Stef.   = *C. raddeanum*

'Cedric's Darkest'   = 'Benton End'

'Cedric's Darkest of All'   = 'Benton End'

'Cedric's Drake'   = 'Benton End' (Anon. 1998)

'Cedric Morris'   = 'Benton End'

chalcedonicum Azn.   **E:** *Bull. Soc. Bot. France* 44: 174 (1897).

chalcedonicum subsp. **punctatum** K. Perss.   **E:** *Candollea* 53: 405 (1998).

'Chequers' (*C. speciosum* Giganteum Group)   **R:** Originated at Chequers, the garden of Jenny Robinson in Boxford, Suffolk. **I:** 2020. **RHS:** AGM (H5) 2017. **D:** See Cultivars chapter.

chimonanthum K. Perss.   **E:** *Pl. Syst. Evol.* 217: 56 (1999).

chionense Haw. ex Kunth   = *C. variegatum*

chlorobasis K. Perss.   **E:** *Edinburgh J. Bot.* 62: 182 (2006).

'Christine' (*C. cupanii*)   **R:** A strain raised by John Walker and named for his wife, originating from blind-collected seed found on the Peloponnese in southern Greece. **I:** Offered by Paul Christian Rare Plants nursery, Wrexham, Wales, in 2016. **D:** Flowers of the same size as the species but of a substantially deeper pink, and leaves about half the size of the typical species.

'Cilician Gates' (*C. cilicicum*)   **R:** Collected in 1960 in Cilicia, Turkey, by Michael and Caryl Baron, and named after a pass in the Taurus Mountains. **D:** A relatively easy garden plant of good substance and colour, but scarcely distinguishable from other forms of the species in cultivation. Flowering late August to September; height in leaf 38–48cm; height in flower 8–14cm. Flowers mid-rose-purple, faintly tessellated, with a median white stripe which is greenish towards the base; outer tepals obovate, 5.5–6.5 × 2.6–2.9cm, subobtuse; inner tepals narrow-obovate, 5.1–2.3(–2.6)cm, subobtuse to subacute; stamens two-thirds the length of the tepals, white with a hint of pink, anthers yellow and 9–10mm long; styles exceed the stamens, almost as long as the tepals, white, stigma punctiform; perianth tube white, 9–10cm long.

cilicicum (Boiss.) Dammer   **E:** *Gard. Chron. ser. 3*, 23: 34 (1898).

clementei Graells   = *C. triphyllum*

clone 1 (*C.* × *alberti*)   = 'Jānis'

clone 3 (*C.* × *alberti*)   = 'Jeanne'

confusum K. Perss.   **E:** *Pl. Syst. Evol.* 217: 60 (1999).

'Constable' (*Colchicum*)   **R:** Unknown. **I:** Unknown. **D:** See Cultivars chapter.

'Conquerer'   = ?'Glory of Heemstede'

'Conquest'   = 'Glory of Heemstede'

cornigerum (Schweinf.) Täckh. & Drar   = *C. schimperi*

corsicum auct., non Baker   = *C. arenasii*

corsicum Baker   **E:** *J. Linn. Soc., Bot.* 17: 431 (1879).

cousturieri Greuter   = *C. cupanii*

'Cretan White' (*C. macrophyllum*)   **R:** Peter Moore. **I:** Late 20th century. **D:** See Cultivars chapter.

cretense Greuter   **E:** *Candollea* 22: 246 (1967).

creticum Turrill   = *C. cupanii*

creticum sensu Rech. f. & P.H. Davis, non Turrill   = *C. cretense*

croaticum Dykes   = *C. hungaricum*

crociflorum Schott & Kotschy   = *C. serpentinum*

crociflorum Sims   = *C. autumnale*

crociflorum (Regel) Regel   = *C. kesselringii*

crociflorum var. *stenosepalum* Regel.   = *C. kesselringii*

crocifolium Boiss.   **E:** *Diagn. Pl. Orient.* ser. 1, 5: 67 (1844).

crocifolium var. *lasiophyllum* Bornm.   = *C. crocifolium*

crocifolium var. *stenanthum* (Bornm.) Stef.   = *C. crocifolium*

cupanii Guss.   **E:** *Fl. Sicul. Prodr.* 1: 452 (1827), as 'cupani'. **RHS:** AM 2005.

cupanii var. *bertolonii* (Steven) Rouy = *C. cupanii*

cupanii subsp. glossophyllum (Heldr.) Rouy   **E:** *Bull. Soc. Bot. France* 52: 646 (1905).

cupanii var. *pulverulentum* Batt. ex Maire & Weiller, nom. inval. = *C. cupanii*

'Daendels' (*Colchicum*)   **R:** Registered by Zocher & Co. **I:** 1918. **D:** See Cultivars chapter.

'Danton' (*Colchicum*)   **R:** Registered by Zocher & Co. **I:** 1900–1905. **D:** Bowles (1924) wrote that he thought it to be identical to, or slightly darker and with a more defined white throat, than 'Conquest' (now 'Glory of Heemstede') and that it was included in a plate in *Gartenschonheit* 1938: 388.

'Darwin' (*Colchicum*)   **R:** R.O. Backhouse of Herefordshire. Registered by R.O. Backhouse. **I:** 1930s. **D:** See Cultivars chapter.

davidovii Stef. = *C. szovitsii*

davisii C.D. Brickell   **E:** *The New Plantsman* 5(1): 15 (1998).

decaisnei sensu Lynch, non Boiss. = *C. cilicicum*

decaisnei Boiss.   **E:** *Fl. Orient. [Boissier]* 5(1): 157 (1882).

decaisnei var. *cilicica* Siehe = *C. polyphyllum*

deserti-syriaci Feinbrun = *C. schimperi*

diampolis Delip. & Cheshm. = *C. szovitsii*

'Dick Trotter' (*Colchicum*)   **R:** Dick Trotter. **I:** 1930s. **RHS:** AM 2008. **D:** See Cultivars chapter.

'Disraeli' (*Colchicum*)   **R:** Registered by Zocher & Co. **I:** 1900–1905. **D:** See Cultivars chapter.

doerfleri Halácsy   **E:** *Denkschr. Kaiserl. Akad. Wiss., Wien. Math.-Naturwiss. Kl.* 64: 739 (1897).

dolichantherum K. Perss.   **E:** *Edinburgh J. Bot.* 56(1): 126 (1999).

'Dombai' (*C. speciosum*)   **R:** Tallinn Botanic Garden, Estonia. Named by by Jānis Rukšāns and registered by C.P.J. Breed in 1999. **I:** Offered by Jānis Rukšāns, Rare Bulb Nursery, Latvia, before 1989. **D:** See Cultivars chapter.

'Dorothee Kersen' (*C. autumnale*)   **E:** Kersen bulb list. **R:** Kersen. **I:** 2019. **D:** See Cultivars chapter.

'Drama Bunch' (*C. autumnale*)   **R:** Collected by Antoine Hoog in 1987 at high altitude in the Falakros Mountains, Drama province, northern Greece.and introduced by Antoine Hoog Authentic Plants. **I:** 2006. **D:** Flowers single (not double as stated in some accounts), mid-pink, up to 17 per corm; leaves considerably shorter than in northern populations of *C. autumnale*.

drenowskii Degen & Rech. f. ex Kitan. in part = *C. autumnale*

drenowskii Degen & Rech. f. ex Kitan. in part = *C. doerfleri*

'Dykes' = 'William Dykes'

'Dykes Seedling' = 'William Dykes'

'E.A. Bowles' (*Colchicum*)   **R:** Named from Myddelton House, the garden of E.A. Bowles in Enfield, Middlesex. **I:** 1980s. **D:** see Cultivars chapter.

'E.K. Balls' (*Colchicum*)   **R:** Unknown. **I:** Unknown. **D:** See Cultivars chapter.

eichleri (Regel) K. Perss. = *C. trigynum*

'Elizabeth' (*C. autumnale*)   **R:** Dick Trotter. **I:** 1996. **N:** Collected by Dick Trotter in the company of E.A. Bowles in the 1930s from the Alpes-Maritimes. It was named during the RHS assessment at Felbrigg in 1996 in honour of Trotter's daughter, Elizabeth Parker-Jervis, who ran P-J Nursery, Longworth, Oxfordshire. **D:** Early-flowering (August to early September) with well-formed, chalice-shaped flowers of pale pink, with broad, oval tepals; height in flower 9–15cm. Leaves late to develop, by early spring often only a few centimetres tall, not fully developed until early summer; height in leaf 30–40cm.

'Emerald Town' (*Colchicum*)   **R:** Leonid Bondarenko, Lithuanian Rare Bulb Garden. **I:** 21st century. **D:** See Cultivars chapter.

'Enigma' (*Colchicum*)   **E:** *Colchicum: The Complete Guide* (2020): 494. **I:** 2020. **D:** see Cultivars chapter.

erdalii Özhatay   **E:** *Istanbul Üniv. Eczac. Fak. Mecm.* 46(2): 126 (2016).

'Erich Pasche' (*C. szovitsii*)   **E:** Paul Christian Rare Plants, catalogue. **I:** 21st century. **D:** Clustered bunches of large, bright, soft-pink flowers in early spring. **N:** Named by Paul Christian for botanist Erich Pasche.

euboeum (Boiss.) K. Perss.   **E:** *Candollea* 53: 400 (1998).

'Fabergé's Silver' (*Colchicum*)   **R:** Leonid Bondarenko, Lithuanian Rare Bulb Garden. **I:** 2017. **D:** See Cultivars chapter.

falcifolium sensu C.D. Brickell, non Stapf = *C. freynii, C. hirsutum, C. serpentinum*.

fasciculare (L.) R. Br.   **E:** *Narr. Travels Africa*

[Denham & Clapperton] 1826: 243 (1826).
fasciculare var. *brachyphyllum* (Boiss. & Hausskn.) Stef.   = *C. szovitsii* subsp. *brachyphyllum*
feinbruniae K. Perss.   **E:** *Israel J. Bot.* 41(2): 75 (1993).
'Felbrigg' (*Colchicum*)   **E:** *The Alpine Gardener: Plant Awards*: 33–34 (2013). **I:** 2010. **RHS:** PC 2010, AGM (H5) 2017. **D:** see Cultivars chapter.
'Felbrigg Violet' (*Colchicum*)   **E:** *Colchicum: The Complete Guide* (2020): 497. **R:** Unknown. **I:** 2020. **D:** See Cultivars chapter.
Felbrigg 20   = 'Felbrigg'
'Ferndown Beauty' (*Colchicum*)   **N:** A name sometimes seen in older nursery catalogues and other publications, but of unknown origin and with a poor or non-existent description. It cannot be reliably traced to a modern equivalent and is probably lost from cultivation.
fharii Fridl.   = *C. lusitanum*
figlalii (Varol) Parolly & Eren   **E:** *Willdenowia* 37: 267 (2007).
filifolium (Cambess.) Stef.   **E:** *Sborn. B'lghar. Akad. Nauk* 22: 58 (1926).
'Flamenco Dance' (*Colchicum*)   **R:** Leonid Bondarenko, Lithuanian Rare Bulb Garden. Registered by Bondarenko in 2010. **I:** 2010. **D:** See Cultivars chapter.
fominii Bordz.   = *C. arenarium*
freynii Bornm.   **E:** *Notizbl. Bot. Gart. Berlin-Dahlem* 7: 172 (1917).
fritillatum Link ex Jahand. & Maire   = *C. lusitanum*
'Fuller's Mill' (*Colchicum*)   **E:** *Colchicum: The Complete Guide* (2020): 499. **R:** Named from Fuller's Mill, West Stow, Suffolk, the garden of Bernard Tickner. **I:** 2020. **D:** see Cultivars chapter.
'Furse'   = 'Polly Furse'
'General Grant'   = 'Glory of Heemstede'. Bowles (1924) wrote that he could not distinguish it from 'Conquest' (now 'Glory of Heemstede') but that it was included in a plate in *Gartenschonheit* 1938: 388.
'Giant' (*Colchicum*)   **R:** Registered by Zocher & Co. **I:** 1900–1905. **D:** See Cultivars chapter.
giganteum hort. ex Stef.   = *C. speciosum* Giganteum Group
Giganteum Group (*C. speciosum*) **E:** *Colchicum: The Complete Guide* (2020): 471. **N:** A cultivar group name established in this book for plants, presumably of Turkish origin, characterised by having larger, broader, more funnel-shaped, pale flowers with a white centre, white perianth tube, pale yellow anthers, and earlier into bloom than *C. speciosum*. Plants showing these characteristics appear in populations of *C. speciosum* in eastern and northeastern Turkey and more commonly in Iranian populations. The group name is based on *C. giganteum* hort. ex Stef. (*Sborn. B'lghar. Akad. Nauk.* 22: 82 (1926); in obs. ex S. Arnott, *Gard. Chron.* ser. 3, 32: 435 (1902); type not traced; syn. *C. speciosum* var. *illyricum* hort.) **D:** See Species chapter and Cultivars chapter, both under *C. speciosum*.
'Giona'   = see Mount Giona
'Glorie van Holland' (*Colchicum*)   **N:** A name sometimes seen in older nursery catalogues and other publications, but of unknown origin and with a poor or non-existent description. It cannot be reliably traced to a modern equivalent and is probably lost from cultivation.
'Glory of Heemstede' (*Colchicum*)   **R:** Registered by J.J. Kerbert of Zocher & Co. **I:** 1900–1905. **RHS:** AM 1928. **D:** See Cultivars chapter.
'Glory of Threave' (*Colchicum*)   **E:** *The Alpine Gardener* 74(4): 460–461, 172 (2006). Registered by the ICRA in 2007. **RHS:** AM 2005, AGM (H5) 2017. **D:** see Cultivars chapter.
glossophyllum Heldr.   = *C. cupanii* subsp. *glossophyllum*
gohariae Gabrieljan   = *C. szovitsii*
'Golden Baby' (*C. luteum*)   **R:** Leonid Bondarenko, Lithuanian Rare Bulb Garden. **I:** 2009. **D:** See Cultivars chapter.
'Golden Elf'   = 'Golden Baby'
gonarei Camarda   **E:** *Boll. Soc. Sarda. Sci Nat.* 17: 227 (1978).
'Gothic Style' (*Colchicum*)   **R:** Leonid Bondarenko, Lithuanian Rare Bulb Garden. **I:** Early 21st century. **N:** 'Looking Up' is a later name for this cultivar. **D:** See Cultivars chapter.
'Gracia' (*Colchicum*)   **R:** Registered by P. Visser Czn. in 1974. **I:** 1974. **D:** See Cultivars chapter.
gracile K. Perss.   **E:** *Bot. Jahrb. Syst.* 127: 489 (2009).
graecum K. Perss.   **E:** *Willdenowia* 18(1): 34 (1988).
'Green Star'   = 'Autumn Queen'
greuteri (Gabrieljan) K. Perss.   = *C. trigynum*
guadarramense Pau   = *C. multiflorum*

***guessfeldtianum* Asch. & Schweinf.** **E:** *Ill. Fl. Égypte, Suppl.* 2: 774 (1889).

'Guizot' (*Colchicum*) **R:** Registered by Zocher & Co. **I:** Pre-1931. **N:** Probably lost from cultivation.

gussonei Lojac. = *C. cupanii*

halepense Freyn = *C. fasciculare*

halophilum Freyn & Bornm. = *C. persicum*

'Hannibal' (*Colchicum*) **R:** Selected by Antoine Hoog in 2011 from seed of 'Spartacus', possibly a hybrid with *C. neapolitanum*. **I:** 2012. **D:** Very dwarf, floriferous, flowers bright pink with large white centres.

'Harlekijn' (*Colchicum*) **R:** Registered by P. Visser Czn. in 1980. **I:** 1988. **D:** See Cultivars chapter.

'Harlequin' = 'Harlekijn'

haussknechtii Boiss. = *C. persicum*

***haynaldii* Heuff.** **E:** *Verh. K.K. Zool.-Bot. Ges. Wien* 8: 213 (1858), as 'haynaldi'.

***heldreichii* K. Perss.** **E:** *Edinburgh J. Bot.* 56: 98 (1999).

'Herbert Kussel' = 'Herbstkugel'

'Herbstkugel' (*Colchicum*) **R:** Hagen Engelmann. **I:** Post 2000. **D:** See Cultivars chapter.

hexapetalum Pourr. ex Lapeyr. = *C. montanum*

'Hidegkut' (*Colchicum*) **N:** A name sometimes seen in older nursery catalogues and other publications, but of unknown origin and with a poor or non-existent description. It cannot be reliably traced to a modern equivalent and is probably lost from cultivation.

hiemale Freyn = *C. pusillum*

hiemale Siehe, nom. nud. = *C. minutum*

***hierosolymitanum* Feinbrun E:** *Palestine J. Bot., Jerusalem ser.* 6: 84 (1953).

'Hirsutum' (*Colchicum*) = *C. hirsutum*

***hirsutum* Stef.** **E:** *Sborn. B'lghar. Akad. Nauk* 22: 34 (1926).

hissaricum (Regel) Stef. = *C. robustum*

hololophum Coss. & Durieu = *C. triphyllum*

'Hora Sfakion' (*C. macrophyllum*) **R:** John Fielding. **I:** 2000. **D:** See Cultivars chapter.

***hungaricum* Janka** **E:** *Természetrajzi Füz.* 10: 75 (1886).

***hungaricum* f. *albiflorum* (K. Maly) Hayek & Markgraf** **E:** *Fedde Rep. Sp. Nov.*, Beihefte 30: iii (1932).

'Huxley' (*Colchicum*) **R:** R.O. Backhouse of Herefordshire. Registered by R.O. Backhouse. **I:** 1930s. **RHS:** AM 1953, AGM (H5) 2017. **D:** See Cultivars chapter.

hydrophilum Siehe = *C. szovitsii* subsp. *brachyphyllum*

***ignescens* K. Perss.** **E:** *Bot. Jahrb. Syst.* 127(2): 191 (2007).

illyricum Stokes = *C. fasciculare*

illyricum Friv. ex Kunth = *C. bivonae*

illyricum auct., non Stokes = *C. speciosum* Giganteum Group

'Illyricum' = ?*C. speciosum*

illyricum superbum **RHS:** AM 1924.

***imperatoris-friderici* Siehe ex K. Perss.** **E:** *Edinburgh J. Bot.* 56: 129 (1999).

'Innocence' (*C.* × *byzantinum*) **E:** *Bull. Alpine Gard. Soc.* 68(2): 201–286 (2000). **R:** Named by Chris Brickell, registered by C.P.J. Breed in 2000. **I:** 1996. **RHS:** PC 1998, AGM (H5) 2017. **D:** See Cultivars chapter.

'Intermediate Dykes' = 'William Dykes'

'Intermediate Woods' = 'Little Woods'

***inundatum* K. Perss.** **E:** *Edinburgh J. Bot.* 56: 99 (1999).

issicum Siehe, nom. nud. = *C. minutum*

'James Pringle' (*Colchicum*) **N:** A name sometimes seen in older nursery catalogues and other publications, but of unknown origin and with a poor or non-existent description. It cannot be reliably traced to a modern equivalent and is probably lost from cultivation.

'Jānis' (*C.* × *alberti*) **R:** Arnis Seisums. **I:** 1989. **N:** Referred to as 'clone 1' in Rolfe (2001). Also described in Rukšāns (2007). **D:** See Cultivars chapter.

jankae Freyn = *C. haynaldii*

'Jarka' (*Colchicum*) **R:** Registered by C.P.J. Breed in 2009. **I:** Offered by Jānis Rukšāns, Rare Bulb Nursery, Latvia, in 1992. **D:** See Cultivars chapter.

'Jarkoslavan' = 'Poseidon'

'Jaroslavna' = 'Poseidon'

'Jaroslawna' = 'Poseidon'

'Jeanne' (*C.* × *alberti*) **R:** Arnis Seisums. **D:** Yellow flowers with prominent reddish-purple stripes externally. **N:** Along with 'Janis' this was found in 1989 by Arnis Seisums growing wild near the airfield of Tovil'-Dora in Tajikistan. It was first offered for sale in 2007 by Paul Christian of Rare Plants nursery, Wrexham, Wales, who obtained it as 'clone 3'.

'Jenny Robinson' (*Colchicum*)   **E:** *Western Morning News* (August 2011). **R:** Originated at Chequers in Boxford, Suffolk, the garden of Jenny Robinson, and named by Matt Bishop in a gardening column he wrote in 2011. **D:** See Cultivars chapter.

jesdianum Czerniak.   = *C. schimperi*

'Jochem Hof'   = 'Poseidon'

'Jochum Hof'   = 'Poseidon'. 'Jochum Hof' was registered by P. Visser Czn. in 1991.

'Jolante'   = 'Jolanthe'

'Jolanthe' (*Colchicum*)   **R:** Unknown. **I:** Unknown.

'Joseph' (*Colchicum*)   **E:** *Colchicum: The Complete Guide* (2020): 512. **R:** Unknown. **I:** Pre-1930s, as *C. bornmuelleri* (*C. bornmuelleri* hort.). **D:** See Cultivars chapter.

'Karin Persson' (*C. autumnale*)   **R:** Selected from a collection (AH 8954) made in 1989 by Antoine Hoog at 1,500m on Katara Pass, Pindus Mountains, Greece. This is the southernmost limit of *C. autumnale* in Greece, primarily because of its preference for moist meadows which only occur further north. Introduced by Antoine Hoog Authentic Plants. **I:** 2009. **D:** Flowering July to early September; height in leaf 28–38cm; height in flower 11–17cm; flowers 4–9 per corm, mid-pink with a hint of lilac and a white throat that merges subtly with the colour zone; tepals 4.5–7cm long; anthers dark yellow. The corms, leaves and flowers are all larger than in other *C. autumnale* cultivars.

kesselringii Regel   **E:** *Trudy Imp. S.-Peterburgsk. Bot. Sada* 8: 646 (1884), as '*kesselringi*'.

'Kiss Me Quick' (*Colchicum*)   **R:** Unknown. **I:** Unknown **D:** See Cultivars chapter.

'Klondike' (*Colchicum*)   **N:** A name sometimes seen in older nursery catalogues and other publications, but of unknown origin and with a poor or non-existent description. It cannot be reliably traced to a modern equivalent and is probably lost from cultivation.

kochii Parl.   = *C. haynaldii*

kotschyi sensu C.D. Brickell, non Boiss.   = *C. imperatoris-friderici*

kotschyi Boiss.   **E:** *Diagn. Pl. Orient. ser. 1* 13: 38 (1854).

kurdicum (Bornm.) Stef.   **E:** *Sborn. B'lghar. Akad. Nauk* 22: 42 (1926).

laetum sensu Baker pro parte, non Steven = *C. balansae, C. kotschyi*.

laetum hort.   = 'Pink Star'

laetum Steven   **E:** *Nouv. Mém. Soc. Imp. Naturalistes Moscou* 1: 262, t. 18 (1829).

lagotum K. Perss.   **E:** *Bot. Jahrb. Syst.* 127(2): 197 (2007).

'Larisa' (*Colchicum*)   **R:** Leonid Bondarenko, Lithuanian Rare Bulb Garden. Registered by Bondarenko in 2010. **I:** 2010. **D:** See Cultivars chapter.

'Larisa Dolina'   = 'Larisa'

latifolium auct., non. Sm.   = *C. macrophyllum*

latifolium Sm.   = *C. bivonae*

latifolium sensu Griseb., non Sm.   = *C. bivonae*

latifolium var. *euboeum* Boiss.   = *C. euboeum*

latifolium var. *longistylum* Pamp.   = *C. macrophyllum*

'Lausanne' (*C. autumnale*)   **R:** Unknown. **I:** Unknown.

'Leithvale'   = 'Giant'

'Leith Vale Seedling'   = 'Giant'

lenkoranicum (Miscz.) Grossh.   = *C. speciosum*

leptanthum K. Perss.   **E:** *Bot. J. Linn. Soc.* 135: 85 (2001).

levieri Janka   = *C. lusitanum*

libanoticum Ehrenb. ex Boiss.   = *C. szovitsii* subsp. *brachyphyllum*

'Lilac Bedder' (*Colchicum*)   **R:** Registered by P. Visser Czn. in 1974. **I:** 1974. **D:** See Cultivars chapter.

'Lilac Wonder' (*Colchicum*)   **R:** Registered by Zocher & Co. **I:** 1900–1905. **D:** See Cultivars chapter.

lingulatum Boiss. & Spruner   **E:** *Diagn. Pl. Orient. ser. 1* 5: 66 (1844).

lingulatum var. *parnassicum* (Sartori, Orph. & Heldr.) Stef.   = *C. parnassicum*

lingulatum subsp. *rigescens* K. Perss.   **E:** *Candollea* 53: 409 (1998).

liparochiadys Woron. ex Czerniak   = *C. woronowii*

liparochlamys Woron.   = *C. woronowii*

'Little Woods' (*Colchicum*)   **R:** Elizabeth Parker-Jervis, P-J Nursery, Longworth, Oxfordshire. **I:** 1980s. **RHS:** AGM (H5) 2017. **D:** See Cultivars chapter.

lockmannii Siehe ex Stef.   = *C. polyphyllum*

longifolium Castagne   **E:** *Cat. Pl. Marseille*: 135 (1845).

longifolium var. *micranthum* Emb. & Maire = *C. longifolium*

'Looking Up'   = 'Gothic Style'

'Lucky Selfmade' (*C.* × *alberti*)   **R:** Leonid Bondarenko, Lithuanian Rare Bulb Garden. **I:** early 21st century. **D:** See Cultivars chapter.

lusitanum **Brot.**   **E:** *Phytogr. Lusitan. Select.* (1816–1827) 2: 211, tt. 173, 174 (1827).

luteum **Baker**   **E:** *Gard. Chron.* n.s. 2: 34 (1874).

'Lysimachus' (*Colchicum*) **R:** Selected by Antoine Hoog in 2006 from seed of *C. autumnale* 'Drama Bunch' raised on his nursery, probably with *C. haynaldii* as the other parent. Introduced by Antoine Hoog Authentic Plants. **I:** 2012. **D:** See Cultivars chapter.

macedonicum **Košanin**   **E:** *Glas Srpske Kraljiveske Akademije* 85: 232, t. 6 (1911).

macrophyllum **B.L. Burtt**   **E:** *Kew Bull.* 5(3): 433 (1951).

'Magnificum' (*Colchicum*)   **R:** Registered by Van Tubergen. **I:** Pre-1927. **D:** Larger flowers and rosy lilac buds. A selection of *C. bornmuelleri* hort. **N:** Probably lost from cultivation.

manissadjianii **(Azn.) K. Perss.**   **E:** *Bot. Jahrb. Syst.* 127(2): 202 (2007).

maraschicum **E. Kaya & Özhatay**   **E:** *Türk. Geofitleri* 3: 539 (2014).

'Maximum' (*C. speciosum*)   **R:** Smith. **I:** Pre-1924. **N:** Mathew *et al.* (1980). **RHS:** AM 1979. **D:** See Cultivars chapter.

'Mells Park' (*Colchicum*)   **R:** Named from Mells Park, Frome, Somerset. **I:** Mid 20th century. **D:** See Cultivars chapter.

'Mellis Park'   = 'Mells Park'

merendera Ker Gawl.   = *C. bulbocodium*

merenderoides E.P. Perrier & Songeon   = *C. alpinum*

micaceum **K. Perss.**   **E:** *Edinburgh J. Bot.* 56: 95 (1999).

micranthum **Boiss.**   **E:** *Fl. Orient. [Boissier]* 5(1): 162 (1882).

'Minion' (*C. luteum*)   **R:** Leonid Bondarenko, Lithuanian Rare Bulb Garden. **I:** 2009.

minutum **K. Perss.**   **E:** *Edinburgh J. Bot.* 56(1): 90 (1999).

mirzoevae (Gabrieljan) K. Perss.   = *C. trigynum*

'Modesty' (*C. kesselringii*)   **R:** Arnis Seisums. **I:** Offered by Jānis Rukšāns, Rare Bulb Nursery, Latvia, in 2005. **D:** Snow-white flowers with a very narrow grey stripe on the outside of the tepals (Rukšāns 2007).

montanum sensu Baker, non. L.   = *C. crocifolium*

***montanum* L.**   **E:** *Sp. Pl.*: 342 (1753).

montanum var. *croaticum* A. Grove   = *C. hungaricum*

montanum var. *pusillum* (Sieber) Fiori   = *C. pusillum*

'Moonlight' (*C.* × *alberti*)   **R:** Leonid Bondarenko, Lithuanian Rare Bulb Garden. **I:** Early 21st century. **D:** The first hybrid created by Bondarenko with pure, pale yellow flowers. Parentage is *C. luteum* × *C. kesselringii* 'Yeti' and the flowers are twice as large as either parent.

'Mount Etna' (*C. bivonae*)   **R:** Unknown. **I:** Unknown. **D:** See Cultivars chapter.

Mount Giona (*C. bivonae*) **R:** Collected by Antoine Hoog (AH 9139) at 1,000m on Mount Giona, central Greece, and introduced by Antoine Hoog Authentic Plants. **I:** 2005. **D:** Not a cultivar or clone, but selected for its extreme variability in corm size, leaf size and flower colour and patterning.

'Mr Kerbert' (*Colchicum*)   **RHS:** [AGM] 1952. **D:** Bowles (1924) wrote that this was very similar to 'Lilac Wonder' but with a significant white throat and white channels extending two-thirds up the petals. **N:** Probably lost from cultivation.

multiflorum sensu Arrigoni, non Brot.   = *C. neapolitanum*

***multiflorum* Brot.**   **E:** *Fl. Lusit.* 1: 597 (1804).

munzurense **K. Perss.**   **E:** *Edinburgh J. Bot.* 56(1): 86 (1999).

'My Choice' (*C. kesselringii*) **R:** Leonid Bondarenko, Lithuanian Rare Bulb Garden. **I:** 21st century. **D:** White flowers with prominent purple stripes externally.

'Myddelton'   = 'Felbrigg'

'Naeisanum' (*C. speciosum*)   **N:** Origin of name unknown, but the cultivar is listed as part of the national collection held by Paweł Kaźmierski in Mileno, Poland.

'Nancy Lindsay' (*C. autumnale* subsp. *pannonicum*)   **R:** Collected by Nancy Lindsay in Romania. **I:** 1936. **RHS:** AGM (H5) 1997. **D:** See Cultivars chapter.

nanum **K. Perss.**   **E:** *Bot. Jahrb. Syst.* 127(2): 206 (2007).

'Natural Firefly' (*Colchicum*)   **E:** *Gardeners' World* 213: 62–63 (2008).

***neapolitanum* (Ten.) Ten.**   **E:** *Fl. Neap. Prod. App.* 5: 11 (1826).

neapolitanum var. *castrense* (De Laramb.) Rouy

= *C. longifolium*

neapolitanum var. *corsicum* (Baker) Fiori = *C. corsicum*

neapolitanum var. *haynaldii* (Heuff.) Asch. & Graebn. = *C. haynaldii*

neapolitanum f. *micranthum* (Emb. & Maire) Maire & Weiller = *C. longifolium*

neapolitanum subsp. *parlatoris* (Orph.) K. Richt. = *C. parlatoris*

'Neptun' (*Colchicum*) **R:** Unknown. **I:** 2017. **D:** See Cultivars chapter.

'Neptune' = 'Neptun'

ninae Sosn. = *C. szovitsii*

nivale (Boiss. & A. Huet) Boiss. & A. Huet ex Stef. = *C. szovitsii*

'Norman Barratt' (*C. montanum*) **E:** Wallis (2003). **R:** Corms of *C. montanum* collected by Norman Barratt in the French Pyrenees (Barratt 1951). **RHS:** AM 2002 (as *Merendera montana* 'Norman Barratt'). **D:** See Cultivars chapter.

'Nutt's Green Star' = 'Autumn Queen'

obtusifolium Siehe ex Hayek = *C. kotschyi*

'October Fest' = 'Oktoberfest'

officinale, Gard. Chron, n.s. 8: 562 (1877) Name of uncertain status and application.

'Oktoberfest' (*Colchicum*) **R:** Rollich, Germany. **I:** Unknown. **D:** See Cultivars chapter.

'Oktober Fest' = 'Oktoberfest'

'Old Bones' = *C. autumnale* 'Album'

'Ordu' (*C. speciosum*) **R:** Collected in the wild by Brian Mathew and John Tomlinson in 1963 under the number 4978. **D:** Flowering late August to September; height in leaf 38–45cm; height in flower 10–16cm. Bright amethyst-violet flowers with a prominent white centre. **N:** No plants under this name appear to be in the horticultural trade at present but may still be in private collections.

orientale Friv. ex Kunth = *C. autumnale*

osmaniyense Özhatay & E. Kaya **E:** Istanbul Üniv. Eczac. Fak. Mecm. 46(2): 126 (2016).

palmetorum Czerniak. = *C. schimperi*

pannonicum **RHS:** AM 2003. **N:** Rolfe (2004)

pannonicum hort. = 'Nancy Lindsay'

pannonicum Griseb. & Schenk (1852) = *C. autumnale* subsp. *pannonicum*

pantocratoris Spreitz. ex Stef. = *C. haynaldii*

'Papa Rema' (*C. bivonae*) **R:** Peter and Penny Watts.

I: 2018. **RHS:** AGM (H5) 2017. **D:** See Cultivars chapter.

parkinsonii Hook. f. = *C. variegatum*

parlatoris Orph. **E:** Atti Congr. Int. Bot. Firenze 1874: 32 (1876).

parnassicum Sartori, Orph. & Heldr. **E:** Diagn. Pl. Orient. ser. 2 4: 122 (1859).

parviflorum Biv. = *C. cupanii*

parvulum Ten. = *C. alpinum*

paschei K. Perss. **E:** Edinburgh J. Bot. 56: 110 (1999).

patens Schultz = *C. × byzantinum*

'Paul Furse' (*C. speciosum*) **R:** Selected from a collection made by Paul and Polly Furse in NW Iran or NE Turkey in the 1950s or 1960s. **D:** See Cultivars chapter.

'Paul Furse Early' (*C. speciosum*) **E:** Mike Salmon, Monocot Nursery catalogue, 1990. **R:** Selected from a collection made by Paul and Polly Furse in NW Iran or NE Turkey in the 1950s or 1960s. **I:** 1990. **D:** See Cultivars chapter.

peloponnesiacum Rech. f. & P.H. Davis **E:** Oesterr. Bot. Z. 95: 427 (1949).

persicum Baker **E:** J. Linn. Soc., Bot. 17: 430 (1879).

'Petrovac' (*C. haynaldii*) **R:** Selected from a collection (AH 8926) made in 1989 by Antoine Hoog from former Yugoslavia and identified by Karin Persson as *C. visianii*. Introduced by Antoine Hoog Authentic Plants. **I:** 2009. **D:** Flowering in early autumn; flowers in small clusters, bright pink, slightly chequered with darker pink; anthers small and yellow on short filaments; style elongated, whitish.

pieperianum Markgr. = *C. macedonicum*

pinatziorum Rech. f. = *C. boissieri*

'Pink Goblet' (*Colchicum*) **R:** Dick Trotter. **I:** 1930s. **RHS:** AGM (H5) 1997. **D:** See Cultivars chapter.

'Pink Star' (*Colchicum*) **R:** Unknown. **I:** 1998. **RHS:** AGM (H5) 2017. **D:** See Cultivars chapter.

'Pleniflorum' (*C. autumnale*) **R:** Unknown. **I:** Pre-1874. **D:** See Cultivars chapter.

'Plenum' = 'Pleniflorum'

'Polly Furse' (*Colchicum*) **R:** Selected from a collection made by Paul and Polly Furse in NW Iran or NE Turkey in the 1950s or 1960s. **D:** Flowers narrow funnel-shaped, deep rose-purple, faintly tessellated, whitish at the base within; outer tepals elliptic-obovate to elliptic-oblanceolate,

6–7.3 × 2.1–2.3cm, subobtuse; inner tepals narrow elliptic-oblanceolate, 5.6–6.1 × 1.5–1.9cm, subacute; stamens half the length of the tepals, the filaments white, greenish towards the base, the anthers creamy yellow, 11–12mm long; styles very slender, almost as long as the tepals, white, rose-purple towards the slightly curved tip; perianth tube whitish with a faint green tinge, 11.5–15.5cm long. Flowering in September.

polymorphum Orph., nom nud.   = *C. sfikasianum*

polyphyllum Boiss. & Heldr.   **E:** *Diagn. Pl. Orient.* ser. 2 **4**: 121 (1859).

'Poseidon' (*Colchicum*)   **R:** Registered by J.J. de Groot in 2007. **D:** See Cultivars chapter.

praecox Spenn.   = *C. autumnale* var. *vernum* L. ex Reichard

'Premier' = 'Lilac Wonder'. 'Premier' was registered by Zocher & Co.

'President Coolidge' (*Colchicum*)   **R:** Registered by Zocher & Co. **I:** 1900–1905. **D:** Bowles (1924) wrote that it differs little from 'Danton', but was redder with less distinct tessellation and the white of the throat extending further up the tepals. **N:** Probably lost from cultivation.

'Pride of Holland' (*Colchicum*)   **E:** *Colchicum: The Complete Guide* (2020): 526. **R:** Unknown. **I:** 2020. **D:** See Cultivars chapter.

'Princess Astrid' = 'Autumn Queen'. 'Princess Astrid' was registered by E. Breed in 2018. As 'Prinses Astrid' it had [AGM] 1952.

procurrens Baker   = *C. boissieri*

'Prosperity' (*C. kesselringii*)   **R:** Arnis Seisums. **I:** Offered by Jānis Rukšāns, Rare Bulb Nursery, Latvia, in 2005. **D:** White flowers with a wide, deep violet-purple stripe on the outside of the tepals that gradually fades to grey at the margin leaving a narrow white edge (Rukšāns 2007).

provinciale Loret   = *C. longifolium*

psaridis sensu C.D. Brickell, non Heldr. ex Halácsy   = *C. minutum*

psaridis Heldr. ex Halácsy   **E:** *Consp. Fl. Graec.* **3**(1): 274 (1904).

pseudoparvulum Lojac.   = *C. alpinum*

pulchellum K. Perss.   **E:** *Willdenowia* 18: 30 (1988).

'Purity' (*Colchicum*)   **N:** A name sometimes seen in older nursery catalogues and other publications, but of unknown origin and with a poor or non-existent description. It cannot be reliably traced to a modern equivalent and is probably lost from cultivation.

'Purple Star' (*C. kesselringii*)   **R:** Leonid Bondarenko, Lithuanian Rare Bulb Garden. **I:** Early 21st century.

'Purpureum' (*C. cilicicum*)   **R:** Unknown. **I:** 1928. **RHS:** AGM (H5) 2017. **D:** See Cultivars chapter.

pusillum Sieber   **E:** *Flora* 5(1): 248 (1822).

pyrenaicum Pourr.   = *C. montanum*

raddeanum (Regel) K. Perss.   **E:** *Fl. Iran. [Rechinger]* 170: 22 (1992).

"ramonensis" in ms. Name proposed by Vasl & Shmida (2015) but not formally published yet.

rausii K. Perss.   **E:** *Pl. Syst. Evol.* 217: 58 (1999).

'Redgrave' (*Colchicum*)   **R:** Found by Christopher Grey-Wilson in a garden in Redgrave, Suffolk. **I:** 2017. **D:** See Cultivars chapter.

regelii Stef.   = *C. kesselringii*

'Revelation' (*C. speciosum* Giganteum Group) **E:** *Colchicum: The Complete Guide* (2020): 473. **R:** Originated from Myddelton House, the garden of E.A. Bowles in Enfield, Middlesex. **I:** 2020. **D:** See Cultivars chapter.

rhodopaeum Kov.   = *C. autumnale*

richii hort.   = *C. ritchii*

ritchii R. Br.   **E:** *Narr. Travels Africa [Denham & Clapperton], App.*: 241 (1826).

ritchii var. *guessfeldtianum* (Asch. & Schweinf.) Stef.   = *C. guessfeldtianum*

robustum (Bunge) Stef.   **E:** *Sborn. B'lghar. Akad. Nauk* 22: 24 (1926).

'Rosa Plenum' = 'Pleniflorum'

'Roseum' (*C. hungaricum*)   **R:** Unknown. **I:** Unknown.

'Rosy Dawn' (*Colchicum*)   **R:** Registered by Barr & Sons. **I:** 1948. **RHS:** AM 1976, FCC 2005, AGM (H5) 1997. **N:** Rolfe (2006). **D:** See Cultivars chapter.

'Rosy Wonder' = 'Huxley'

'Rubens'   **D:** Bowles (1924) wrote that it flowers in early August with 'Princess Astrid' and 'Autumn Queen'. **N:** Probably lost from cultivation.

'Rubrum' (*C. speciosum*)   **R:** Unknown. **I:** Pre-1900. **D:** See Cultivars chapter.

'Ruby Queen' (*Colchicum*)   **D:** Bowles (1924) wrote that it has large flowers opening widely on long tubes. **N:** Probably lost from cultivation.

sanguicolle K. Perss.   **E:** *Edinburgh J. Bot.* 56(1): 92 (1999).

schimperi Janka ex Stef.   **E:** *Sborn. B'lghar. Akad.*

Nauk. 22: 31 (1926).

'Sentinel' = 'Rosy Dawn'

serpentinum Woronow ex Miscz.   **E:** *Fl. Caucas. Crit.* 3(4): 114 (1912).

sfikasianum Kit Tan & Iatroú   **E:** *Rock Gard.* 24(3): 255 (1995).

sibthorpii Baker   = *C. bivonae*

sibthorpii sensu B.L. Burtt, non Baker   = *C. lingulatum*

sieheanum Hausskn. ex Stef.   **E:** *Sborn. B'lghar. Akad. Nauk.* 22: 47 (1926).

'Snow of Highland' (*C. kesselringii*)   **R:** Leonid Bondarenko, Lithuanian Rare Bulb Garden. **I:** Early 21st century. **D:** See Cultivars chapter.

'Snowwhite' (*C. szovitsii*)   **R:** Arnis Seisums. **I:** Pre-1997. **N:** Also described in Rukšāns (2007). **D:** See Cultivars chapter.

'Snow White' = 'Snowwhite'

soboliferum (C.A. Mey.) Stef.   **E:** *Sborn. B'lghar. Akad. Nauk* 22: 44 (1926).

'Spartacus' (*Colchicum*)   **R:** Selected by Antoine Hoog in 2007 from seed of *C. autumnale* 'Drama Bunch' raised on his nursery, with *C. bivonae* as the other parent. Introduced by Antoine Hoog Authentic Plants. **I:** 2012. **D:** See Cultivars chapter.

'Spartacus' hort.   = 'Enigma'

speciosum sensu Stef., non Steven   = *C. balansae*

speciosum Steven   **E:** *Nouv. Mém. Soc. Imp. Naturalistes Moscou* 1: 265, t. 15 (1829).

speciosum var. *illyricum* hort.   = *C. speciosum* Giganteum Group

stenanthum Bornm.   = *C. crocifolium*

'Steveni' = *C. stevenii*

stevenii auct., non Kunth   = *C. freynii*

stevenii Kunth   **E:** *Enum. Pl. [Kunth]* 4: 144 (1843), as 'steveni'.

stevenii subsp. *taurii* (Stef.) Thiébaut, comb. inval. = *C. serpentinum*

stevenii var. *vernale* Bornm.   = *C. freynii*

'Surprise' (*Colchicum*)   **R:** Registered by P. Visser Czn. in 1985. **I:** 1985. **N:** Probably lost from cultivation.

syriacum Siehe ex Stef.   = *C. szovitsii*

szovitsii Fisch. & C.A. Mey.   **E:** *Index Seminum [St. Petersburg (Petropolitanus)]* 1: 24 (1835).

szovitsii var. *bifolium* (Freyn & Sint.) Bordz. = *C. szovitsii*

szovitsii subsp. **brachyphyllum** (Boiss. & Hausskn.) K. Perss.   **E:** *Bot. Jahrb. Syst.* 127(2): 221 (2007). **RHS:** PC 1959 (as *C. fasciculare* var. *brachyphyllum*).

szovitsii var. *cornigerum* Schweinf.   = *C. schimperi*

szovitsii var. *nivale* Boiss. & A. Huet   = *C. szovitsii*

tauri Siehe ex Stef.   = *C. serpentinum*

tauri sensu Mouterde ex Feinbrun, non Siehe ex Stef.   = *C. antilibanoticum*

taygeteum Heldr., in sched.   = *C. graecum*

tenorei sensu J.G. Baker   = *C.* × *ambiguum*

tenorei Parl.   = *C. cilicicum*

tessellatum hort. Angl. ex Baker   = *C.* × *agrippinum*

tessellatum Salisb.   = *C. variegatum*

tessulatum Mill.   = *C.* × *agrippinum*

'Teufelskralle' (*Colchicum*)   **R:** Hagen Engelmann. **I:** 1990s. **D:** Purplish pink flowers with white, constricted tips to the tepals and lacking styles. It resembles 'Harlekijn' and 'Jarka'. Found by the raiser growing near *C. bornmuelleri* hort. (now 'Joseph').

texedense Pau   = *C. lusitanum*

'The Giant' = 'Giant'

'The Premier' = 'Lilac Wonder'

'Tivi' (*C. szovitsii*)   **R:** Arnis Seisums. **I:** Late 20th century. **RHS:** AM 2008. **N:** Also described in Rukšāns (2007) and Rolfe (2009). **D:** See Cultivars chapter.

transsilvanicum Schur   = *C. autumnale*

trapezunticum Boiss. ex Baker   = *C. umbrosum*

trigynum (Steven ex Adams) Stearn   **E:** *J. Bot.* 72: 344 (1934).

triphyllum Kunze   **E:** *Flora* 29: 755 (1846)

troodi sensu C.D. Brickell, non Kotschy   = *C. decaisnei*

troodi Kotschy   **E:** in Unger, F. & Kotschy, C.G.T. *Ins. Cypern*: 190 (1865).

tunicatum Feinbrun   **E:** *Palestine J. Bot., Jerusalem Ser.* 6: 87 (1953).

turcicum sensu Bornm., non Janka   = *C. haynaldii*

turcicum Janka   **E** *Oesterr. Bot. Z.* 23: 242 (1873).

tuviae Feinbrun   **E:** *Palestine J. Bot., Jerusalem Ser.* 6: 79 (1953).

umbrosum Steven   **E:** *Nouv. Mém. Soc. Imp. Naturalistes Moscou* 1: 264, t. 14 (1829).

'Vahsh' = 'Vakhsh'

'Vakhsh' (*C. luteum*)   **R:** Leonid Bondarenko, Lithuanian Rare Bulb Garden. **I:** Early 21st century. **N:** Originally named 'Vahsh' but the raiser corrected it to 'Vakhsh' to better reflect the orthography. **D:** See

Cultivars chapter.

'Valentine' (*C. doerfleri*)   **R:** A clone grown in the Netherlands and named by Antoine Hoog Authentic Plants when supplied in 2007 to overseas nurseries. **D:** See Cultivars chapter.

valeryi Tineo   = *C. cupanii*

'Van Tubergen's' (*C. speciosum*)   **R:** van Tubergen. **N:** In the late 20th century the Dutch firm of Van Tubergen marketed a selection of *C. speciosum* under this name. Plants similar to it are still offered by Dutch wholesalers. **D:** Height in leaf 36–44cm; outer leaves elliptic-oblong, 32–35 × 6.5–8cm; inner leaves more lancolate-elliptic, 33–35 × 4–5cm; corms quite large, narrow ovate-oblong; height in flower 10–16cm; flowers relatively slim, mid-reddish purple, appearing mid-September into October.

'Vardahovit' (*C. szovitsii*)   **R:** Arnis Seisums. **I:** Late 20th century. **N:** Also described in Rukšāns (2007). **D:** See Cultivars chapter.

***varians* (Freyn & Bornm.) Dyer**   **E:** *Index Kew.*, Suppl. 2: 45 (1904).

***variegatum* L.**   **E:** *Sp. Pl.*: 342 (1753).

variegatum var. *desii* Pamp.   = *C. variegatum*

variegatum subsp. *parkinsonii* (Hook. f.) K. Richt.   = *C. variegatum*

'Velebit Star' (*C. hungaricum*)   **R:** Collected in the Velebit Mountains, Croatia, by Antoine Hoog. Introduced by Antoine Hoog Authentic Plants and registered by C.P.J. Breed in 2014. **I:** 2003. **D:** A seed-raised selection and, as a result, shows some variability. Flowering in January and February, white (which can be assigned to f. *albiflorum*) or palest pink flowers with contrasting black anthers.

velutinum Bornm. & Kneuck.   = *C. schimperi*

veratrifolium S. Arn.   = *C. × byzantinum*

verlaqueae Fridl.   = *C. corsicum*

vernale Hoffm.   = *C. autumnale* subsp. *vernum*

vernum Ker Gawl. ex Stef.   = *C. bulbocodium*

vernum (Reichard) Georgi   = *C. autumnale* subsp. *vernum*

versicolor Ker Gawl.   = *C. bulbocodium* subsp. *versicolor*

'Vesta' (*C. bivonae*)   **R:** A clone grown in the Netherlands and received by Antoine Hoog in 1991 from Jaap Zweeris of 't Zand as *C. sibthorpii*. Hoog named and registered it before introducing it through Hoog & Dix Export. **I:** 1995. **D:** Flowering mid-August to early September; height in leaf 38–45cm; height in flower 10–14cm. Very similar to 'Apollo' but flowers more cup-shaped, more lightly tessellated and with a less prominent white centre; anthers purple-brown; styles slender with prominent curved tips.

'Violet Queen' (*Colchicum*)   **R:** Registered by Zocher & Co. **I:** 1900–1905. **D:** Probably extinct in cultivation but there are enough details to provide a description, should it ever be found persisting in a garden. It was mentioned briefly by Synge (1961) as 'Deep purplish violet with pointed petals and conspicuous white throat, moderately tessellated. Mid-September'. Height in leaf 34–37cm; height in flower 19–23cm; flowering September to early October. Corm broadly ovoid, 4.4–6.2 × 4.5–5.5cm; foot rather obscure; neck 5.5–8.5cm long. Leaves 3–4, grey-green, matt or with a slight shine, strongly ridged longitudinally; lowermost leaf elliptic, 20–23.5 × 7.2–8.2cm, apex subobtuse; uppermost leaf elliptic-oblanceolate to oblanceolate, 22.5–28.4 × 6–7.7cm, apex subacute. Flowers chalice-shaped with pointed tepals, deep purplish violet, strongly tessellated, with a conspicuous white throat, extended up as a thin white stripe along the centre of each tepal for a half to two-thirds from the base; outer tepals oval-oblanceolate, 8.4–8.7 × 3.3–3.5cm; inner tepals elliptic-oblanceolate, 8.1–8.5 × 2.4–3cm; stamens with purplish filaments and purplish pink anthers; styles pale pink-purple with a darker tip, relatively short; perianth tube pale yellowish green, stained purple towards the top, 11.5–14cm long. Some plants in UK private collections grown as 'Violet Queen' differ from the original Zocher description and are now named 'Felbrigg Violet'. Plants sold under the name 'Violet Queen', often from Dutch sources, with non-tessellated, pale pinkish purple flowers without a hint of violet, and which are smaller overall, are now named 'Pride of Holland'.

'Violet Queen' hort., ex Netherlands   = 'Pride of Holland'

'Violet Queen' hort., ex UK private collections   = 'Felbrigg Violet'

visianii Parl.   = *C. haynaldii*

vranjanum Adamović ex Stef.   = *C. autumnale*

'W. Kerbert' (*Colchicum*)   **R:** Zocher & Co. **I:** 1900–1905. **RHS:** AM 1952. **N:** Probably lost from cultivation.

'W.A. Constable' = 'Constable'

'W.R. Dykes' = 'William Dykes'

'Waterlily' (*Colchicum*) **R:** Registered by Zocher & Co. **I:** 1900–1905. **RHS:** AM 1928, AGM (H5) 1997. **D:** See Cultivars chapter.

wendelboi K. Perss. **E:** *Fl. Iranica [Rechinger]* 170: 19 (1992).

'White Wonder' (*C. kotschyi*) **R:** Collected and named by Arnis Seisums, Latvia, who found it north of Tehran, Iran, on the 1998 SLIZE expedition. **I:** 2011 without cultivar name by Jānis Rukšāns, Rare Bulb Nursery, Latvia; then with cultivar name in 2014. **D:** Pure white flowers that appear in August to September in Latvia (Rukšāns 2007).

'Whitton Globe' (*Colchicum*) **R:** Richard Hobbs. **I:** late 20th century. **D:** See Cultivars chapter.

'William Dykes' (*Colchicum*) **R:** Registered by Zocher & Co. **I:** 1900–1905. **D:** See Cultivars chapter.

'Wine Cup' (*C. speciosum*) **R:** Werner Wolf, Berlin. **I:** Pre-2011. **D:** Deep purplish pink flowers with white centre.

'Woods' = 'Little Woods'

woronowii M.R. Bokeriya **E:** *Bot. Zhurn. (Moscow & Leningrad)* 75(2): 201 (1990)

'World Champion's Cup' (*Colchicum*) **R:** Leonid Bondarenko, Lithuanian Rare Bulb Garden. Registered by Bondarenko in 2018. **I:** 2018. **D:** See Cultivars chapter.

'World Cup Champion' = 'World Champion's Cup'

'Yellow Empress' (*C. luteum*) **R:** Arnis Seisums. **I:** Offered by Jānis Rukšāns, Rare Bulb Nursery, Latvia, before 1997. **D:** Large, elegant, elongated, bright yellow flowers with no brown speckling (Rukšāns 2007).

'Yeti' (*C. kesselringii*) **R:** Leonid Bondarenko, Lithuanian Rare Bulb Garden. **I:** Early 21st century. **D:** See Cultivars chapter.

zahnii Heldr. = *C. psaridis*

zangezurum Grossh. = *C. freynii*

'Zephyr' (*Colchicum*) **R:** Registered by P. Visser Czn. **I:** 1985. **D:** See Cultivars chapter.

'Zigana' (*C. szovitsii*) **I:** Offered by Paul Christian Rare Plants nursery, Wrexham, Wales, in 2016. **D:** Later flowering and more intensely coloured than other selections of the species; flowers relatively small, deep pinkish-purple.

Protologues and synonyms in genera other than *Colchicum*

Androcymbium Willd., *Ges. Naturf. Freunde Berlin Mag. Neuesten Entdeck. Gesammten Naturk.* 2: 21 (1808) **Type:** *Androcymbium melanthioides* Willd., designated by Muller-Doblies & Muller Doblies, *Feddes Repert.* 109: 553 (1998)

Bulbocodium L., *Sp. Pl.* 1: 294 (1753)
 Type: *Bulbocodium vernum* L., designated by Steudel, *Nom.* ed. 2, 1: 236 (1840)

Bulbocodium atticum (Spruner ex Tommas.) Nyman = *Colchicum atticum*

Bulbocodium autumnale (L.) Lapeyr. = *Colchicum montanum*

Bulbocodium balearicum Nyman = *Colchicum filifolium*

Bulbocodium broteroi Welw. ex Baker = *Colchicum montanum*

Bulbocodium colchicoides Nyman = *Colchicum montanum*

Bulbocodium dioszegianum Rapaics = *Colchicum bulbocodium* subsp. *versicolor*

Bulbocodium edentatum Schur = *Colchicum bulbocodium* subsp. *versicolor* var. *edentatum*

Bulbocodium lusitanicum Heynh. = *Colchicum montanum*

Bulbocodium montanum Fisch. = *Colchicum bulbocodium* subsp. *versicolor*

Bulbocodium montanum (L.) Heynh. = *Colchicum montanum*

Bulbocodium persicum var. *turkestanicum* Regel = *Colchicum robustum*

Bulbocodium ruthenicum Bunge = *Colchicum bulbocodium* subsp. *versicolor*

Bulbocodium trigynum Steven ex Adams = *Colchicum trigynum*

Bulbocodium vernum L. = *Colchicum bulbocodium*

Bulbocodium vernum f. *dioszegianum* (Rapaics) Soó = *Colchicum bulbocodium* subsp. *versicolor*

Bulbocodium versicolor (Ker Gawl.) Spreng. = *Colchicum bulbocodium* subsp. *versicolor*

Erythrostictus Schltdl., *Linnaea* 1: 90 (1826)
 Type: *Erythrostictus gramineus* (Cav.) Schltdl.

Geophila pyrenaica Bergeret = *Colchicum montanum*

Merendera Ramond, *Bull. Sci. Soc. Philom. Paris* 2: 178

(1801) **Type**: *Merendera bulbocodium* Ramond, nom. illeg. (= *M. montana* (L.) Lange, based on *Colchicum montanum* L., *Sp. Pl.* 1: 342 (1753)

Merendera aitchisonii Hook. f. = *Colchicum robustum*

Merendera androcymbioides Valdés = *Colchicum androcymbioides*

Merendera attica (Spruner ex Tommas.) Boiss. & Spruner = *Colchicum atticum*

Merendera badghyi Korsh. = *Colchicum robustum*

Merendera bulbocodioides Willd. = *Colchicum montanum*

Merendera bulbocodium Ramond = *Colchicum montanum*

Merendera candidissima Miscz. ex Grossh. = *Colchicum trigynum*

Merendera caucasica M. Bieb. = *Colchicum trigynum*

Merendera eichleri (Regel) Boiss. = *Colchicum trigynum*

Merendera figlalii Varol = *Colchicum figalii*

Merendera filifolia Cambess. = *Colchicum filifolium*

Merendera ghalghana Otsch. = *Colchicum trigynum*

Merendera greuteri Gabrieljan = *Colchicum trigynum*

Merendera jolantae Czerniak. = *Colchicum robustum*

Merendera kurdica Bornm. = *Colchicum kurdicum*

Merendera manissadjianii Azn. = *Colchicum manissadjianii*

Merendera mirzoevae Gabrieljan = *Colchicum mirzoevae*

Merendera montana (L.) Lange = *Colchicum montanum*

Merendera nivalis Stapf = *Colchicum szovitsii*

Merendera persica Boiss. = *Colchicum robustum*

Merendera pyrenaica (Pourr.) P. Fourn. = *Colchicum montanum*

Merendera raddeana Regel = *Colchicum raddeanum*

Merendera robusta Bunge = *Colchicum robustum*

Merendera schimperiana Hochst. = *Androcymbium schimperianum*

Merendera sobolifera Fisch. & C.A. Mey - *Colchicum soboliferum*

Merendera trigyna (Steven ex Adams) Stapf = *Colchicum trigynum*

Merendera verna (L.) Bubani = *Colchicum bulbocodium*

Merendera wendelboi (K. Perss.) Oganezova = *Colchicum wendelboi*

Synsiphon crociflorus Regel = *Colchicum kesselringii*

Excluded species

Colchicum albanense = *Androcymbium albanense*
Colchicum albofenestratum = A South African species for which there is no combination in *Androcymbium*
Colchicum albomarginatum = *Androcymbium albomarginatum*
Colchicum amphigaripense = *Androcymbium amphigaripense*
Colchicum asteroides = *Androcymbium asteroides*
Colchicum austrocapense = *Androcymbium austrocapense*
Colchicum bellum = *Androcymbium bellum*
Colchicum buchubergense = *Androcymbium buchubergense*
Colchicum burkei = *Androcymbium burkei*
Colchicum capense = *Androcymbium capense*
Colchicum cedarbergense = *Androcymbium cedarbergense*
Colchicum circinatum = *Androcymbium circinatum*
Colchicum clanwilliamense = *Androcymbium albanense* subsp. *clanwilliamense*
Colchicum coloratum = *Androcymbium burchellii*
Colchicum crenulatum = *Androcymbium crenulatum*
Colchicum crispum = *Androcymbium crispum*
Colchicum cruciatum = *Androcymbium cruciatum*
Colchicum cuspidatum = *Androcymbium cuspidatum*
Colchicum decipiens = *Androcymbium decipiens*
Colchicum dregei = *Androcymbium dregei*
Colchicum eghimocymbion = *Androcymbium eghimocymbion*
Colchicum etesionamibense = *Androcymbium etesionamibense*
Colchicum eucomoides = *Androcymbium eucomoides*
Colchicum europaeum = *Androcymbium*

europaeum
Colchicum exiguum = *Androcymbium exiguum*
Colchicum falcifolium = *Iris caucasica*
Colchicum gramineum = *Androcymbium gramineum*
Colchicum greuterocymbium = *Androcymbium greuterocymbium*
Colchicum hantamense = *Androcymbium hantamense*
Colchicum henssenianum = *Androcymbium henssenianum*
Colchicum hierrense = *Androcymbium hierrense*
Colchicum hughocymbion = *Androcymbium hughocymbion*
Colchicum huntleyi = *Androcymbium huntleyi*
Colchicum irroratum = *Androcymbium irroratum*
Colchicum jordanicola = *Androcymbium palaestinum*
Colchicum karooparkense = *Androcymbium karooparkense*
Colchicum knersvlaktense = *Androcymbium knersvlaktense*
Colchicum kunkelianum = *Androcymbium kunkelianum*
Colchicum leistneri = *Androcymbium leistneri*
Colchicum longipes = *Androcymbium longipes*
Colchicum melanthioides = *Androcymbium melanthioides*
Colchicum natalense = *Androcymbium natalense*
Colchicum orienticapense = *Androcymbium orienticapense*
Colchicum palaestinum = *Androcymbium palaestinum*
Colchicum poeltianum = *Androcymbium poeltianum*
Colchicum praeirroratum = *Androcymbium praeirroratum*
Colchicum psammophilum = *Androcymbium psammophilum*
Colchicum rechingeri = *Androcymbium rechingeri*
Colchicum roseum = *Androcymbium roseum*
Colchicum scabromarginatum = *Androcymbium scabromarginatum*
Colchicum schimperianum = *Androcymbium schimperianum*
Colchicum stirtonii = *Androcymbium stirtonii*
Colchicum striatum = *Androcymbium striatum*
Colchicum swazicum = *Androcymbium swazicum*
Colchicum undulatum = *Androcymbium undulatum*
Colchicum vanjaarsveldii = *Androcymbium vanjaarsveldii*
Colchicum villosum = *Androcymbium villosum*
Colchicum walteri = *Androcymbium walteri*
Colchicum worsonense = *Androcymbium worsonense*
Colchicum wyssianum = *Androcymbium gramineum*

Glossary

Acaulescent – stemless.

Actinomorphic (of the flower) – regular, with radial symmetry.

Amplexicaul – stem-clasping, with reference to expanded leaf or bract bases.

Anther – the pollen-producing part of the stamen.

Anthesis – the point at which the flowers open and become ready for pollination.

Auriculate – possessing an ear-like projection (auricle).

Axil – the angle between the upper side of a petiole or branch and the stem or trunk to which it is attached.

Axillary bud – a bud that is formed in the axil, or sometimes a bulbil or proto-corm formed in the lower leaf bases.

Base plate – the interface between the roots and the corm.

Basifixed (of the anther) – attached at the base to the filament (see also dorsifixed).

Bract – a modified leaf, usually associated with the inflorescence. Bracts range from leaf-like to scale-like on different plants.

Bracteole – secondary bract.

Calyx – a collective term for the sepals.

Campanulate – bell-shaped. In *Colchicum* often referred to as goblet-shaped.

Capsule – a dry, many-seeded, dehiscent fruit. In *Colchicum* fruit capsules are 3-chambered and located at ground level or elevated above on a short pseudostem to 30cm.

Carpel – the basic unit of the female part of a flower made up of the ovary (containing ovules), stigma and style.

Cartilaginous – flexible but firm or tough, in relation to the corm tunic or leaf margin.

Cataphyll – a modified leaf generally lacking in chlorophyll, and fleshy, cylindrical, white, whitish or rarely green to purplish. They are usually produced from the base plate and encased wholly or partly within the tunic, serving to protect the emerging leaves and flowers, extending up to ground level or sometimes beyond and generally withering away after anthesis.

Ciliate – the presence of fine, marginal hairs fringing the leaves, sepals or other organs.

Claw – the narrowed, lower part of the tepals (perianth lobes) in species that do not form a fused perianth tube.

Colchicine – a toxic alkaloid extracted from *Colchicum* seeds and corms, used mainly in medicine and plant breeding.

Convolute – rolled up longitudinally, often wrapped around itself.

Coriaceous – leathery in texture.

Corm – a solid, bulb-like, subterranean storage organ. In *Colchicum* normally upright and asymmetric, but occasionally horizontal and in effect stoloniferous. The main structure is typically referred to by gardeners as the 'mother' corm; in certain species this produces several cormlets from side buds, known as 'daughter' offsets.

Corolla – a collective term for the fused petals of the flower.

Corymb – a raceme in which the lower flowers have progressively longer stalks (pedicels) bringing all the flowers to one level, flat-topped. Sometimes called a pseudo-umbel.

Cucullate – hooded.

Cultivar – a clear-cut variant of a species or hybrid, named and maintained in horticulture.

Cuneate – wedge-shaped, typically with reference to a leaf, bract, petal or tepal.

Cyme – an inflorescence involving repeated lateral branching, each branch terminating in a flower.

Decurrent (of the stigma) – continuing as a wing or flange down a main structure. In some *Colchicum* species the stigma is decurrent down the tip of the style for up to 6mm.

Dehiscence – the release of the contents of a plant structure, usually an anther or fruit, that splits along a line of weakness at maturity.

Dorsifixed (of the anther) – attached by their middle to the filament of the stamen (see also basifixed).

Elliptical – narrowly oval in outline.

Emarginate – notched at the top, generally referring to sepals, petals or tepals, or leaves.

Endosperm – the tissue produced inside a seed.

Entire (of the leaf margin) – without teeth, indentations or lobes.

Exserted – protruding.

Extrorse – anthers that release pollen through a fissure on their outer side.

Fascicled – clustered; a bunch or bundle.

Filament – the slender stalk of the stamen which bears the anther.

Filament channels – pronounced ridges that sometimes form either side of the filament base, where the filament attaches to the tepals.

Filiform – thread-like.

Fimbriae – a fringe, sometimes interlocking, of short, sometimes hooked, hair-like projections at the base of the tepal lobes. Fimbriae can be present on leaves, filaments and other organs.

Garrigue – a more open, often degraded plant community than maquis, consisting of low, open shrubs and many aromatic plants. Widespread in the Mediterranean region and given different names in different places, including phrygana.

Geophyte – a perennial plant that resides below ground for part of the year in the form of a bulb, corm or tuber.

Glabrous – hairless.

Gynoecium – the ovary and styles; the female organs of the flower.

Hermaphrodite – fertile stamens (male) and ovaries (female) present in the same flower.

Hispid – coarsely, stiffly hairy, or bristly.

Hypopodium – a downward-pointing projection at the base of the corm, sometimes referred to as a foot.

Hysteranthous – flowering in advance of the emergence of the leaves. Subhysteranthous means flowering with just the tips of the leaves present.

Idioblasts – an isolated plant cell that differs from neighbouring tissues; in *Androcymbium* they are seen as black or brown dots.

Inflorescence – a group of flowers with or without a branching system arranged on a single stalk (peduncle).

Infundibuliform – funnel-shaped.

Introrse – anthers that release pollen through a fissure on their inner side.

Keeled – a thickened, central ridge or flange on the underside of the leaf, as in the keel of a boat.

Lamellae – narrow plates of tissue.

Lanceolate – shaped like a lance, the broadest part towards the base (linear-lanceolate is slenderer still).

Ligulate – strap-shaped.

Littoral – growing on or close to the shore, especially that of the sea.

Loculicidal – a form of dehiscence in which the fruit capsule splits along the midrib of each ovary, not at the septa (see also septicidal).

Locule (of the ovary and fruit) – a chamber of the ovary (and then fruit) that contains ovules. In *Colchicum* the ovary and fruit are 3-locular.

Maquis (macchie) – a plant community typical of the Mediterranean consisting of dense, evergreen small trees and shrubs and often very aromatic.

Membranous – like a membrane; parchment-like, thin, dry and flexible.

Molecular phylogeny – the science of analysing evolutionary relationships by means of their hereditary, molecular differences.

Naturalised – a non-native species that has become established and is self-perpetuating.

Nectary – a nectar-secreting organ.

Oblanceolate – reverse of lanceolate, i.e. lance-shaped but broadest in the upper third.

Obovate – reverse of ovate, i.e. egg-shaped with the broadest part towards the top, in particular a leaf, petal or tepal attached at the narrow end.

Ovary – the female part of a flower that holds the ovules and develops into the seedpod. In *Colchicum* the ovary is always superior, i.e. with the perianth tube formed from beneath, not attached to the top as in *Crocus*.

Ovate – an outline akin to a hen's egg, broadest towards the base.

Ovoid – egg-shaped and three-dimensional.

Pedicel – the stalk of an individual flower.

Peduncle – the main stalk of an inflorescence.

Pedunculate – possessing a peduncle.

Perfoliate – leaves or bracts whose bases fuse, making it appear that the stem grows through them, or, on occasion, the base of a single leaf or bract that entirely encircles the stem.

Perianth – the outer, non-sexual, whorls of the flower (the calyx and corolla together). In *Colchicum* this consists of six similar looking tepals, arranged in two series of three.

Perianth tube – a tube formed at the base of the tepals by their fusion. In many *Colchicum* species and cultivars there is a very long perianth tube, in some the tube is split and the tepals not united to one another.

Phrygana – see garrigue.

Pilose – covered in spreading, long hairs.

Ploidy – the number of sets of chromosomes in the nucleus of a plant cell. Normal cells have two sets and are known as diploid; a triploid cell has three, a tetraploid has four etc.

Polje – a large, steep-sided basin, found in karstic regions of eroded limestone.

Procumbent – spreading on the ground.

Pseudostem – an arrangement of the leaf bases, tightly wrapped around one another to form a stem-like cylindrical structure. Banana plants are the best example but on a much smaller scale the structure is present in many of the large-flowered, autumn-flowering species and cultivars of *Colchicum*.

Pubescent – covered with short, downy hairs.

Puberulous – covered in fine down.

Punctiform – with a dot-like stigma (see also decurrent).

Recurved – curved down or back.

Rotate (of the flower) – star- or wheel-shaped, the petals or tepals spreading at right angles to the axis.

Scabrid – rough to the touch, minutely bristly.

Scorpioid – with reference to cymes, coiled in the manner of a scorpion's tail.

Sensu lato (s.l.) – in the broad sense, usually in terms of the application of a scientific name..

Sensu stricto (s.s.) – in the narrow sense, usually in terms of the application of a scientific name.

Septicidal – a form of dehiscence in which the fruit capsule splits down the partitions (septa) of the locule walls, dividing into its component carpels.

Sessile – lacking a stalk; this can be the lack of a petiole or pedicel.

Soboliferous – refers to species in which the corm is elongate and somewhat grub-like, behaving like a horizontal rhizome (a sobole).

Stamen – the male component of the flower, comprising a stalk or filament and a pollen-bearing anther.

Stellate – star-like.

Stigma – the receptive area at the top of the style that receives the pollen.

Stolon – a modified, creeping stem, either subterranean, as in some *Colchicum* species, or occurring above ground.

Style – the slender, stalk-like appendage linking the ovary to the receptive tip or stigma.

Subglobose – not quite globe-shaped.

Superior (of the ovary) – with the perianth tube formed from beneath, not attached to the top (as in *Crocus*).

Subtend – positioned below but in close proximity, particularly a bract relating to a flower in its axil.

Synanthous – flowering when the leaves have emerged and are part or wholly developed.

Taxon – any group or rank (such as a species or a family) in a biological classification into which related organisms are classified. The plural is taxa.

Tendrillate – a spirally coiled, thread-like outgrowth derived from a leaf or a stem, and able to twine round a support.

Tepal – the undifferentiated outer part, or perianth segment of a flower; often used when sepals and petals all look similar and are difficult to differentiate.

Terete – a stem or leaf which is smooth and rounded in cross-section

Terra rossa – a reddish, clayey soil formed from the weathering of limestone and characteristic of Mediterranean regions.

Tessellated – with a chequered patterning. In *Colchicum* this is in reference to the tepals.

Tunic – the protective outer layer(s) surrounding corms and bulbs, formed from successive leaf bases and, in *Colchicum*, varying in shades of brown, reddish-brown or orange-brown from one species to another.

Tunicated – possessing a tunic, consisting often of concentric layers. In *Colchicum* corms it often extends upwards into a neck.

Ultramafic – dark-coloured, igneous rocks, mainly ferromagnesian.

Umbel – an inflorescence in which all the flowers, usually stalked, arise from a single point at the end of the stem, like the spokes of an umbrella.

Undulate – waved, in a plane at right angles to the surface.

Variety – a variant differing in one or several small details from the norm, often found within the general population of a particular species or taxon.

Versatile (of the anther) – the ability of a dorsifixed anther to pivot (see also basifixed).

Zygomorphic (of the flower) – divisible in one plane only into two equal halves.

Appendix

New taxon

Colchicum × ambiguum Grey-Wilson, hybr. nov.

Hybrid with certain features of *C. cilicicum*, flowering in advance of leaf production at the same time as that species, with faintly tessellated, rose-purple tepals, yellow anthers and hooked reddish purple styles bearing punctiform stigmas. *Colchicum × ambiguum* has inherited various features of *C. cilicicum* in its overall morphology, yet the two can be separated on a combination of characters. *Colchicum cilicicum* generally produces a lot of flowers per corm, commonly 3–15, sometimes more, whereas *C. × ambiguum* produces 1–4, although its corms tend to clump together to give the impression of many more flowers per corm. The flowers of *C. cilicicum* are rather larger, the tepals 5–7.5cm long (versus 3.9–6.6cm in *C. × ambiguum*), while the stamens are two-thirds to four-fifths the length of the tepals with a noticeable yellow swollen base to the filaments (versus no more than half the length of the tepals and with a scarcely swollen greenish base in *C. × ambiguum*). In *C. cilicicum* the stigmas are often, not always, shortly decurrent, while in *C. × ambiguum* they are always punctiform. Marked differences in chromosome number are indicated below. This hybrid has been cultivated since at least the mid-19th century and has not been found in the wild. One parent is *C. cilicicum* (2n=54) and the other is likely to be *C. neapolitanum* (2n=90).

Type: Cultivated plant (accession no. W853856A) collected at RHS Garden Wisley, Surrey, UK, 2 Sep. 1997.

Holotype: WSY (WSY0002294).

Synonym: *C. tenorei* (as *C. tenorii*) sensu J.G. Baker, *J. Linn. Soc. Bot.* 17: 427 (1879) & sensu E.A. Bowles, *A Handbook of Crocus and Colchicum for Gardeners*: 184 (1952), non Parl.

Etymology: The name refers to Baker's (1879) ambiguous description which cites purple anthers, although he was almost certainly referring to this plant and was one of the first to recognise it.

Cytology: 2n=72.

References & bibliography

Abbott, R.D. et al. (2013) Hybridization and speciation. *J. Evol. Biol.* 26: 229–246

Aeschimann, D., Lauber, K., Moser, D.M. & Theurillat, J.-P. (2004) *Flora Alpina*. Haupt, Bern

Akan, H. & Eker, I. (2005a) Checklist of the genus *Colchicum* in the *Flora of Turkey*. *Turk. J. Bot.* 29: 327–331

Akan, H. & Eker, I. (2005b) A new record for Turkey: *Colchicum crocifolium* Boiss., with a contribution to the taxonomy of the species. *Belgian J. Bot.* 138: 93–96

Alexiou, S. The genus *Colchicum* L. (*Colchicaceae*) in Greece. *Parnassiana Arch.* 1: 59–73

Allen, M. (1973) *E.A. Bowles and his Garden at Myddelton House 1865–1954*. Faber & Faber, London

Anderson, E.B. (1973) *Seven Gardens or Sixty Years of Gardening*. Michael Joseph, London

Anon. (1954) The Award of Garden Merit LXXXVII *J. Roy. Hort. Soc.* 79(3): 128–133

Anon. (1998) Minutes of the Joint Rock Garden Plant Committee. *Proc. Roy. Hort. Soc.* 1998: 50–52

Arrigoni, P.V. (2015) *Flora dell'Isola di Sardegna Vol. 5*. Carlo Delfino, Sassari, Sardinia

Aznavour, G.V. (1908) Un nouveau *Merendera* d'Anatolie. *Bull. Herb. Boissier Ser.* 2: 249–250

Baillon, H.E. (1894) *Liliacées*. In Baillon, H.E. (ed.) *Histoire des Plantes Vol. 12*. Hachette et Cie, Paris

Baker, J.G. (1879) A synopsis of *Colchicaceae* and the aberrant tribes of *Liliaceae*. *J. Linn. Soc. Lond. Bot.* 17: 405–510

Baliousis, E. & Tan, K. (2012) In Vladimirov, V., Dane, F. & Tan. K. (eds.) New floristic records in the Balkans: 20. *Phytol. Balcan.* 18(3): 333–373

Barina, Z., Pifkó, D. & Mesterházy, A. (2011) Contributions to the flora of Albania, 3. *Willdenowia* 41: 329–339

Barratt, T.N. (1951) Return to Gavarnie. *Quart. Bull. Alpine Gard. Soc.* 19: 173–176

Bentham, G. & Hooker, J.D. (1883) *Genera Plantarum Vol. 3*. L. Reeve & Co., London

Biel, B. & Tan, K. (2009) In Vladimirov, V. et al. New floristic records in the Balkans: 12. *Phytol. Balcan.* 15(3): 431–452

Birkinshaw, D. (1971) Lebanese colchicums and others. *Lily Year Book* 34: 140–148

Blakeslee, A.F. & Avery, A.G. (1937) Methods of inducing chromosome doubling in plants by treatment with colchicine. *J. Heredity* 28: 393–411

Blamey, M., & Grey-Wilson, C. (1993) *Mediterranean Wild Flowers*. HarperCollins, London

Boissier, P.E. (1882) *Flora Orientalis Vol. 5*. Basel

Bokeriya, M.R. (1990) On the forgotten species of *Colchicum* (*Colchicaceae*) from the Abkhazian automonous Soviet Socialist Republic. *Bot. Zhurn.* 75: 199–203 [in Russian]

Bowles, E.A. (1915, revised 1972) *My Garden in Autumn and Winter*. David & Charles, Newton Abbot

Bowles, E.A. (1924, reprinted 1952) *A Handbook of Crocus and Colchicum for Gardeners*. The Bodley Head, London

Brickell, C.D. (1980) *Colchicum* L. In Tutin, T.G. et al. (eds.) *Flora Europaea Vol. 5*. Cambridge University Press, Cambridge

Brickell, C.D. (1983) *Colchicum* L. In Davis, P.H. (ed.) Materials for a flora of Turkey 38. *Notes Roy. Bot. Gard. Edinburgh* 41(1): 49–51

Brickell, C.D. (1984) *Colchicum* L., *Merendera* Ramond. In Davis, P.H. (ed.) *Flora of Turkey and the East Aegean Islands. Vol. 8*. Edinburgh University Press, Edinburgh

Brickell, C.D. (1986a) *Colchicum* L. In Cullen, J. et al. (eds.) *The European Garden Flora Vol. 1*. Cambridge University Press.

Brickell, C.D. (1986b) *The Vanishing Garden*. John Murray.

Brickell, C.D. (1998) *Colchicum davisii* C.D. Brickell; a new species from Turkey. *The New Plantsman* 5 (1): 15–22

Briquet, J. (1910) *Prodrome de la Flore Corse*. Paul Lechevalier, Paris

Brotero, F.A. (1816) *Phytographia Lusitaniae Selectior* (edn. 3). Typographia Regia, Lisbon

Buchan, U. (2007) *Garden People*. Thames & Hudson, London

Burtt, B.L. (1955) Notes on *Colchicum*. *Notes Roy. Bot. Gard. Edinburgh* 21: 296–300

Burtt. B.L. (1968) *Colchicum* and *Merendera*. A Lily Group Discussion. *Lily Year Book* 31: 90–105

Burtt, B.L. (1970) The evolution and taxonomic significance of a subterranean ovary in certain monocotyledons. *Israel J. Bot.* 19: 77–90

Burtt, B.L. (1981) The name of the Spanish *Merendera*. *Taxon* 30: 299–300

Butcher, R.W. (1954). *Colchicum* L. *J. Ecol.* 42(1): 249–257

Buxbaum, F. (1925) Vergleichende anatomie der *Melanthoideae*. *Repert. Spec. Nov. Regni Veg. Beih.* 29: 1–80

Buxbaum, F. (1936) Der entwicklungslinien der *Liliodeae* 1, *Wurmbaeioideae*. *Bot. Arch.* 38: 213–293

Camarda, I. (1978) Le piante endemiche della Sardegna 21–23: *Colchicum gonarei* species nova. *Boll. Soc. Sarda Sci. Nat.* 17: 227–242

Camarda, I. (1979) Actuelles connaissances du genre *Colchicum* en Sardaigne. *Webbia* 34(1): 481–485

Camarda I. (1990) Le piante endemiche della Sardegna 198: *Colchicum corsicum* Baker (1879). *Boll. Soc. Sarda Sc. Nat.* 27: 283–287

Ceschmedjiev, I.V. (1994) Reports 313–366. In Kamari, G., Felber, F., Garbari, F. (eds.) Mediterranean chromosome number reports 4. *Fl. Medit.* 4: 269–279

Chacón, J., Cusimano, N., & Renner, S.S. (2014) The evolution of *Colchicaceae*, with a focus on chromosome number. *Syst. Bot.* 39(2): 415–427

Chittenden, F.J. (1927) The Award of Garden Merit IX. *J. Roy. Hort. Soc.* 52(1): 82–85

Clusius, C. (1601) *Rariorum Plantarum Historia*. Ioannem Moretum, Antwerp

Dahlgren, R.M.T., Clifford, H.T. & Yeo, P.F. (1985) *The Families of the Monocotyledons*. Springer-Verlag, Berlin

D'Amato, F. (1955) Revisione citosistematica del genere *Colchicum*. I. *C. autumnale* L., *C. lusitanum* Brot. e *C. neapolitanum* Ten. *Caryologia* 7: 292–349

D'Amato, F. (1957a) Revisione citosistematica del genere *Colchicum*. II. Nuove località di *C. autumnale* L., *C. lusitanum* Brot. e *C. neapolitanum* Ten. e delimitazione dell'areale delle tre specie nella penisola Italiana. *Caryologia* 9: 315–339

D'Amato, F. (1957b) Revisione citosistematica del genere *Colchicum*. III: *C. alpinum* Lam. & DC., *C. cupanii* Guss., *C. bivonae* Guss. e chiave analitica per la determinazione delle specie di *Colchicum* della flora italiana. *Caryologia* 10: 111–151

Davis, P.H. (1939) Some plants of the Eastern Mediterranean. *Quart. Bull. Alpine Gard. Soc.* 7: 25–64

Davis, P.H. (1971) Distribution patterns in Anatolia with particular reference to endemism. In Davis, P.H., Harper, P.C. & Hedge. I.C. (eds.). *Plant life of South West Asia*. The Botanical Society of Edinburgh, Aberdeen

de Passe the Younger, C. (1614) *Hortus Floridus*. Arnheim

Del Hoyo, A. & Pedrola-Montfort, J. (2006) Missing links between disjunct populations of *Androcymbium* (*Colchicaceae*) in Africa using chloroplast DNA noncoding sequences. *Aliso* 22: 606–618

Dimitrellos, G. & Christodoulakis, D. (1995) The Flora of Mt Timfristos (NW Sterea Ellas, Greece). *Fl. Medit.* 5: 9–51

Dimopoulos, P. (1993) *Chloridiki kai Phitokinoniki Ereuna tou Orous Killini-Oikologiki Prosegisi*. Ph.D. thesis, University of Patras, Greece

Dimopoulos, P., Raus, Th., Bergmeier, E., Constantinidis, Th., Iatrou, G., Kokkini S., Strid, A. & Tzanoudakis, D. (2013) *Vascular Plants of Greece: An Annotated Checklist*. Botanischer Garten und Botanisches Museum Berlin-Dahlem, Berlin & Hellenic Botanical Society, Athens

Dobeś, C. & Hahn, B. (1997) IOPB chromosome data 11. *IOPB Newsletter* 26–27: 15–18

Domac, R. (1973) *Mala Flora Hrvatske i Susjednih Područja*. Školska Knjiga, Zagreb

Domac, R. (1994) *Flora Hrvatske, Priručnik za Oderedivanje Bilja*. Školska Knjiga, Zagreb

Dusen, O. & Sumbul, H. (2007) A morphological investigation of *Colchicum* L. (*Liliaceae*) species in the Mediterranean region in Turkey. *Turk. J. Bot.* 31: 373–419

Evans, S., Henrici, A. & Ing, B. (2006) *Red Data List of*

Threatened British Fungi. British Mycological Society

Fakas, G., Korakis, A. & Alexiou, S. (2014) *Colchicum atticum (Colchicaceae), a new taxon for Samos island, Greece. Parnassiana Arch.* 2: 55–59

Farrer, R. (1928) *The English Rock-Garden. Vol. 1.* T.C. & E.C. Jack Ltd., London

Feinbrun, N (1953) The genus *Colchicum* in Palestine and neighboring countries. *Palestine J. Bot. Jerusalem Ser.* 6: 71–95

Feinbrun, N. (1958) Chromosome numbers and evolution in the genus *Colchicum. Evol.* 12: 173–18

Feinbrun-Dothan, N. (1986) *Flora Palaestina.* Israel Academy of Sciences and Humanities, Jerusalem

Fernandes, A. & Franca, F. (1977) Le genre *Colchicum* L. au Portugal. *Bol. Soc. Brot., Sér.* 2 51: 5–36

Franco, J.A. & Afonso, M.L.R. (1994) *Colchicum* & *Merendera. Nova Flora de Portugal, Vol. 3.* Escolar Editora, Lisbon

Fridlender, A. (1999a) Sur l'identité des colchiques (*Colchicum* L, Liliaceae) a feuilles hystéranthées du Maroc: I. *Colchicum fharii* sp. nov. *Bull. Mens. Soc. Linn. Lyon* 68: 251–278

Fridlender, A. (1999b) *Originalités Biologiques et Systématiques des Especes Rares. Quelques Exemples Choisis dans la Flore Tyrrhénienne.* Thesis, Muséum National d'Histoire Naturelle, Paris.

Fridlender, A. (1999c) Une nouvelle espece corse de colchique: *C. arenasii* sp. nov. (Liliaceae). *Acta Bot. Gallica* 146(2): 157–167

Fridlender, A. (1999d) Description d'une espece nouvelle de colchique (*Colchicum*, Liliaceae) en Sardaigne: *Colchicum actupii* Fridlender. *Bull. Mens. Soc. Linn. Lyon* 68: 193–200

Fridlender, A., Brown, S., Verlaque, R., Crosnier, M.T. & Pech, N. (2002) Cytometric determination of genome size in *Colchicum* species (*Liliales, Colchicaceae*) of the western Mediterranean area. *Plant Cell Rep.* 21: 347–352

Fridlender, A. & Pignal, M. (2013) Les colchiques de Provence: état de conservation et nouveautés taxonomiques. *Nature de Provence* 2: 19–35

Furse, P. (1963) Iran and Turkey, 1962. *J. Roy. Hort. Soc.* 88: 166–176, 199–211, 247–251

Furse, P. (1968) *Colchicum* and *Merendera* in Turkey, Iran and Afghanistan. *Lily Year Book* 31: 106–115

Gabrieljan, E.C. (2000) The genus *Colchicum* in southern Transcaucasus. *Bot. Chron. (Patras)* 13: 229–239

Galil, J. (1969) On the laterally-contracting root of *Colchicum stevenii. Beitr. Biol. Pflanz.* 46: 315–322

Giannopoulos, K., Tan, K. & Vold, G. (2011) In Vladimirov, V. *et al.* New floristic records in the Balkans: 16. *Phytol. Balcanica* 17(2): 247–264

Goula, K. & Konsoulas, G. (2016) *Colchicum chimonanthum* (Colchicaceae): confirmation of old reports and discovery of new populations. *Parnassiana Arch.* 4: 23–26

Govaerts, R. & Persson, K. (2008) Proposals to reject the names *Colchicum tenorei* and *Colchicum todaroi* (Colchicaceae). *Taxon* 57: 995–996

Greuter, W. (1971) Betrachtungen zur pflanzengeographie der Südägäis. *Opera Bot.* 30: 49–64

Grey, C.H. (1937) *Hardy Bulbs.* Williams & Norgate, London

Grey-Wilson, C. & Mathew, B. (1981) *Bulbs: The Bulbous Plants of Europe and their Allies.* Collins, London

Grey-Wilson C. (2010) *A Field Guide to the Bulbs of Greece.* Alpine Garden Society, Pershore

Grey-Wilson C. (2019) Colchicums on trial. *The Garden* 144(10): 57

Grintescu, I. (1966) *Colchicum, Merendera.* In Săvulescu, T & Nyárády, E.I. (eds) *Flora Republicii Socialiste România. Vol 11.* Editura Academiei Republicii Socialiste România, Bucharest

Guinochet, M. & de Vilmorin, R. (1978) *Flore de France, Vol. 3.* Centre National de la Recherche Scientifique, Paris

Halácsy, E.V. (1904) *Conspectus Florae Graecae, Vol. 3.* Sumptibus Guilelmi Engelmann, Leipzig

Havas, L. (1937) Effects of colchicine and of *Viscum album* preparations upon germination of seeds and growth of seedlings. *Nature* 139: 371

Hedge, I. & Wendelbo, P. (1978) Patterns of distribution and endemism in Iran. *Notes Roy. Bot. Gard. Edinburgh* 36: 441–464

Hegi, G. (1990) *Illustrierte Flora von Mitteleuropa, Vol. 2* (3rd edn). Parey, Berlin & Hamburg

Heldreich, T. von (1882) *Flore de l'Île de Céphalonie.* Bridel, Lausanne

Hess, H.E., Landolt, E., Hirzel, R. (1976) *Flora der Schweiz und Angrenzender Gebiete, Vol. 1* (2nd edn.). Birkhäuser, Basel

Holubec, V. & Krivka, P. (2006) *The Caucasus and its Flowers*. Loxia, Prague

Horvat, I., Glavač, V., Ellenberg, H. (1974) *Vegetation Südosteuropas*. G. Fischer, Stuttgart

Hršak, V. (2001) Notulae ad indicem Florae Croaticae, 3. *Natura Croatica* 10(1): 67–72

Huxley, A., Griffiths, M. & Levy, M. (1992) *The New Royal Horticultural Society Dictionary of Gardening*. Macmillan, London

Jarvis, C.E., Barrie, F.R., Allan, D.M., Reveal, J.L. (eds.) (1993) A list of Linnaean generic names and their types. *Regn. Veg.* 127

Jung, L.S., Winter, S., Eckstein, R.L., Kriechbaum, M., Karrer, G., Welk, E., Elsässer, M., Donath, T.W. & Otte, A. (2011) *Colchicum autumnale*, Biological Flora of Central Europe. *Perspect. Plant Ecol. Evol. Syst.* 13: 227–244

Karidas, A., Giannakis, T. & Antonopoulos, Z. (2017) A new locality of *Colchicum soboliferum* (*Colchicaceae*) from Macedonia, Greece. *Parnassiana Arch.* 5: 57–59

Kit, T. & Vold, G. (2006) In Vladimirov, V. *et al.* New floristic records in the Balkans: 1. *Phytol. Balcanica* 12(1): 107–128

Kit, T., Vold, G. & Sfikas, G. (2006) In Vladimirov, V. *et al.* New floristic records in the Balkans: 3. *Phytol. Balcanica* 12(3): 413–440

Kitanov, B. (1950) Kritische bemerkungen über zwei pflanzenarten Mazedoniens. *Izv. Bot. Inst. (Sofia)* 1: 377–392

Kleizen, C., Midgley, J. & Johnson, S.D. (2008) Pollination systems of *Colchicum* (*Colchicaceae*) in southern Africa: evidence for rodent pollination. *Ann. Bot.* 102(5):747–55

Kokmotos, E. & Georgiadis, Th. (2005) The flora of Mountain Elikon, Xerovouni and Neraidolakkoma (Boeotia, Sterea Ellas, Greece). *Fl. Medit.* 15: 403–451

Komarov, V.L. (1935, translated 1968) *Colchicum*. In Komarov, V.L. (ed.) *Flora of the USSR, Vol. IV*. Israel Program for Scientific Translations, Jerusalem.

Koninklijke Algemeene Vereeniging voor Bloembollencultuur (continually updated) Siergewassen cultivar registration database. www.kavb.nl/siergewassen

Kougioumoutzis, K., Tiniakou, A., Georgiadis, T. & Georgiou, O. (2012) Contribution to the flora of the south Aegean Volcanic Arc: the Methana peninsula (Saronic Gulf, Greece). *Edinburgh J. Bot.* 69 (1): 53–81

Küçüker, O. (1995) Contribution to the knowledge of some endangered *Colchicum* species of Turkey. *Fl. Medit.* 5: 211–219

Lafranchis, T. & Sfikas, G. (2009) *Flowers of Greece, Vol. 2*. Diatheo, France

Leeds, R. (2005) *Autumn Bulbs*. B.T. Batsford Ltd, London

Lentini, F. & Raimondo, F.M. (1984) Contribution à la connaissance du genre *Colchicum* L. en Sicile. *Webbia* 38: 745–755

Levin, D.A. (1983) Polyploidy and novelty in flowering plants. *Amer. Naturalist* 122: 1–25

Linnaeus, C. (1753) *Species Plantarum*. Laurentius Salvius, Stockholm

Loudon, J.C. (1839) *Hortus Britannicus with second additional supplement*. Longman, Orme, Brown, Green & Longmans, London

Lovka, M. (1975) Prispevek k citologiji jugoslovanskih semenovk (*Spermatophyta*), I: *Liliaceae* s. lat. *Biol. Vestn.* 23: 25–40

Mabey, R. (1996) *Flora Britannica*. Sinclair-Stevenson, London

Maire, R. (1958) *Flore de l'Afrique du Nord, Vol.* 5. Paul Lechevalier, Paris

Májovsky, J., *et al.* (1978) Index of chromosome numbers of Slovakian flora (part 6). *Acta Fac. Rerum Nat. Univ. Comenianae, Bot.* 26: 1–42

Malo, S. & Shuka, L. (2013) Distribution of *Colchicum doerfleri* Halácsy, *Colchicum triphyllum* Kunze and *Colchicum bivonae* Guss., in Albania. *Int. J. Ecosystems Ecol. Sci.* 3(2): 273–278

Manning, J., Forest, F. & Vinnersten, A. (2007) The genus *Colchicum* L. redefined to include *Androcymbium* Willd. based on molecular evidence. *Taxon* 56 (3): 872–882

Marie, D. & Brown, S. (1993) A cytometric exercise in plant DNA histograms, with 2C values for 70 species. *Biol. Cell* 78: 41–51

Markgraf, F. (1975) Der übergang der vegetationsstufen Albaniens in die östliche Balkanhalbinsel. In Jordanov, D. (ed.) *Problems of Balkan Flora and Vegetation*. Bulgarian Academy of Sciences, Sofia

Maroulis, G. & Artelari, R. (2001) New records to the Flora of Mt. Erimanthos (NW Peloponnisos, Greece). *Fl. Medit.* 11: 311–331

Mathew, B (1965) The Bowles Scholarship Botanical Expedition to Iran, 1963. *J. Roy. Hort. Soc.* 90(1): 5–18

Mathew, B., Starling, B.N. et al. (1980) Plant Awards 1979–80. *Quart. Bull. Alpine Gard. Soc.* 48: 285-320

Mathew, B. (1982) *J.P. Redouté: Lilies and Related Flowers.* Michael Joseph, London

Mathew, B. (1987) *The Smaller Bulbs.* B.T. Batsford, London

Mathew, B. (2000) Colchicum 'Waterlily'. *The Garden* 125(9): 684

Mathew, B. & Baytop, T. (1984) *The Bulbous Plants of Turkey.* B.T. Batsford, London.

Meikle, R.D. (1985) *Flora of Cyprus,* Vol. 2. Royal Botanic Gardens, Kew

Meusel, H., Jäger, E. & Weinert, E. (1965) *Vergleichende Chorologie der Zentraleuropäischen Flora.* Vol. 1. G. Fischer, Jena

Mouterde, P. (1966) *Nouvelle Flore du Liban et de la Syrie* Vol. 1. Editions de l'Imprimerie Catholique, Beyrouth

Murín, A. & Májovsky, J. (1979) Karyological study of Slovakian flora. I. *Acta Fac. Rerum Nat. Univ. Comenianae, Bot.* 27: 127–133

Nerantzis, X., Koudros, V. & Alexiou, S. (2014) New localities for *Colchicum atticum* (Colchicaceae) from NE Greece. *Parnassiana Arch.* 2: 53–54

Nikolić, T. (ed.) (2000) Flora Croatica index Florae Croaticae, Pars 3. *Natura Croatica* 9, suppl. 1

Nguyen, T.P.A., Kim, J.S. & Kim, J.-H. (2013) Molecular phylogenetic relationships and implications for the circumscription of Colchicaceae (Liliales). *Bot. J. Linn. Soc.* 172: 255–269

Nordenstam, B. (1982) A monograph of the genus *Ornithoglossum* (Liliaceae). *Opera Bot.* 64: 1–51

Nordenstam, B. (1998) Colchicaceae. In Kubitzki, K. (ed.) *The Families and Genera of Vascular Plants, Vol. 3 (Flowering Plants, Monocotyledons).* Springer-Verlag, Berlin

Nutt, R.D. (1971) Snowdrop cultivars and colchicums in cultivation. *J. Scott. Rock Gard. Club* 12(3): 190–209

Oganezova, G.G. (2011) Osovennosti geografii i napravlenii evolyutsii histerantnykh i sinantnykh vidov roda *Colchicum* s.str. (Colchicaceae). *Takhtajania* 1: 87–97

Özhatay, N., Kültür, Ş. & Aslan, S. (2009) Check-list of additional taxa to the Supplement Flora of Turkey IV. *Turk. J. Bot.* 33: 191–226

Paradis, G. & Alphand, J. (1994) *Colchicum corsicum* Baker. Notes et contributions a la flore de Corse. *Candollea* 49: 576

Parkinson, J. (1629) *Paradisi in Sole Paradisus Terrestris.* Humfrey Lownes & Robert Young, London

Parlatore, F. (1860) *Flora Italiana,* Vol. 3. Tipografia Le Monnier, Florence

Peri, O. (2015) *Bulbs of the Eastern Mediterranean.* Alpine Garden Society, Pershore

Perrenoud, R. & Favarger, C. (1971) Sur l'existence d'hybrides entre le colchique des alpes (*Colchicum alpinum* DC.) et le colchique d'automne (*C. autumnale* L.) dans les Alpes Françaises. *Bull. Soc. Neuchâtel. Sci. Nat.* 94: 21–27

Persson, K. (1988) New species of *Colchicum* (Colchicaceae) from the Greek mountains. *Willdenowia* 18: 29–46

Persson, K. (1991) Colchicum. In Strid, A. & Tan, K. (eds) *Mountain Flora of Greece.* Vol. 2. Edinburgh University Press, Edinburgh

Persson, K. (1992) Liliaceae III. In Rechinger, K.H. (ed.) *Flora Iranica.* Vol. 170. Akademische Druck-u-Verlagsanstalt, Graz

Persson, K. (1993a) *Colchicum feinbruniae* sp. nov. and allied species in the Middle East. *Israel J. Bot.* 41: 75–86

Persson, K. (1993b) Reproductive strategies and evolution in *Colchicum.* In Demiriz, H. & Özhatay, N. (eds) *Proceedings of the Fifth Optima Meeting, Istanbul, 8–15 September 1986.* Istanbul University, Istanbul

Persson, K. (1998a) Comments on some tessellated *Colchicum* species in the East Mediterranean area. *Candollea* 53: 399–418

Persson, K. (1998b) The genus *Colchicum* in Turkey. I. New species. *Edinburgh J. Bot.* 56 (1): 85–102

Persson, K. (1999a) New and revised species of *Colchicum* (Colchicaceae) from the Balkan Peninsula. *Pl. Syst. Evol.* 217(1–2): 55–80

Persson, K. (1999b) The genus *Colchicum* in Turkey. II. Revision of the large-leaved autumnal species.

Edinburgh J. Bot. 56 (1): 103–142

Persson, K. (2001a) *Colchicum* L. In Güner, A. *et al.* (eds) *Flora of Turkey and the East Aegean Islands, Vol. 11, Suppl. 2.* Edinburgh University Press, Edinburgh

Persson, K. (2001b) A new soboliferous species of *Colchicum* in Turkey. *Bot. J. Linn. Soc.* 135: 85–88

Persson, K. (2006) A new Turkish species of *Colchicum* (*Colchicaceae*) related to *C. boissieri*. *Edinburgh J. Bot.* 62(3): 181–192

Persson, K. (2007) Nomenclatural synopsis of the genus *Colchicum* (*Colchicaceae*), with some new species and combinations. *Bot. Jahrb. Syst.* 127(2): 165–242

Persson, K. (2008) A new species of *Colchicum* (*Colchicaceae*) from southern Italy. *Bot. Jahrb. Syst.* 127(3): 283–288

Persson, K. (2009) In Marhold, K. (ed.) IAPT/IOPB chromosome data 7. *Taxon* 58(1): 181–183

Persson, K., Petersen, G., Hoyo del, A., Seberg, O. & Jørgensen, T. (2011) A phylogenetic analysis of the genus *Colchicum* L. (*Colchicaceae*) based on sequences from six plastid regions. *Taxon* 60(5): 1349–1365

Phillips, R. & Rix, M (1989) *Bulbs*. Pan Books, London

Phitos, D., Constantinidis, T. & Kamari, G. (eds.) (2009) *The Red Data Book of Rare and Threatened plants of Greece*. Hellenic Botanical Society, Patras

Pignatti, S. (1982) *Flora d'Italia*. Edagricole, Bologne

Pils, G. (2006) *Flowers of Turkey, A Photo Guide*. Eigenverlag G. Pils, Austria

Polunin, O. (1980) *Flowers of Greece and the Balkans*. Oxford University Press, Oxford

Polymenakos, K. & Tan, K. (2016) In Vladimirov, V. *et al.* New floristic records in the Balkans: 31. *Phytol. Balcanica* 22(3): 429–467

Popova, M.T., Češchmedjiev, I.V. (1978) Reports. In Löve, A. (ed.) IOPB chromosome number reports LXI. *Taxon* 27: 384–385

Post, G.E. & Dinsmore, J.E. (1933) *Flora of Syria, Palestine and Sinai, Vol. 2* (2nd edn.). American Press, Beirut

Radić, J. (1976) *Bilje Biokova*. Institut Planina I More Malakološki Muzej, Makarska

Rechinger, K.H. (1950) Grundzüge der pflanzenverbreitung in der Ägäis (mit 30 karten). *Vegetatio* 2: 55–119

Redouté, P.-J. (1802–1816) *Les Liliacées*. Imprimerie de Didot Jeune, Paris

Rix, M. (1983) *Growing Bulbs*. Croom Helm, London

Rolfe, R. (2000) Plant awards 1998–9. *Quart. Bull. Alpine Gard. Soc.* 68(2): 201–286

Rolfe, R. (2001) Plant awards 2000–2001. *The Alpine Gardener* 69(4): 482–550

Rolfe, R. (2004) Plant awards 2003–4. *The Alpine Gardener* 72(4): 385–450

Rolfe, R. (2006) Plant awards 2005–6. *The Alpine Gardener* 74(4): 454–522

Rolfe, R. (2009) Plant awards 2008–9. *The Alpine Gardener* 77(4): 461–510

Rolfe, R. (2013) Plant awards 2010–11. *The Alpine Gardener: Plant Awards* 33–34

Rukšāns, J. (2007) *Buried Treasures*. Timber Press, Portland, Oregon

Sevgi, E. & Küçüker, O. (2011) Morpho-anatomical observations on *Colchicum boissieri* Orph. in Turkey. *IUFS J. Biol.* 70 (2): 53–61

Sheasby, P. (2007) *Bulbous Plants of Turkey and Iran*. Alpine Garden Society, England

Skalińska, M., Banach-Pogan, E., & Wcislo, H. *et al.* (1957) Further studies in chromosome numbers of Polish angiosperms. *Acta Soc. Bot. Polon.* 26: 215–245

Šopova, M. (1969) Cytological study in the genus *Colchicum* from Macedonia. *Godisen. Zborn. Biol. Fak. Skopje Univ. Prir.-Mat.* 21: 119–130

Stearn, W.T. (1996) *Stearn's Dictionary of Plant Names for Gardeners*. Cassell, London

Stefanov, B. (1926) Monografiya na roda *Colchicum* L. *Sborn. Bălg. Akad. Nauk. Sofiya* 22: 1–100 [English translation at RBG Kew]

Stefanović, V. (1996) Analysis of the Central European and Mediterranean orophytic element on the mountains of the W. and Central Balkan Peninsula, with special reference to endemics. *Bocconea* 5: 77–97

Stevanović, V., Tatić, B., Janković, M., Diklić, N., Jovanović, B., Vasić, O., Jovanović, S., Niketić, M., Butorac, B. & Boza, P. (1999) *Crvena Knjiga Flore Srbije 1*. Ministry of Environmental Protection, Belgrade

Strid, A. (1996) The Greek mountain flora, with special reference to the Central European element. *Bocconea* 5: 99–112

Strid, A. (2016) Atlas of the Aegean flora. *Englera* 33 (1 & 2)

Svešnikova, L.I., Kričfalušij, V.V. (1985) Čisla hromosom nekotoryh predstavitelej semejstv *Amaryllidaceae* i *Liliaceae* flory USSR i GSSR. *Bot. Zhurn.* 70: 1130–1131

Synge, P.M. (1961) *Collins Guide to Bulbs.* Collins

Tan, K. & Iatrou G. (2001) *Endemic Plants of Greece, The Peloponnese.* Gads Forlag, Copenhagen

Thi, N.P.I, Kim, J.S. & Kim, J.-H. (2013) Molecular phylogenetic relationships and implications for the circumscription of *Colchicaceae* (*Liliales*). *Bot. J. Linn. Soc.* 172: 255–269

Trigas, P. & Iatrou, G. (2006) The local endemic flora of Evvia (W Aegean, Greece). *Willdenowia* 36: 257–270

Turland, N., Chilton, L. & Press, J. (1993) *Flora of the Cretan Area. Annotated Checklist & Atlas.* H.M.S.O., London

Turrill, W.B. (1929) *The Plant-life of the Balkan Peninsula.* Clarendon Press, Oxford

van Scheepen, J. (1991) *KAVB International Checklist for Hyacinths and Miscellaneous Bulbs.* Royal General Bulbgrower's Association, Hillegom

Vasl, A & Shmida, A. (2015) The adaptive role of nectarial appendages in *Colchicum. Pl. Syst. Evol.* 301(6): 1713–1723

Vassiliades, D. & Persson, K. (2002) A new winter-flowering species of *Colchicum* from Greece. *Preslia* 74: 57–65

Velenovský, J. (1891) *Flora Bulgarica.* Caroli Bellmanni, Prague

Vinnersten, A. & Larsson, S. (2010) Phylogenetic relationships within *Colchicaceae. Am. J. Bot.* 90: 1455–1462

Vinnersten, A. & Manning, J. (2007) A new classification of *Colchicaceae. Taxon* 56: 171–178

Vinnersten, A. & Reeves, G. (2003) Phylogenetic relationships within *Colchicaceae. Am. J. Bot.* 90 (10): 1455–1462

Visiani, R. de (1842) *Flora Dalmatica, Vol. 1.* F. Hofmeister, Leipzig

Visiani, R. de (1877) *Florae Dalmaticae supplementum alterum, Part 1.* Mem. Inst. Veneto, Venice

Wallis, R. (2003) *Merendera montana* 'Norman Barratt'. *Quart. Bull. Alpine Gard. Soc.* 71: 164, 382–383

Wendelbo, P. (1985) *Colchicum* L. In Townsend, C.C. & Guest, E. (eds) *Flora of Iraq, Vol. 8.* Ministry Of Agriculture & Agrarian Reform, Baghdad

Weston, R. (1771) *The Universal Botanist and Nurseryman.* London

Yakovlev, S., Pustahija, F., Solic, E.M., Bogunic, F., Muratovic, E., Basic, N., Catrice, O. & Brown, S.C. (2010) Towards a genome size and chromosome number database of Balkan flora: C-values in 343 taxa with novel values for 242. *Adv. Sci. Lett.* 3: 190–213

Zarkos, G., Christodoulou, V., Kit, T. & Vold, G. (2015) In Vladimirov, V. *et al.* New floristic records in the Balkans: 28. *Phytol. Balcanica* 21(3): 367–399

Zarkos, G., Christodoulou, V., Kit, T. & Vold, G. (2016) In Vladimirov, V. *et al.* New floristic records in the Balkans: 31. *Phytol. Balcanica* 22(3): 429–467

Zervou, S., Raus, T. & Yannitsaros, A. (2009) Additions to the flora of the island of Kalimnos (SE Aegean, Greece). *Willdenowia* 39: 165–177

Zaharijeva, O.I., Makušenko, L.M. (1969) Hromosomnye čisla odnodolnyh rastenij iz sem. *Liliaceae, Iridaceae, Amaryllidaceae* i *Araceae. Bot. Zhurn.* 54: 1213–1227

Zohary, M. (1938) On the vegetative reproduction of some oriental geophytes. *Palestine J. Bot. (Jerusalem series)* 1(1): 35–41

Index

Subjects in the **Checklist of epithets** chapter are not indexed.

Species and cultivar names in the tables are not indexed.

Synonyms and their cross-references can be found in the **Checklist of epithets** chapter.

Main entries, with illustration/s, are indicated in **bold**.

Illustrations are indicated in *italics*.

AB&S 32, 33
AC&W 31, 32
Acaena inermis 85
   microphylla 85
Aitchison, James 21
Ajuga pyramidalis 85
   reptans 85
     'Atropurpurea' *84*
AH 30
Alexander of Tralles 74
Alliaceae 41
Allium 36, 102
   regelii 28
   ursinum 75
Alpine Garden Society 33, 103, 402, 466
Alstroemeriaceae 41
Amand, John 95
Anderson, Edward B. 357, 394, 402
Androcymbium 41, 42, 43, 44, 45, 50, 51, 70
   europaeum 49
   palaestinum 47
   psammophilsum 48
   rechingeri 49
   vanjaarsveldii 48
Anemone nemorosa 26, 165
Angiosperm Phylogeny Group 52
aphids 92
Archibald, Jim and Jenny *32*, 33, 34, 35, 154
Arnott, Sam 471
Asparagaceae 41
Asparagales 41
Asphodelaceae 41
Avicenna 74
Backhouse Nursery, York 23
Backhouse, Robert Ormston 24
Baeometra 42, 44
Baker, Stuart 28, *28*
Bankier, Keith 95
Barbarossa, Frederick (Holy Roman Emperor I) 256
Barr and Sons 24
Barter, David 28, *28*
BATMAN 99
Baytop, Asuman 158
Baytop, Turhan 158, 377
Behçets disease 73, 75
Bellevalia glauca 27
Birch Farm Nursery, Sussex 516
Bird, Peter 32
Bishop, Joy 93
Blackthorn Nursery, Hampshire 458
Blanchard, John 32
B&M 32
Bondarenko, Leonid 36, 80, 97, 450, 462, 463, 480, 493, 495, 498, 504, 515, 534
Bornmüller, Joseph 382, 469, 512
Botrytis cinerea 92, 103
Bowles, Edward A. 21, 23, 24, 25, *26*, 39, 150, 436, 438, 473, 474, 491, 517, 530
Bowles Scholarship Botanical Expedition (BSBE) 28, *28*, 29
Brickell, Chris 31, 52, 93, 95, 157, 304, 340, 369, 384, 459, 484, 502, 524
Brueghel the Elder, Jan 39
Brunnera 85
Bulbocodium 41, 42, 43, 44, 45, 47, 48, 50, 52, 105, 109, 110, 111, 121
   eichleri 398
   vernum 47
Bullock, Edward 93
Burchardia 34
Burtt, Bill 171
Bydžovský, Jiři 35
Cambridge University Botanic Garden 103, 148
Camptorrhiza 42, 44
Campynemataceae 41
cataphylls *62*, *63*
CE&H 30
CH 30
Charybdis maritima 291
Chittenden, Frederick J. 97
Christian, Paul 30
Clusius, Carl 20, 441
Colchicaceae 16, 42, 43, 48, 52, 53
Colchicine 73, 74, 75, 76, 77
   tablets *75*

Colchicos 74
Colchicum
  × *agrippinum* *1*, 38, *57*, 61, 82, 123, 422, *430*, **431**, *444*, 446
  × *alberti* 36, 38, 123, 262, *432*, **433**, 445, 450
    'Jānis' 36, 433, **450**
    'Jeanne' 36, 433, 450
    'Lucky Selfmade' 36, 433, **450**
    'Moonlight' 36, 433, 450
  alpinum 54, 55, 115, 122, *128*, **129**, 313
    var. *parvulum* 129
  × *ambiguum* 38, 123, *434*, **435**, 436, *437*, 438, *439*, *444*, 446, 524
  androcymbioides 121, 123, *130*, **131**
  'Antares' *24*, **479**
  antepense 54, 102, 120 123, *132*, **133**
  antilibanoticum *34*, 53, 106, 114, 115, 122, *134*, **135**
  arenarium 54, *70*, 116, 122, 129, *136*, **137**, 321, 415
  arenasii 116, 122, *138*, **139**, 189
  'Artur Klark' 384, **480**, 512
  asteranthum 54, 122, *140*, **141**
  atropurpureum 21, 148, 150
  'Atropurpureum' 150
  atticum 54, *55*, *67*, 102, 121, 123, 131, *142*, **143**, 379
  autumnale 14, 15, *18*, *19*, 20, 37, 39, *40*, 54, *64*, *68*, *72*, 74, 75, 80, 81, 82, 93, 105, 107, 116, 122, 123, 141, *144*, **145**, *146*, 148, *149*, 150, 185, 186, 207, 240, 245, 279, 281, 282, 309, 315, 316, 402, 431, 442, *444*, 445, 451, 476, 483, 485, 518, 520, 530
    'Alboplenum' 148, **451**, 455, 530
    'Album' 23, 24, 25, 37, 83, 148, 382, **452**, 467, 468, 484, 489, 523
    var. *algeriense* AB&S 4353 32
    'Annecy' 146
    var. *atropurpureum* 150, *151*
    'Dorothee Kersen' 146, **453**
    'Drama Bunch' 146, 519, 529
    'Nancy Lindsay' *71*, 148, *444*, *447*, 453, **454**
    'Old Bones' 452
    var. *pannonicum* 146
    subsp. *pannonicum* 113, *147*, **148**, 453, 454
    'Pleniforum' 146, 451, **455**, 530
    subsp. *vernum* 146
    var. *vernum* 146
    × *C. haynaldi* 38
    × *C. variegatum* 38, 431
  'Autumn Herald' *69*, **481**
  'Autumn Queen' *8*, *17*, 23, 38, *56*, *80*, 93, 162, *444*, **482**, 501
  balansae 54, 55, 62, 69, 116, 122, *152*, **153**, 154, *155*, 256, 265, 297, 436
  baytopiorum 59, 115, 119, 122, *156*, **157**, *159*, 179, 199, 200
  'Beaconsfield' **483**
  'Benton End' 37, 39, 83, *86*, 411, *444*, **484**, 491, 502, 523
  biflorum 19
  bivonae 21, 23, 24, 25, 30, 37, 38, 54, 64, 67, 70, 112, 123, 141, *160*, **161**, *162*, 214, 220, 231, 291, 374, 435, 438, 445, 456, 457, 458, 482, 483, 490, 501, 529, 535
    AH 9139 from Mount Giona 30
    'Apollo' 162, **456**
    × *C. lusitanum* 231
    'Mount Etna' 162, *457*
    'Papa Rema' 162, **458**
  boissieri 54, 59, 71, 81, 115, 122, 157 *164*, **165**, 179
    AC&W 2352 31
  bornmuelleri 21, 23, 25, 382, 384, 442, 469, 473, 480, 506, 517
  bornmuelleri Freyn 382, 469, 512
  bornmuelleri hort. 512, 527, 534
  bowlesianum 25, 37, 220
  'Boxford' 37, **485**, 511
  bulbocodium 46, 69, 121, 123, *166*, **167**, 168, *169*, 262
    subsp. *versicolor* **168**, *169*
  burttii 54, 65, 119, 123, *170*, **171**
  byzantinum 20, 435, 438
  × *byzantinum* *20*, 38, 74, 80, 123, 256, *440*, **441**, 442, *443*, *444*, 445
    'Album' 94, 442, 459
    'Innocence' *84*, 94, 442, *444*, **459**
  callicymbium 240
  candidum 153
  'Cedric Drake' 484
  'Cedric Morris' 484
  chalcedonicum 54, 113, 123, *172*, **173**, 275, 276, 299

subsp. *chalcedonicum* 174
subsp. *punctatum* 174, *175*
chimonanthum 54, 102, 119, 122, *176*, **177**, 193, 227
chionense 422
chlorobasis 54, 116, 122, *178*, **179**, 377
cilicicum 37, 54, 68, 94, 106, 112, 116, 122, 174, *180*, **181**, 182, *183*, 199, 363, 435, 436, 438, 441, 442, 445, 460, 496, 512, 524
  'Cilician Gates' 182
  × ?*C. autumnale* 38
  MS&CL 541 32
  × ?*C. neapolitanum* 38
  subsp. *punctatum* 54
  'Purpureum' 182, *444*, **460**
confusum 54, 122, 123, 145, *184*, **185**, 186, *187*, 240
'Conquest' **501**
'Constable' **486**
corsicum 30, 55, 108, 116, 122, 139, *188*, **189**
  Watt s.n. 32
cousturieri 196
cretense 54, *65*, 114, 122, 141, *190*, 191, 329, 330, 347, 348
crocifolium 27, 28, 53, 118, 120, 122, 177, *192*, **193**, 227, 262
cupanii 54, *66*, 105, 118, 123, *194*, **195**, 196, *197*, 340
  B&M 10149 32
  subsp. *glossophyllum* 54, *109*, 195, **196**, *197*
'Daendals' 23, 446, **487**
'Danton' 23, 39
'Darwin' *2*, 24, 63, 95, **488**
davisii *31*, 54, 54, 112, 123, *198*, **199**, 200, *201*
decaisnei 33, 54, 117, 455, *202*, **203**, 204, *205*, 245, 405, 406
diampolis 392
'Dick Trotter' 25, 39, **489**, 498, 504, 508, 515, 523
'Disraeli' 23, 38, 39, 162, **490**
doefleri rear jacket, 34, 54, 99 120, 123, *206*, **207**, 249, 461
  'Valentine' 34, **461**
dolichantherum 54, 117, 119, 122, 182, *208*, **209**, 256
drenowskii 207
'E.A. Bowles' 25, 37, 39, 83, 94, *97*, 411, 484, **491**, 499, 502
'E.K. Balls' **492**, 505
eichleri 398
eichleri f. *rossa* aff. hybrid 99
'Elizabeth' 146
'Emerald Town' **493**
'Enigma' **494**, 529
erdalii 53, *210*, **211**, 369, 370
'Eric Pasche' 392
euboeum 54, 112, 123, *212*, **213**, 214, *215*
'Fabergé's Silver' 384, **495**, 512
falcifolium 369
fasciculare 53, 106, 118, 123, *216*, **217**, 247, 347, 348
feinbruniae 53, 111, 112, 123, *218*, **219**, 220, *221*
  J&JA 314.250 33
'Felbrigg' 37, 94, *444*, **496**
'Felbrigg Violet' **497**, 526
'Ferndown Beauty' 39
figlalii 54, 122, 123, *222*, **223**
filifolium 54, 121, 123, *224*, **225**
'Flamenco Dance' **498**, 515
fominii 137
freynii 54, 119, 122, *226*, **227**
'Fuller's Mill' *36*, 37, *78*, **499**

'Giant' *rear jacket*, 23, 93, **500**
giganteum 21, 23, 25, 37, 82, 382, 384, 471, 512
'Glorie van Holland' 39
'Glory of Heemstede' *front jacket*, *23*, 38, 39, 162, *444*, 490, **501**
'Glory of Threave' *rear jacket*, *2*, 39, *70*, *82*, *84*, *86*, *94*, 446, 499, **502**, 525
gonarei 55, 67, 68, 117, 122, 129, *228*, **229**, 313
'Gothic Style' **504**
'Gracia' 492, **505**
gracile 112, 122, *230*, **231**
graecum 54, 113, 119, 122, 214, *232*, **233**, 234, *235*, 325
greuteri 108
guessfeldtianum 106, 114, 123, *236*, **237**, 413
'Guizot' 39
hadriaticus 17
'Harlekijn' 24, 384, **506**, 510, 512, 534
haynaldii 55, 113, 123, 185, *238*, **239**, 240, *241*, 519
  CH 871 30
heldreichii 54, 122, *242*, **243**, 265,
'Herbstkugel' **507**
'Hidegkut' 39
hiemale 304
hierosolymitanum 113, 116, 122, 204, 220, *244*, **245**, 273, 337
hirsutum 27, 35, 53, 123, *246*, **247**
hissaricum 360
hungaricum 34, 54, 99, 118, 120, 207, *248*, **249**, 251, 461

f. *albiflorum* **250**, *251*
    'Velebit Star' 34, 250
'Huxley' 24, *38*, 39, 489, 504, 507, **508**, 523
ignescens 53, 123, *252*, **253**, 398
illyricum 471
imperatoris-friderici 55, 114, 122, 182, *254*, **255**, 256, *257*
'Intermediate Dykes' 533
inundatum 55, 114, 122, *258*, **259**
'James Pringle' 39
'Jarka' 384, 506, **510**, 512, 534
'Jarkoslavan' 525
'Jaroslavna' 525
'Jaroslawna' 525
'Jenny Robinson' **511**
'Jochem Hof' 525
'Jochum Hof' 525
'Joseph' 384, 469, 480, 487, 495, 506, 510, **512**, 517, 527, 534
'Karin Persson' 146
kesselringii 13, 21, 28, 36, 46, 47, 55, 67, 68, 69, 70, 106, 111, 123, 167, *260*, **261**, 262, *263*, 286, 360, 433, 450, 462
    'Boldness' 262
    × *C. luteum* 36, 38
    'Modesty' 262
    'My Choice' 262
    'Prosperity' 262
    'Purple Star' 262
    'Snow of Highland' 36, 262, 286, 450, **462**
    'Yeti' 36, 262, 450, **462**
'Kiss Me Quick' **514**
'Klondike' 39
kotschyanus 17
kotschyi 30, 33, 53, 114, 122, 243, *264*, **265**, 297, 327
kurdicum 46, 53, 107, 121, 123, 253, *266*, **267**
    J&JA 314.789 33
laetum 35, 69, 70, 82, 94, 117, 122, 153, *268*, **269**, 524
laetum hort. 524
lagotum 53, *71*, *99*, 120, 123, *270*, **271**, 392
'Larisa' 498, **515**
latifolium 214, 442
'Lausanne' 146
leptanthum 53, 108, 122, 141, *272*, **273**
levieri MS 937 32
'Lilac Bedder' *2*, **516**
'Lilac Wonder' 23, 39, 82, 486, **517**, 533
lingulatum 55, 68, 71, 81, 113, 116, 117, 122, 174, *274*, **275**, 276, *277*, 325
    subsp. *lingulatum* 276
    subsp. *rigescens* **276**, *277*
liparochiadys 429
liparochlamys 382, 429
'Little Woods' *444*, **518**
longifolium 122, *278*, **279**
'Looking Up' 504
lusitanum 32, 55, 114, 122, 231, 279, 280, **281**, *282*, *283*, 309, 316, 436, 438
luteum 13, *21*, 22, 36, 46, 47, 55, 67, 68, 106, 111, 123, 167, 262, *284*, **285**, 286, *287*, 360, 433, 450, 463
    'Carrot Line' 286
    'Golden Baby' 286, **463**
    'Golden Elf' 463
    'Minion' 286
    'Vakhsh' 286, **463**
    'Yellow Empress' 286
'Lysimachus' 38, **519**
macedonicum 55, 108, 111, 115, 122, *288*, **289**, 343, 344,
macrophyllum *60*, *65*, 67, *70*, 112, 123, *290*, **291**, 292, *293*, 374, 464, 465, 55
    AH 9806 30
    'Anopolis' 292, **464**
    'Cretan White' 37, 292, **465**
    'Hora Sfakion' 292, **465**
manissadjianii 55, 122, 123, *294*, **295**
maraschicum 115, 122, *296*, **297**
meadow saffron 19
'Mells Park' **520**
merenderoides 129
micaceum 55, 122, *298*, **299**, 301
micranthum 30, 55, 117, 122, 299, *300*, **301**, 355, 415
minutum 22, 35, 54, 59, 119, 122, 141, 273, *302*, **303**, 304, *305*, 311, 340
montanum 21, 55, 121, 123, 168, 225, *306*, **307**, 466
    'Norman Barratt' 307, **466**
multiflorum 122, *308*, **309**, 316
munzurense 54, 99, 103, 120, 122, 273, 304, *310*, **311**
    KPPZ 208 35
'Myddelton' 496
'Naeisanum' 39, 382
nanum 55, 118, 122, 129, 229, *312*, **313**
neopolitanum 32, 55, 116, 122, *217*, 231, 279, 309, *314*, **315**, 316, 436
    var. *micranthum* 32
'Neptun' **521**
ninae 392
'Nutt's Green Star' 482

'Oktoberfest' 162, **522**
osmaniyense 55, 119, 123, *318*, **319**, 388
palatoris 55, 115, 122, *320*, **321**, *323*, 329, 330, 348
pannonicum 148
parkinsonii 422
parnassicum 55, 112, 116, 122, 214, 233, 234, *324*, **325**
parvulum 129
paschei 54, 117, 122, *326*, **327**
peloponnesiacum 31, 54, **65**, 119, 123, *324*, *325*, *327*, 329, 330, 387, 388
   MS&CL 195 32
persicum 55, 117, 122, 259, *332*, **333**, 334, *335*, 409
   J&JA 316.707 33
'Petrovac' 39
pinatziorum 165
'Pink Goblet' *12*, *25*, 39, 484, 489, 508, **523**
'Pink Star' 82, 94, 269, 438, *444*, **524**
polymorphum 374
polyphyllum 33, 53, 64, 115, 122, 220, *336*, **337**
'Poseidon' 80, 446, **525**
praecox 146
'Premier' 517
'President Collidge' 23, 39
'Pride of Holland' **526**
procumbens 85
procurrens M&T 1862 32
psaridis 55, 59, 81, 105, 118, 122, 196, 304, 321, 330, *338*, **339**, 340, *341*
   MS&CL 198 32
pseudoparvulum 129
pulchellum 55, **68**, 115, 117, 122, 276, 289, 355, *342*, **343**, 344, *345*

'Purity' 39
pusillum 32, 55, *104*, 105, 119, 123, 143, 191, 304, 321, 329, 330, *346*, **347**, 348, *349*, 387
raddeanum *46*, 53, 121, 122, 123, 253, 295, *350*, **351**, 398
"ramonensis" 68, 106, 118, 123, 351, *352*, **353**
rausii 55, 59, 115, 122, 344, *354*, **355**
'Redgrave' 37, 384, 487, 495, 506, 510, 512, **527**, 534
rhodopaeum 146
ritchii *53*, 68, 106, 118, 119, 123, 237, *356*, **357**, 366, 413
robustum 28, 36, 55, 68, 121, 123, 225 262, *358*, **359**, *361*
'Rosy Dawn' *10*, 24, *84*, 93, 162, *444*, 446, 511, 522, **528**
'Rosy Wonder' 508
'Ruby Queen' 39
sanguicolle 30, 53, *62*, 114, 122, *362*, **363**
schimperi 53, 118, 120, 123, 217, 237, 353, *364*, **365**, *367*, 425
serpentinum 30, 35, 53, 65, *100*, 120, 123, 133, 171, 211, 247, 273, *368*, **369**, 370, *371*
sfikasianum 55, 113, 123, 321, *372*, **373**, *375*
sibthorpii 25, 37
sieheanum 59, 122, 179, *376*, **377**
soboliferum 55, 59, *61*, *103*, 121, 123, *378*, **379**
'Spartacus' 494, **529**

speciosum 19, 21, 23, 24, 25, 30, 33, 37, 38, 39, 54, 64, 70 ,74, 80, 82, *105*, 107, 114, 122, 146, 153, 200, 269, 334, *380*, **381**, 382, *383*, 384, *385*,429, 442, *444*, 445, 467, 468, 469, 470, 471, 472, 473, 474, 475, 476, 477, 484, 491, 493, 502, 508, 512
   'Album' *81*, *84*, *90*, *91*, **467**
   'Atrorubens' 23, 47, 83, 382, **468**
   Bornmuelleri Group 54, 114, 384, **469**, 500, 512
   'Chequers' 37, *59*, 384, **472**, 473, 511, 530
   'Dombai' 382, **470**
   Giganteum Group 384, **471**, 500, 512
   'Maximum' 382, **474**
   'Ordu' 382, 498, 515
   'Paul Furse' 382, **475**, 476
   'Paul Furse Early' 382, **476**
   'Revelation' 384, **473**
   'Rubrum' 382, **477**
stevenii 55, 71, 119, 123, 143, 319, 329, 330, 347, 348, *386*, 388, **387**, *389*
strangularium 20
'Surprise' 39
szovitsii 19, 27, *29*, *30*, 33, 34, 35, 46, *54*, *63*, 99, 101, 107, 120, 123, 141, 271, 273, *390*, **391**, 392, *393*, *395*, 417, 418, 478
   subsp. brachyphyllum 53, 102, **392**, 394, *395*
   JCA 1147 33
   × C. luteum 99
   M&T 4530 29
   'Snowwhite' 34, 392, **478**
   subsp. szovitsii 392

'Tivi' 34, 392, **478**
'Vardahovit' 34, *35*, 392, **478**
taygeteum 234
tenorei 30, 435, 436, 438
× *tenorei* 438
trigynum *15*, 28, 33, 53, 66, 100, 108, 123, 168, 223, 253, 273, 351, *396*, **397**, 398, *399*, 425
  Palandöken Group 100
triphyllum 54, *66*, 100, 120, 123, 171, 145, *400*, **401**, 402, *403*
  JCA 825 33
troodi 30, 55, 115, 117, 122, 204, *404*, **405**, 406, *407*
tunicatum 55, 115, 122, 334, *408*, **409**
turcicum 21, 30, 37, 39, 55, 68, 112, 150, *410*, **411**, 445, 484, 491,502
tuviae 53, 66, 68, 106, 118, 123, 237, *412*, **413**
umbrosum 54, 117, 123, 137, 269, 301, *414*, **415**
varians 54, 120, 123, 247, *416*, **417**, 418, *419*
  J&JA 3178.700 33
variegatum 21, 30, 55, *65*, *67*, 71, 81, 112, 123, 291, 374, *420*, **421**, 422, *423*, 431
  PB 408 32
verlaqueae 189
vernale 146
vernum 146
'Vesta' 162
'Violet Queen' 23, 24, 39, 497, 526
'W. Kerbert' 23, 39
'Waterlily' *5*, 23, 39, *444*, 451, 455, **530**

wendelboi 33, 54, 98, 121, 122, 123, *424*, **425**, 426, *427*
'Whitton Globe' **532**
'William Dykes' *13*, **533**
'World Champion's Cup' 384, 506, 512, **534**
woronowii 54, 114, 123, 382, *428*, **429**
zahnii 340
'Zephyr' 24, **535**
'Zigana' 392
colchicum around tree *83*
Colchis 15
Convention on International Trade in Endangered Species of Wild Fauna and Flora (CITES) 108
corms *16*, *56*, *57*, *59*, *60*, *61*, *62*, *63*, *80*, *87*
Corsiaceae 41
Corydalis 36, 41, 100
Cotoneaster dammeri 85
Crocus 13, *16*, 17, 22, 27, 70, 103, 171
  aerius 29
  boryi 321
  hermoneus 135
  longiflorus 17
  niveus 17, 321
  nudiflorus 17
  sativus 15, 17, 20
  speciosus 70
Cucurbita 76
cut flowers *76*, 86, *86*
Daisy Hill Nursery, Newry 474
Datura 76
Davis, Peter 30, 31, 200, 330
Dianthus webbianus 31
Dioscorides, Pedanius 73
diseases 91
Disporum 42, 43
Dix Export BV 494

Dryden, Kath 93
Dykes, William Rickatson 533
East Ruston Old Vicarage Garden, Norfolk 96, *97*, 481, 494
Ebers, Georg 73
Ebers papyrus *73*
Edwards, Alan 93
Elliott, David 30
Elwes, Henry 436
Engelmann, Hagen 507
Eranthis 22, 36
Erskine, Peter 93
Erythronium 102
Evans, Alf 93
Farrer, Reginald 85, 422, 442
Felbrigg Hall, Norfolk 93, 95, 97, 438, 459, 473, 484, 496, 497, 501, 502, 524
Fielding, John 464, 465
Fığlalı, Ethem Ruhi 223
filament bases *71*
filius ante patrem 20
Fillan, Mark 32
Flowers in a Vase 39
Franklin, Benjamin 74
Fritillaria 29, 36, 103
  alburyana 31
fruits *70*, *88*, *89*
Fuchs, Leonhart 20
Fullerton, Alice 454
Furse, Paul 26, 28, 39
Furse, Paul and Polly 26, *27*, 28, 475, 476
G.C. Meeuwen nursery 39
Gagea 28
Galanthus 26
Garden Museum, London 39
Gerard, John 19, 20, *74*
Gladiolus 16, 77
Gloriosa 42,43
Gothenburg Botanical Garden, Sweden 79, 90, *98*, *101*,

102, 133, 179, 186, 220, 223, 234, 271, 282, 322, 330, 344
Grey, Alan 95, 97
Grey, Charles 150
grey mould (*Botrytis cinerea*) 92, 103
Grey-Wilson, Christine 96
Grey-Wilson, Christopher 95, 96, 384, 418, 426, 447, 496, 527
Gussone, Giovanni 435
Hall, Tony 95
Handbook of Crocus and Colchicum for Gardeners, A 24, *25*
Hedera 85
Hemerocallis 77
Hermodactylus tuberosus 74
Heuchera 85
　'Palace Purple' *84*
Hexacyrtis 42, 44
Hobbs, Richard 95, 497, 532
Hodgkin, Eliot 392
Hoog & Dix Export 456
Hoog, Antoine 30, 34, 456, 461, 519, 529
Hyacinthaceae 41
Ingerwersen nursery 31, 516
International Union for Conservation of Nature's Red List of Threatened Species (IUCN) 108, 189
Iphigenia 42, 44
Iridaceae 16, 42
Iris 16, 36, 69, 533
　bakeriana 29
　lazicus 69
　pseudocaucasica 369
　reticulata 28, 33
　tuberosa 74

unguicularis 69
Janaki Ammal, E.K. 77
JCA 33
Jekyll, Gertrude 520
Jevremovac Botanical Garden, Belgrade 382
Jilich, Jan 36
J&JA 33
Jost, Vaclav 35
Joyce, Doug 95
Kaźmierski, Paweł 97
Kehr, August 77
Kerbert, Jacobus Johannes 23
Kerndorff & Pasche 179
KPPZ 36, 99
Kreutzberger, Sibylle 93
Kunkeliocymbium 51
Kuntheria 42, 44, 53
Laborde and Houde 74
Lagotis stolonifera 259
leaf shapes *65*
Leeds, Rod 93, 95,
Leeds, Rod and Jane 96
Leichtlin, Max 240
Leptinella pusilla 85
Leslie, Alan 93
Lilium 77
Liliaceae 41
Lindsay, Norah 454
Linnaeus, Carl 20, 21
Lithuanian Rare Bulb Garden 450, 462, 463, 480, 493, 495, 498, 504, 515, 534
Littonia 43
Lord, Tony 93
Lovell, Chris 32
Lutyens, Edwin 520
Magnolia 77
　kobus 'Janaki Ammal' 77
　stellata 'Two Stones' 77, *77*
　'Sun Ray' 77

Mathew, Brian 28, *28*, 29, 32, 150, 530
McSeveney, Andrew 95
Melanthiaceae 41
Merendera 28, 36, 41, 42, 43, 44, 45, 46, 47, 48, 50, 52, 53, 67, 105, 109, 110, 111, 121, 143, 168, 225, 253, 262, 267, 295, 307, 360, 392
　androcymbioides 131
　attica 121, 143
　longifolia 49
　montana 21, 466
　robusta 49
　schimperiana 49
　trigyna 29
Monocot Nursery, Somerset 32
Moore, Peter 465
Morley, John 93, 95, 96, 189
Morley, John and Diana 497, 532
Morris, Cedric 39, 484, 528
MS 32
MS&CL 32, 33
M&T 33
Myddelton House, Enfield 24, 25, 473, 491
Narcissus 102
National Plant Collection 25, 96, 97, 494
Neodregea 44
Nutt, Richard 26, 93, 482
Onixotis 44
Orchidaceae 42
Origanum munzurense 311
Ornithoglossum 42, 43
Parker-Jervis, Elizabeth 25, 93, 95, 489, 491, 518, 533
Parker-Jervis, Johnnie 25, 26
Pasche, Eric 179, 327
PB 33

Pelletier and Caventou 74
Per Wendelbo Memorial Garden 99
Pernice, B 76
Persson, Jimmy 99
Persson, Jimmy and Karin 344
Persson, Karin 36, 46, 48, 49, 52, 99, 105, 150, 154, 177, 185, 204, 214, 231, 234, 256, 276, 340, 360, 369, 382, 384, 406, 426, 435, 436, 438, 442, 469
pests 91
Petermanniaceae 41
PF 27, 28
Philesiaceae 41
PJC 30
P-J Nursery, Oxfordshire 25, 489, 491, 501, 518
Plant Heritage 25, 97
Plantaginaceae 259
Polunin, Oleg 31
Polypodium 85
Portulaca 76
Primula capitellata 29
propagation 87
Psarides, Elias 340
pseudostem *64*
Pulmonaria 26, 85
Pycraft, David 28, *28*
Randall, Mary 93, 95
Ransome, Mrs 21
Ranunculus munzurensis 311
Rare Bulb Nursery, Latvia 35, 470, 510
Rare Plants, Wrexham 30
RHS Award of Garden Merit (AGM) 23, 431, 438, 442, 446
RHS Garden Hyde Hall, Essex 37, 93, 94, *95*, *96*, 97, 442,,447, 469, 471, 475, 491, 499, 508, 511, 517, 522
RHS Garden Wisley, Surrey 27, 77, 96, 97
RHS Lindley Library 27
RHS plant trials 93
Ripongonaceae 41
Rix, Martyn 30
Robinson, Jenny 472, 485, 511, 528
Royal Botanic Garden Edinburgh 171, 200, 282, 384, 411
Royal Botanic Gardens, Kew 21, 27, 148, 186, 234, 322, 377, 436
Rukšāns, Jānis 34, 35, 36, 80, 97 470, 510
S&Z 99
Salmon, Michael 32, 33, 476
Sandersonia 42, 43
Schelhammera 42, 43, 53
Schwerdt, Pam 93
Scottish Rock Garden Club 26, 33
seeds *70*, *71*, *89*
Seisums, Arnis 34, 36, 450, 478
Sfikas, George 373
Shmida, Avi 353
Siehe, Walter 256, 304
slugs 91
Smilacaceae 41
Smith, Tom 474
smut (*Urocystis colchici*) *92*, 93
Species Plantarum 20
Stachys munzurdagensis 311
Stapf, Otto 148
Stearn, William 15, 148, 150
Sternbergia 286
  lutea 292
Stevens, Norman 33
styles *68*, *69*
Styrax 33
Synge, Patrick 27, 37, 39
T4Z 99
Tallinn Botanic Garden, Estonia 470
Tenore, Michele 435
Threave Garden, Dumfries and Galloway 502
Tickner, Bernard 96, 499
Tomlinson, John 29
Tripladenia 42, 43, 53
Triticale 77
Triticum 77
Trotter, Richard Durant (Dick) 25, *26*, 489, 491, 523
University of Edinburgh 30, 31
University of Leipzig, Germany 73
Urocystis colchici 92, 93
Uvularia 42, 43, 53
van Eeden, William 21
Van Tubergen 24, 150, 456, 459, 382
Vasl, Amiel 353
Veratrum 65, 291
viruses *92*, 93
Visser, Kees 24
von Heldreich, Theodor 340
von Regel, Eduard August 360
Wallis, Bob and Rannveig 34, 35, 103, 466
Watson, John 31
Watts, Peter and Penny 458
Wendelbo, Per 98
West, W.A. 392
White, Robin 458
Wurmbea 42, 44
Young, Ian 36
Zetterlund, Henrik 98
Zocher & Co 23, 24
Zubov, Dimitri 35

Photography credits

Numbers refer to pages on which photographs appear

B = bottom, C = centre, L = left, R = right, T = top

Sotiris Alexiou 212, 215TL, 215TR, 277TR, 345TL

Jacques Amand 79

Sergey Banketov 127T, 268

Ilan Biel 364, 367TL, 367B

Biodiversity Heritage Library 168TR

Leonid Bondarenko 450B, 462T, 462B, 463B, 493, 515

Razvan Chisu 49T, 65BL, 109, 149T, 149B, 341TR, 342

Yiannis Christofides 346

Howard Clase 90

Wendy Copage 345B

Jacqui Dracup / Garden World Images front jacket

Hagen Engelmann 507

Jon Evans 46B, 72, 97R, 159BL, 160, 168TL, 172, 184, 198, 224, 238, 241TL, 241B, 242, 248, 251B, 293B, 349TL, 378, 383B, 389BR, 393TR, 407TR, 407B, 430, 452, 463T

Demetrios Fallieras 277TL, 277BR

John Fielding 60, 65TL, 66TL, 67L, 68T, 69, 70BL, 71TL, 104, 159TR, 235T, 283TR, 323B, 323T, 341TL, 372, 375TR, 447, 465B, 466, 531

Valerie Finnis / Lindley Library / RHS 26R, 27

Thomas Giannakis / Flora of Greece Web 176

Philippa Gibson / RHS 501

Katerina Goula 140, 354

Mike Grant 25R, 91, 95

Christine Grey-Wilson 10

Christopher Grey-Wilson rear jacket L, rear jacket TR, 2C, 2R, 9, 11, 13, 17, 31, 36, 48L, 49B, 56, 57, 59, 61BL, 61BR, 62L, 64, 65BR, 66TR, 68B, 70T, 76, 78, 80T, 81, 82, 84TL, 86TR, 86B, 88, 89, 94, 96, 97L, 128, 144, 147L, 163TL, 163B, 164, 180, 188, 194, 196TL, 196TR, 196B, 200TL, 200TR, 222, 260, 283B, 290, 293TL, 293TR, 320, 331TL, 338, 375TR, 383TL, 385BL, 410, 437T, 437B, 439B, 443TL, 443TR, 443B, 454, 454 inset, 455, 458, 459, 460, 464, 465T, 467, 469, 471, 472, 472 inset, 473, 474, 475, 479, 480, 482, 482 inset, 483, 483 inset, 484, 485, 486, 488, 491, 491 inset, 492, 494, 495, 496, 496 inset, 497, 498, 499, 502, 503, 504, 505, 506, 508, 509, 511, 513, 514, 518, 518 inset, 520, 520 inset, 522, 523, 523 inset, 524, 524 inset, 525, 525 inset, 527, 528, 528 inset, 529 inset, 532, 533, 533 inset, 534

Wilf Halliday / RHS 19

Yvette Harvey / RHS 519

Herbarium, Botanic Garden and Botanical Museum Berlin-Dahlem 288

Herbarium, RHS 23

Herbarium, Royal Botanic Gardens, Kew 151

Herbarium, University of Gothenburg 230

Antoine Hoog 187B, 232, 235B, 241TR, 305TL, 317TL, 317TR, 331TR,

Doug Joyce 2L, 25L, 38, 71R, 92R, 147TL, 147TR, 163TR, 221TR, 283TL, 385T, 434, 439T, 456, 457, 468, 476, 477, 481 inset, 484 inset, 487, 487 inset, 489, 490, 517 inset, 526, 529

Erdal Kaya 210, 252, 296

Tristan Lafranchis 331B

Lindley Library / RHS 1, 18, 20, 26L, 28, 29, 40, 74

Mihail Lubinsky 70BR

Giuliano Mereu 228, 312

Nikos Nikitidis / Flora of Greece Web 187TL

Johan Nilsson 100, 130, 270, 318, 371TR, 371B, 416

Oron Peri 47, 53, 134, 202, 216, 218, 221TL, 244, 280, 356, 389TR, 389TL, 395TL, 404, 407TL, 408, 412, 420

Andrew Radgick 419T

Jon Richfield / Wikimedia Commons / CC BY-SA 3.0 42

Robert Rolfe 46T, 63T, 66B, 98, 101, 102, 103, 136, 140, 142, 156, 166, 168B, 170, 226, 228 inset, 251TL, 251TR, 258 inset, 263TL, 263TR, 294, 298 inset, 305TR, 310, 335TR, 345TR, 349B, 354 inset, 358, 361B, 393TL, 396, 399BL, 399BR, 403TL, 410 inset, 427TR, 432, 440, 453, 478R

Jānis Rukšāns 21, 30, 35, 54, 62R, 63B, 65TR, 67R, 71BL, 99, 132, 159TL, 183T, 190, 205TL, 246, 257B, 258, 266, 272, 278, 287TR, 314, 317B, 326, 328, 335TL, 349TR, 361TL, 361TR, 372, 375TL, 389BL, 403TR, 423T, 423BL, 450T, 461, 470, 478TL, 478BL, 481, 510, 516, 517, 521, 535

Tim Sandall / RHS 16, 41, 80B, 83, 444, 451

Luc Scheldeman 187TR, 215BL, 215BR, 375B

Science History Images / Alamy Stock Photo 73

Peter Sheasby 55, 155TL, 155TR, 159BR, 175T, 175B, 178, 183B, 192, 205TR, 264, 274, 300, 305B, 335B, 336, 350, 362, 371TL, 383TR, 385BR, 386, 395B, 403B, 414, 427TL

Carol Sheppard / RHS 24, 77, 324

Avi Shmida 236, 352

Mike Sleigh / RHS 500

Mark Smyth 61T, 87

Tigerente / Wikimedia Commons / CC BY-SA 3.0 306

Rogier Vugt 48R

Bob Wallis rear jacket BR, 15, 34, 105, 152, 155B, 200B, 205B, 206, 208, 254, 257T, 263B, 277BL, 284, 287TL, 287B, 298, 302, 308, 341B, 368, 380, 390, 393B, 395TR, 399T, 400, 423BR, 424, 427B, 428

Bobby Ward 32

Christine Whitehead / Alamy Stock Photo 75

Richard Wilford 221B, 332

Mark Winwood / RHS / Dorling Kindersley 5

Ray Woods 14, 92L

Dimitri Zubov 127B, 136, 376, 419BL, 419BR

Publisher's acknowledgements

In additon to those thanked in the authors' acknowledgements (p11) and the photography credits (pp574–575, many of whom have also helped with other aspects of the book), the publisher acknowledges the assistance of Heather Anderson, James Armitage, Tim Berry, John Birks, Rob Brett, Charlotte Brooks, Martin Cheek, Jan Conway, Matthew Cromey, John David, Panayotis Dimopoulos, Crestina Forcina, Vikky Furse, Alistair Griffiths, Yvette Harvey, Debora Hodgson, Fiona Hood, Carel Hoog, Michelle Housden, Vanessa Invernon, Hayley Jones, Jeremy Kirk, Diana Levy, Rosalyn Marshall, Florent Martos, Anthony Masi, Christine McGregor, Andrew McSeveney, Charlotte Olver, Neriman Özhatay, Ally Page, Matt Pottage, Kristallenia Pougadaki, Debbie Roe, Sonya Roebuck, Johan van Scheepen, Rae Spencer-Jones, Melanie Steel, Lou Tee, Mark Timothy, Tim Upson, Julian Weigall, Amy Williams and Chris Young.